John F. Helliwell, Richard Layard, Jeffrey D. Sachs,
Jan-Emmanuel De Neve, Lara B. Aknin, and Shun Wang

This publication may be reproduced using the following reference: Helliwell, J. F., Layard, R., Sachs, J. D., De Neve, J.-E., Aknin, L. B., & Wang, S. (Eds.). (2025). *World Happiness Report 2025.* University of Oxford: Wellbeing Research Centre.

Full text and supporting documentation can be downloaded from the website: **worldhappiness.report**

ISBN 978-1-7348080-8-7

The *World Happiness Report* is a publication of the Wellbeing Research Centre at the University of Oxford in partnership with Gallup, the UN Sustainable Development Solutions Network, and the WHR's Editorial Board.

Table of Contents

World Happiness Report 2025

This version was last updated on 13 March 2025. Please see worldhappiness.report for the latest data.

Chapter 1

Executive summary

John F. Helliwell
Vancouver School of Economics,
University of British Columbia

Richard Layard
Centre for Economic Performance,
London School of Economics and Political Science

Jeffrey D. Sachs
Center for Sustainable Development,
Columbia University

Jan-Emmanuel De Neve
Wellbeing Research Centre,
University of Oxford

Lara B. Aknin
Department of Psychology,
Simon Fraser University

Shun Wang
International Business School Suzhou,
Xi'an Jiaotong-Liverpool University

doi.org/10.18724/whr-nkm9-5m31

This version was last updated on 13 March 2025. Please see worldhappiness.report for the latest data.

Like 'mercy' in Shakespeare's *Merchant of Venice*, caring is "twice-blessed" — it blesses those who give and those who receive.

In this year's issue, we focus on the impact of caring and sharing on people's happiness. Like 'mercy' in Shakespeare's *Merchant of Venice*, caring is "twice-blessed" – it blesses those who give and those who receive. In this report, we investigate both of these effects: the benefits to the recipients of caring behaviour and the benefits to those who care for others.

There is a wealth of evidence about the extent of caring behaviour around the world. In the Gallup World Poll, people are asked if, in the last month, they gave money to charity, if they volunteered, and if they helped a stranger. They were also asked, in 2019, if they think other people would help them by returning their lost wallet.

Some key findings jump out of the data.

First, people are much too pessimistic about the benevolence of others. For example, when wallets were dropped in the street by researchers, the proportion of returned wallets was far higher than people expected. This is hugely encouraging.

Second, our wellbeing depends on our *perceptions* of others' benevolence, as well as their actual benevolence. Since we underestimate the kindness of others, our wellbeing can be improved by receiving information about their true benevolence (see Chapter 5).

Third, when society is more benevolent, the people who benefit most are those who are least happy. As a result, happiness is more equally distributed in countries with higher levels of expected benevolence (see Chapter 2).

Finally, benevolence increased during COVID-19 in every region of the world. People needed more help and others responded. This 'benevolence bump' has been sustained since then. Despite a fall from 2023 to 2024, benevolent acts are still about 10% above their pre-pandemic levels (see Chapter 2).

Benevolence also brings benefits to those doing the caring and sharing. This works best if the motivation is to help others (rather than to feel good yourself), if the act is voluntary, and if it has an obvious positive impact on the beneficiary. All this is shown in Chapter 2, where the usual country rankings of happiness are supplemented by rankings for benevolent acts and expected wallet return.

There are many ways in which we care and share with each other. Perhaps the most universal example is sharing meals. As Chapter 3 shows, dining alone is not good for your wellbeing. People who eat frequently with others are a lot happier and this effect holds even taking into account household size. The increasing number of people who eat alone is one reason for declining wellbeing in the United States.

Another important form of caring and sharing is the family. Latin American societies, characterised by larger household sizes and strong family bonds, offer valuable lessons for other societies that seek higher and sustainable wellbeing. In Chapter 4, we see that happiness rises with household size up to four people, but above that happiness declines. Notably, people living alone are much less happy than people who live with others.

Trends towards increased loneliness are most evident among young people. In 2023, 19% of young adults across the world reported having no one they could count on for social support, a 39% increase compared to 2006. However, as we have said, they often underestimate the benevolence of other people. After a powerful intervention, students at Stanford University became much happier when given evidence of the kindness of their peers (see Chapter 5).

The opposite of happiness is despair, which can lead to death by suicide or substance abuse – also known as 'deaths of despair'. Fortunately, deaths of this kind are falling in the majority of countries, though not in the United States or Republic of Korea. As Chapter 6 shows, deaths of despair are significantly lower in countries when more people report donating, volunteering, or helping strangers.

The degree of benevolence in a country also has a profound impact on its politics (see Chapter 7). Populism is largely due to unhappiness. But whether populists are on the left or the right depends on trust. People who trust others veer to the left, those who do not veer to the right.

For many people, how to express their benevolence is a serious question. Where should you donate your money? The logical answer is to generate as much extra happiness (or reduced unhappiness) as possible. This means choosing charities that yield the most happiness per dollar. Chapter 8 explains this method and illustrates it across a range of interventions. Even in low-income countries, mental health treatments emerge as an especially effective way of spending money.

In what follows, we summarise the key insights from each chapter and encourage you to dig deeper into this year's report.

Chapter 2
Caring and sharing: global analysis of happiness and kindness

- People are too pessimistic about the kindness of their communities. The return rate of lost wallets is far more than people expect.
- In 2024, benevolent acts continue to be 10% more frequent than in 2017–19 in all generations and almost all global regions, despite evidence of a return towards pre-COVID levels.
- Benevolent acts and expected kindness both matter for individual happiness levels (Figure 2.4).
- Within-country inequality of happiness has been growing over the past 15 years, while international inequality of happiness has remained roughly constant (Figure 2.5).
- Expected and actual kindness both reduce the inequality of wellbeing (Figure 2.6).
- The wellbeing benefits of benevolent acts depend on why and how people engage in them. Both helpers and recipients experience greater happiness from caring and sharing when they do so in the context of caring connections, choice, and clear positive impact.
- Untied foreign aid is positively related to national happiness in the donor countries. But, on average, countries with high refugee shares are less happy, since refugee flows are more often based on location than invitation.

Chapter 3
Sharing meals with others: how sharing meals supports happiness and social connections

- This chapter presents new Gallup evidence on an understudied measure of social connection – sharing meals. Given the relatively objective way in which it is measured, sharing meals is uniquely comparable across countries and cultures, between individuals, and over time, unlike many other social indicators.
- There are stark differences in rates of meal sharing around the world. While residents of some countries share almost all of their meals with other people, residents of other countries eat almost all of their meals alone. These differences are not fully explained by differences in income, education, or employment.
- Sharing meals has a strong impact on subjective wellbeing – on par with the influence of income and unemployment. Those who share more meals with others report significantly higher levels of life satisfaction and positive affect, and lower levels of negative affect. This is true across ages, genders, countries, cultures, and regions.
- In the United States, using data from the American Time Use Survey, the authors find clear evidence that Americans are spending more and more time dining alone. In 2023, roughly 1 in 4 Americans reported eating all of their meals alone the previous day – an increase of 53% since 2003. Dining alone has become more prevalent for every age group, but especially for young people.
- Meal sharing also appears to be closely related to some, but not all, measures of social connectedness. Most notably, countries where people share relatively more meals tend to display higher levels of social support and positive reciprocity, and lower levels of loneliness.
- Nevertheless, there remain vast gaps in our understanding of the causal dynamics of meal sharing, subjective wellbeing, and social connections. The authors point to a number of promising avenues for future research.

Chapter 4

Living with others: how household size and family bonds relate to happiness

- For most people in the world, family is a source of joy and support. This chapter explores how the size and configuration of households affect people's happiness.
- In Mexico and Europe, a household size of four to five predicts the highest levels of happiness. Couples who live with at least one child, or couples who live with children and members of their extended family, have especially high average life satisfaction.
- People living on their own often experience lower levels of happiness. People in very large households can also experience lower happiness, probably linked to diminished economic satisfaction.
- Latin American societies, characterised by larger household sizes and strong family bonds, offer valuable lessons for other societies that seek to enrich relational satisfaction and improve overall happiness metrics and research approaches.
- Understanding the drivers of family happiness requires surveys that measure their dynamics, interactions, processes, and outcomes. National statistical offices should prioritise the development of metrics that assess the quantity and quality of interpersonal relationships and the bonds that underpin them.
- Public policies should consider how economic decisions may have secondary effects on relationships, hence affecting the wellbeing of families.

Chapter 5
Connecting with others: how social connections improve the happiness of young adults

- Social connections are vital for the wellbeing of young adults as they provide a buffer from the toxic effects of stress.
- However, social disconnection is quite prevalent among young adults. In 2023, 19% of young adults across the world reported having no one that they could count on for social support, representing a 39% increase compared to 2006.
- Early social ties during young adulthood have long-lasting effects. For university students, friendships formed in the first few weeks of college increase the likelihood of flourishing and reduce the likelihood of developing depressive symptoms over the subsequent years.
- Many young adults underestimate their peers' empathy, leading them to avoid connecting with others and missing out on opportunities for meaningful relationships.
- Fortunately, there are interventions that can bridge this 'empathy perception gap' by informing young adults about the empathy of their peers. Undergraduate students who were exposed to these interventions saw others as more empathic and were more likely to make new connections and build larger social networks.

Chapter 6
Supporting others: how prosocial behaviour reduces deaths of despair

- Increasing prosocial behaviour (donating, volunteering, and helping strangers) is connected to decreasing deaths of despair around the world. Regression results indicate that a ten percentage-point increase in the share of people engaging in prosocial behaviour is associated with approximately 1 fewer death per year per 100,000.
- Deaths of despair have declined since 2000 in 75% of 59 countries. The largest declines occurred in northeastern Europe, from very high initial levels, but deaths of despair are still high and rising in a few countries including the United States and Republic of Korea. In 2019, Slovenia had the highest level, with more than 50 deaths per 100,000.
- Deaths of despair are nearly four times higher among men than women, and more than double among those aged 60+ compared with those aged 15–29. Three-quarters are due to suicide, followed by deaths due to alcohol and drug abuse.
- Previous research indicates that prosocial behaviour contributes to individual wellbeing. This chapter further demonstrates that increasing prosocial behaviour is reliably connected to decreasing deaths of despair. Societies could benefit from investing in the conditions supporting prosocial behaviour.

Chapter 7
Trusting others: how unhappiness and social distrust explain populism

- Subjective experiences like life satisfaction and trust play a much greater role in shaping values and voting behaviour than traditional ideologies or class struggle.
- In Europe and the United States, the decline in happiness and social trust explains a large share of the rise in political polarisation and votes against 'the system'.
- The decline in life satisfaction explains the overall rise in anti-system votes but trust in others then comes into play. Among unhappy people attracted by the extremes of the political spectrum, low-trust people are more often found on the far right, whereas high-trust people are more inclined to vote for the far left.

Chapter 8
Giving to others: how to convert your money into greater happiness for others

- The authors estimate how much happiness per dollar is created by specific forms of charitable expenditure. Happiness is measured in well-being-years (WELLBYs).
- They find that the wellbeing cost-effectiveness of charities varies dramatically. The best charities in their sample are hundreds of times better at increasing happiness than others. This implies that donors can multiply their impact, *at no extra cost*, by funding the most cost-effective charities.
- A key gap in the evidence is the lack of well-being evaluations for large, well-known charities. The authors discuss the challenges in evaluating large charities and explain why they have questions about the impact of these organisations.
- To conclude, the authors set out directions for how to improve the new discipline of wellbeing cost-effectiveness analysis.

Acknowledgements

We have had a remarkable range of contributing authors and expert reviewers this year. We are deeply grateful for their willingness to share their knowledge with our readers. Although the editors and authors of the World Happiness Report are volunteers, there are administrative and research support costs that are generously funded by our partners: Glico, Katsuiku Foundation, Blue Zones, Ajinomoto, illycaffè, and Fondazione Ernesto Illy.

We greatly value our special relationship with Gallup who provide us with early access to the World Poll data that underpins so much of the report. We very much appreciate the continued work by Stislow Design and Ryan Swaney for their skills in design and web development. New this year, we are joined by Barry Grimes as our Production Editor and benefit from his passion and commitment to producing the best possible report. We also continue to benefit from the professional support of Jonathan Whitney and Leoni Boyle and thank them for their extraordinary efforts.

This year, thanks to support from the Robert Wood Johnson Foundation (RWJF), we ran our first-ever open call for chapter proposals and convened the authors in Rome for a productive workshop which helped to improve the report and build the scholarly community. RWJF also funded the development of our new data dashboard to bring Gallup's data to life and improve people's understanding of how their country is doing compared to others.

All of these contributions together are what makes the World Happiness Report the go-to source for so many people around the world who seek the latest evidence and analysis on the global state of happiness.

John Helliwell, Richard Layard, Jeffrey D. Sachs, Jan-Emmanuel De Neve, Lara B. Aknin, and Shun Wang.

Chapter 2

Caring and sharing

Global analysis of happiness and kindness

John F. Helliwell
Vancouver School of Economics, University of British Columbia

Lara B. Aknin
Department of Psychology, Simon Fraser University

Haifang Huang
Department of Economics, University of Alberta

Max Norton
Vancouver School of Economics, University of British Columbia

Shun Wang
International Business School Suzhou, Xi'an Jiaotong-Liverpool University

Felix Cheung
Department of Psychology, University of Toronto

Yeeun Archer Lee
Department of Psychology, Trinity Western University

Sara Konrath
Lilly Family School of Philanthropy, Indiana University
Institute for Social Research, University of Michigan

The authors are grateful for assistance and advice from Kenith Chan, Anat Noa Fanti, Jessica Mahoney, Abigail Marsh, Julie Ray, Mariano Rojas, Marwan Saleh, Meik Wiking, and especially Len Goff. We also continue to be grateful for the efficiency and kindness of the Gallup World Poll data team, who deliver important data to us in time for our 20 March launch.

doi.org/10.18724/whr-31zp-ga37

This version was last updated on 13 March 2025. Please see worldhappiness.report for the latest data.

Generous behaviours are associated with — and often directly cause — increased wellbeing among givers across the world.

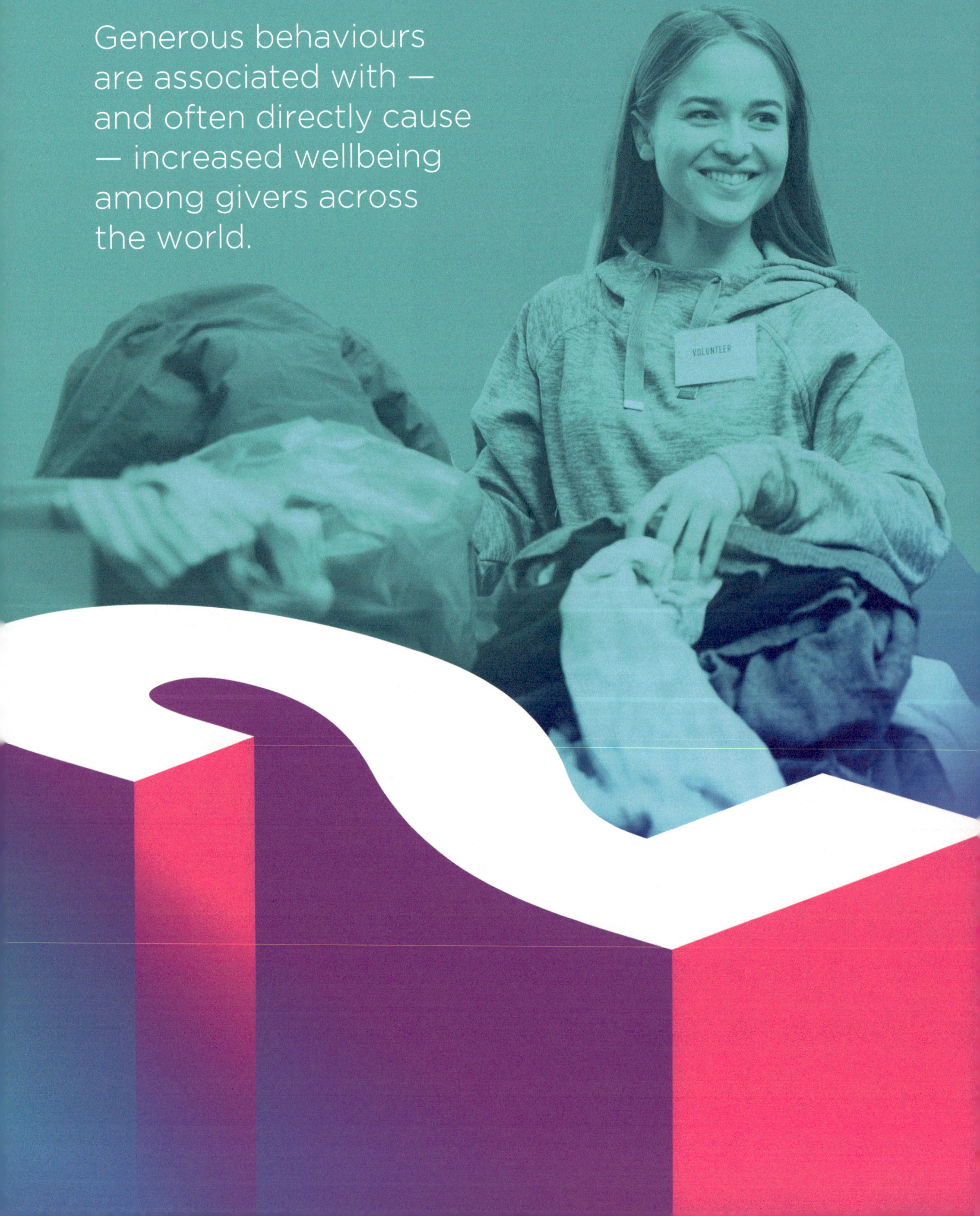

Key Insights

Happiness and benevolence rankings

- Our annual happiness ranking is, once again, led by Nordic countries, with Finland still first among them.
- The Nordic countries also rank among the top places for expected and actual return of lost wallets.
- Country rankings for the three benevolent acts covered by the Gallup World Poll – donating, volunteering, and helping strangers – vary depending on cultural and institutional differences.

Review of previous findings

- Research shows that the wellbeing benefits of benevolent acts depend on why and how people do things for others.
- Both helpers and recipients experience greater happiness from caring and sharing in the context of three Cs: caring connections, choice, and clear positive impact.

Our new results on caring and sharing

- During 2024, the COVID-era surge in benevolent acts fell significantly but remains more than 10% higher than 2017–19 levels almost everywhere.
- In 2024, helping strangers remains significantly higher than in 2017-19 in all global regions, by a global average of 18%.
- Expecting kindness from others is a stronger predictor of happiness than major actual or expected harms (Figure 2.4).
- People are too pessimistic about the kindness of their communities. The return rate of lost wallets is much higher than people expect, especially in the Nordic countries, which have the highest rates of both expected and actual wallet returns.
- Engaging in benevolent acts and expecting kindness from others both matter for individual happiness levels. The effect of expected wallet return is almost twice as large as for the frequency of benevolent acts (Figure 2.4).
- Across countries, expected wallet return significantly predicts the return of wallets dropped in experiments.
- Happiness inequality within countries has increased by about one-quarter over the past two decades, while happiness inequality between countries has remained roughly constant (Figure 2.5).
- Expected fairness and kindness reduce happiness inequality and add to the benefits of trust and social connections (Figures 2.6 and 2.7).
- On average, countries that provide more untied official development assistance (ODA) are happier. Countries with high refugee population shares are less happy, as refugee flows are more often based on location than invitation.

Setting the stage

In *World Happiness Report 2024*, we studied happiness by age group and birth cohort, finding contrasting patterns in different parts of the world. We also continued to show how three types of benevolent acts - donating, volunteering, and helping strangers - changed during the COVID-19 years. We previously found a global surge in benevolent acts during 2020, led by the helping of strangers, which continued through subsequent years. Last year, we found these acts to be prevalent in all generations, especially among Millennials and Gen Z. We suggested that this upsurge of benevolent acts might have led people to feel better about themselves and their neighbours. These positive wellbeing effects appear to have offset the negative effects felt by many of those whose lives were changed, endangered, and sometimes harmed during the pandemic.

We also found that feelings of social support were twice as common as feelings of loneliness and had larger connections to life evaluations.[1] In *World Happiness Report 2021*, we found that people who felt others in their communities were watching their backs and would return a lost wallet were far happier with their lives. Seeing kindness in one's community provides a greater happiness boost than the absence of violent crime, mental illness, or having a much higher income. These findings, along with the relative stability of life evaluations during COVID-19, led us to dig deeper into these questions.

This year, we present several strands of evidence on our theme of caring and sharing. First, we deliver our usual rankings of nations in terms of the average life evaluations of their residents, along with our modelling of how differences across countries and over time are connected to a variety of life circumstances and the prevalence of positive and negative emotions. The main life circumstances we consider continue to be GDP per capita, healthy life expectancy, having someone to count on, having a sense of freedom to make key life decisions, average frequency of donations,[2] and perceptions of corruption in government and business.

Next, we present rankings (using 2022–2024 data) for the three benevolent acts covered in every Gallup World Poll plus national perceptions, from the 2019 Lloyd's Register Foundation World Risk Poll, of the likelihood of the return of a lost wallet if found by: (a) a neighbour, (b) a stranger, or (c) a police officer. As we shall see, each of these three wallet questions captures different aspects of society. The benevolence of neighbours represents the local social context, while expected wallet return by strangers reflects the broader social fabric. The expectation of wallet return by a police officer captures the perceived honesty and benevolence of an important public institution. We later track time trends of the three Gallup World Poll benevolence measures from 2006 through 2024.

We then survey the large literature that considers the conditions under which benevolent acts are likely to be of more benefit to givers and receivers. In doing so, we present evidence on the link between individual benevolent actions, wellbeing, and the social context.

Another central social issue relates to the distribution of wellbeing. We have previously shown that within-country inequality of wellbeing has been increasing over the past twenty years and that this inequality is less in high-trust environments. This year, we show how living in a society believed to be benevolent mitigates the harmful effects of unfortunate circumstances and thereby reduces the inequality of wellbeing.

Finally, we report on the extent of international caring and sharing and its possible linkages to national happiness. We consider both the levels of official development assistance and each country's population share of refugees.

Box 2.1: Measuring subjective wellbeing

Our measurement of subjective wellbeing continues to rely on three main wellbeing indicators: life evaluations, positive emotions, and negative emotions (with the last two often referred to as positive and negative affect). Our happiness rankings are based on life evaluations, as the more stable measure of the quality of people's lives.

Life evaluations

The Gallup World Poll, which remains the principal source of data in this report, asks respondents to evaluate their current life as a whole using the image of a ladder, with the best possible life for them as a 10 and the worst possible as a 0. Each respondent provides a numerical response on this scale, referred to as the Cantril Ladder. Typically, around 1,000 responses are gathered annually for each country. Weights are used to construct population-representative national averages for each year in each country. We base our happiness ranking on a three-year average of these life evaluations since the larger sample size enables more precise estimates.

Positive emotions

Positive affect is given by the average of individual yes or no answers about three positive emotions: laughter, enjoyment, and interest (for details see Box 2.2).

Negative emotions

Negative affect is given by the average of individual yes or no answers about three negative emotions: worry, sadness, and anger.

Comparing life evaluations and emotions

Life evaluations provide the most informative measure for international comparisons because they capture quality of life in a more complete and stable way than emotional reports based on daily experiences.

Life evaluations vary more between countries than emotions and are better explained by the diverse life experiences in different countries. Emotions yesterday are well explained by events of the day being asked about, while life evaluations more closely reflect the circumstances of life as a whole. In Table 2.1, we show that emotions are significant supports for life evaluations.

Positive emotions are still more than twice as frequent as negative emotions, even during the years since the onset of COVID-19.

Happiness ranking

Countries are ranked according to their self-assessed life evaluations averaged over the years 2022–2024.[3] The overall length of each country bar in Figure 2.1 represents the average response to the Cantril Ladder question in the Gallup World Poll. The confidence intervals for each country's average life evaluation are shown by horizontal whiskers at the right-hand end of each country bar. Confidence intervals for each country's rank are displayed in brackets to the right of the rank number.[4] These ranking ranges are wider where there are many countries with similar averages and for countries with smaller sample sizes.[5]

Figure 2.1 includes colour-coded sub-bars in each country row representing the extent to which six key variables contribute to explaining life evaluations. These variables (described in more detail in Box 2.2) are log GDP per capita, social support, healthy life expectancy, freedom, generosity, and corruption. As already noted, our happiness rankings are *not* based on any index of these six factors. Rather, rankings are based on individuals' assessments of their own lives, in particular their answers to the single-item Cantril Ladder life evaluation question. We use observed data on the six variables and estimates of their associations with life evaluations to help explain the variation of life evaluations across countries, much as epidemiologists estimate the extent to which life expectancy is affected by factors such as smoking, exercise, and diet.

Each of the country bars in Figure 2.1 is divided into seven segments, showing our research efforts to find possible sources for the ladder levels. The first six sub-bars show how much each of the six key variables is calculated to contribute to that country's ladder score, relative to that in a hypothetical country called 'Dystopia', so named because it has values equal to the world's lowest national averages for 2022–2024 for each of the six key variables used in Table 2.1. We use Dystopia as a benchmark against which to compare contributions from each of the six factors. The choice of Dystopia as a benchmark permits every real country to have a positive (or at least zero) contribution from each of the six factors. We calculate, based on the estimates in the first column of Table 2.1, that Dystopia had a 2022–2024 ladder score equal to 1.37 on the 0–10 scale. The final sub-bar is the sum of two components: the calculated average 2022–2024 life evaluation in Dystopia (= 1.37) and each country's own prediction error, which measures the extent to which life evaluations are higher or lower than predicted by our equation in the first column of Table 2.1. These residuals are as likely to be negative as positive.

Consistency and change in happiness rankings

Two features carry over from previous editions of the World Happiness Report. First, there is still a lot of year-to-year consistency in the way people rate their lives in different countries. Since our rankings are based on a three-year average, there is information carried forward from one year to the next. The effects of cataclysmic events depend on when the survey took place and are muted by the three-year averaging.

Second, there remains a large gap between the top and bottom countries – more than six points (on a 0–10 scale) between Finland at the top and Afghanistan at the bottom. The top countries are more tightly grouped than the bottom ones. The top twenty have a spread of less than one point on the 0–10 scale, with the corresponding spread among the bottom twenty being three times as great. The remaining 100-odd countries cover the remaining 2.3 points of the total range. This means that relatively modest changes in a national average can lead to a large shift in rank, as illustrated by 95% confidence regions of more than 25 ranks for several countries in the middle of the global list.

Happiness scores are based on the resident populations in each country, rather than their citizenship or place of birth. In *World Happiness Report 2018*, we split the responses between the locally- and foreign-born populations in each country and found the happiness rankings to be essentially the same for the two groups.[6] There was some source-country effect after migration and some tendency for migrants to move to happier countries so that, among the 20 happiest countries in that report, the average happiness for the locally-born was about 0.2 points higher than the foreign-born.

Nordic countries once again lead the happiness rankings. Finland, Denmark, Iceland and Sweden are still the top four and in the same order. The confidence intervals for the rankings show Finland still in a group by itself, with Denmark and Iceland following in a group of two, and Sweden in a range that runs from 4 to 8.

If we compare this year's top-ranking countries with those in the 2013 report – the first to assign rankings based on three-year averages – we find 14 western industrial countries in the top 20 in both years. In 2013, these countries were accompanied by four from Latin America and one from the Middle East. Reflecting the long-term convergence between Eastern and Western Europe, three of the top 20 countries in 2025 are from Central and Eastern Europe (Lithuania at 16, Slovenia at 19, and Czechia at 20). This year's top 20 also includes two countries from Latin America (Costa Rica at 6 and Mexico at 10) and one from the Middle East (Israel at 8). In 2013, the top ten countries were all western industrial countries but now only seven are. As a group,

Figure 2.1: Country rankings by life evaluations (part 1)

Gallup World Poll (2022–2024)

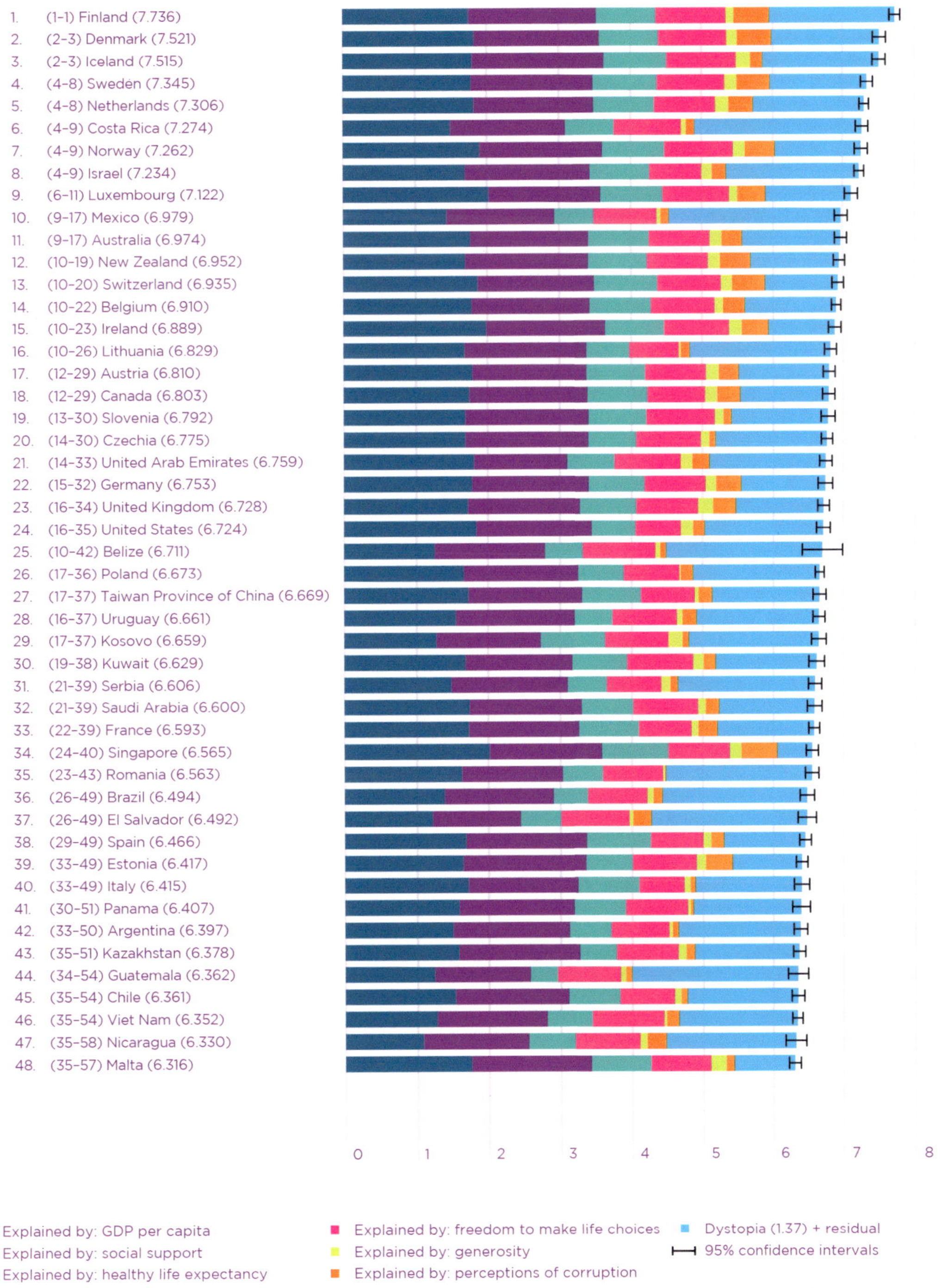

Figure 2.1: Country rankings by life evaluations (part 2)

Gallup World Poll (2022-2024)

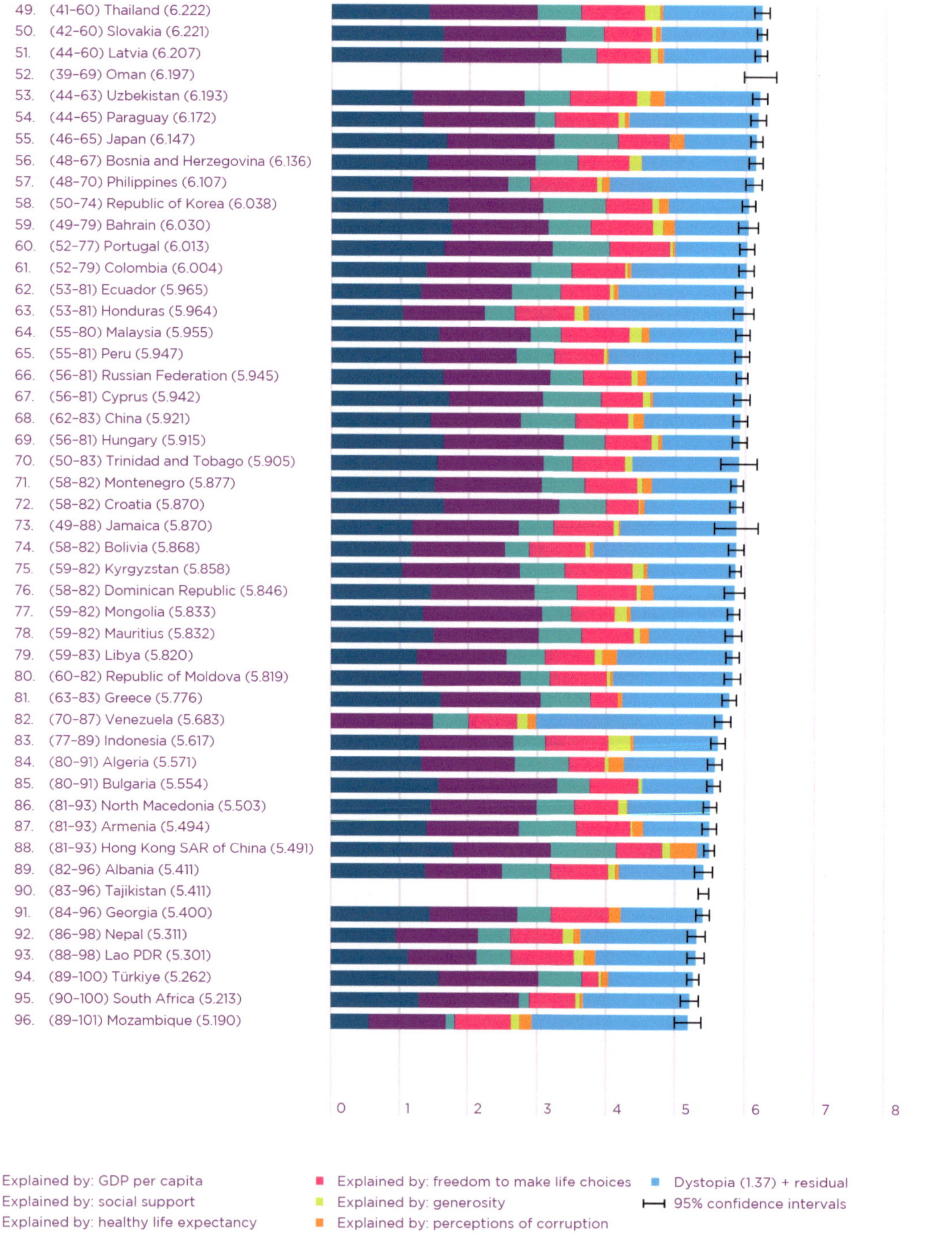

Note: Countries with empty columns have life evaluation scores but are missing one or more of the supporting factors for decomposition analysis

Figure 2.1: Country rankings by life evaluations (part 3)

Gallup World Poll (2022–2024)

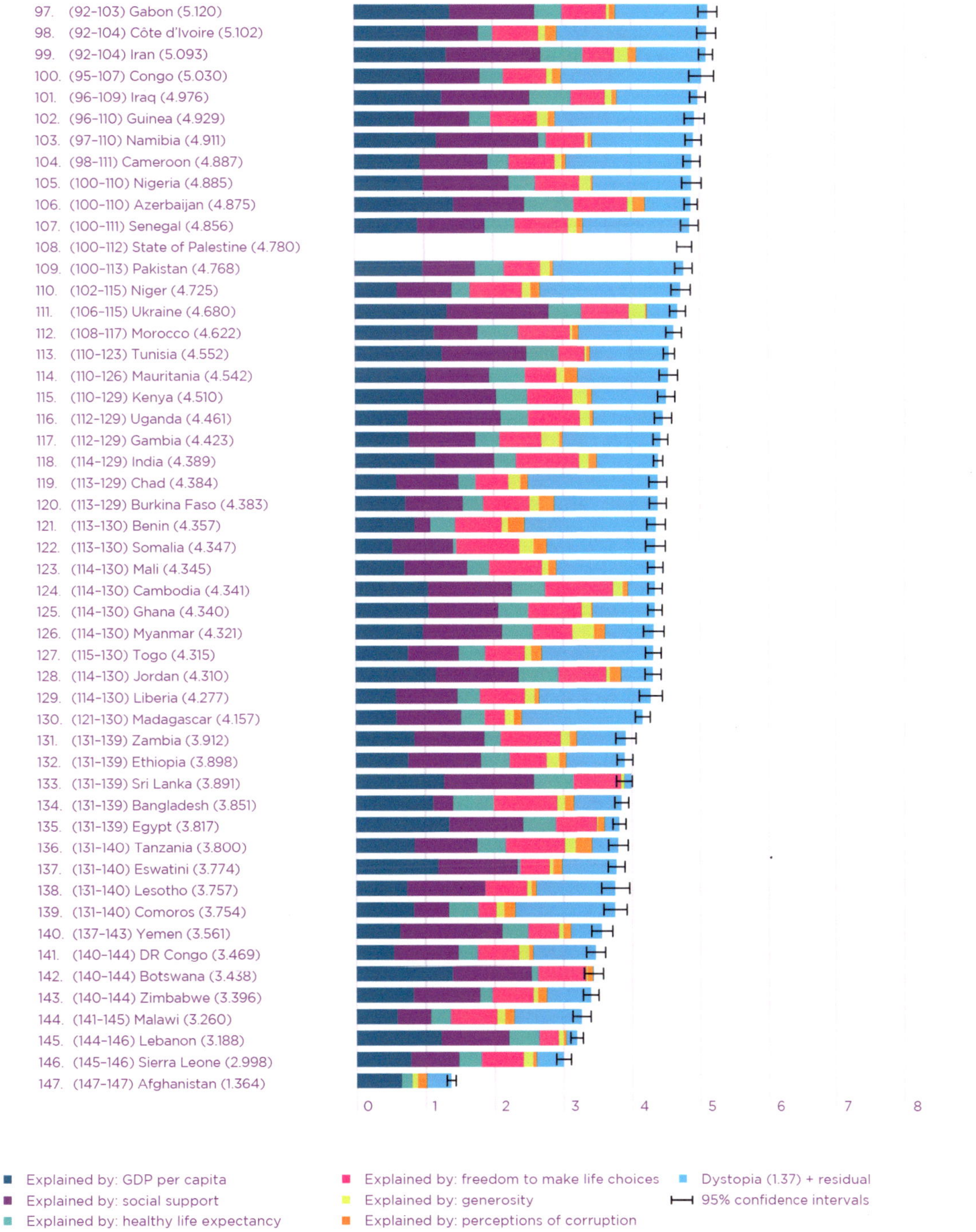

Note: Countries with empty columns have life evaluation scores but are missing one or more of the supporting factors for decomposition analysis

the five Nordic countries have improved their positions in the top ten, with an average rank of 4.8 in 2013 rising to 3.4 in 2025. This increase is driven mainly by Finland (from 7 to 1) and Iceland (from 9 to 3). The industrial countries pushed out of the top ten between 2013 and 2025 include Switzerland (3 in 2013, 1 in 2015, and 13 in 2025), Canada (6 in 2013 and 18 in 2025), and Australia (10 in 2013 and 11 in 2025). This year, for the first time, none of the large industrial powers ranked in the top 20.

For the least happy countries, ranks are not so easily compared since there were 156 countries ranked in 2013 compared to 147 this year. Togo was the least happy country in 2013 and has since risen twenty places, with an average life evaluation almost 1.4 points higher now than then. Afghanistan has gone in the reverse direction with a drop of almost 2.7 points between 2013 and 2025. The average life evaluation is now 1.36, by far the lowest average score ever seen in all our reports. Furthermore, life is especially difficult for Afghan women, as their average is only 1.16 points.[7]

In the middle and lower sections of the rankings, it is more meaningful to look at average life evaluations, because a country's rank can change many places with only a small change in average life evaluation. That is why, when we consider changes in happiness, we consider how current average life evaluations compare with those during the first years of the Gallup World Poll (2005–2010). Tables 24 to 26 in the online statistical appendix measure the change in average life evaluations from the 2005–2010 base period to the current ranking period, 2022–2024. The top five gainers are all in Central and Eastern Europe: Serbia, Bulgaria, Georgia, Latvia, and Romania. Of the 19 countries that have gained a point or more on the 0–10 scale, 12 are Central and Eastern Europe, reflecting the European happiness convergence that has been clear for more than a decade. Other big gainers include Togo and Congo in Africa; China, Mongolia, the Philippines and Viet Nam in Asia; and Nicaragua in Latin America.

There are, fortunately, fewer countries whose life evaluations have fallen by more than one point on the 0–10 scale. Going from the largest to the smallest drops in life evaluations, these seven countries are Afghanistan, Lebanon, Jordan, Malawi, Venezuela, Egypt and Botswana. These are mainly countries in or near zones of major conflict.

In general, the western industrial countries are now less happy than they were between 2005 and 2010.

In general, the western industrial countries are now less happy than they were between 2005 and 2010. Fifteen of them have had significant drops, compared to four with significant increases.[8] Three western countries had drops exceeding 0.5 on the 0–10 scale (the United States, Switzerland, and Canada) putting them among the fifteen largest losers.

Among the 136 countries included in the 2005–2010 and 2022–2024 data, there are 67 with statistically significant gains[9] and 42 with significant drops in their life evaluations.[10] Those with significant drops include western industrialised countries with previously, and even currently, high rankings.

The rankings for positive emotions are shown in Tables 66–68 of the online statistical appendix. The top ten include six from Latin America, three from Southeast Asia, and one from Africa. The lowest frequency of positive emotions is in Afghanistan. It also has the most frequent negative emotions. In the top ten for negative emotions it is joined by three Middle Eastern countries, five African countries, and Armenia (see Tables 69–71 of the online statistical appendix).

Why do happiness levels differ?

In Table 2.1, we present our latest modelling of national average life evaluations and measures of positive and negative emotions (often referred to as positive and negative affect) by country and year.[11] The results in the first column explain national average life evaluations in terms of six key variables: log GDP per capita, healthy life expectancy, having someone to count on, freedom to make life choices, generosity, and freedom from corruption.[12] Taken together, these six variables explain more than three-quarters of the variation in life evaluations across countries and years, using data from 2005 through 2024.[13]

The six variables were originally chosen as the best available measures of factors established in both experimental and survey data as having significant links to subjective wellbeing, and especially to life evaluations.[14] The explanatory power of the unchanged model has gradually increased as we have added more years to the sample, which is now almost three times as large as when the equation was first introduced in *World Happiness Report 2013*. We keep looking for possible improvements when and if new evidence becomes available.[15] The number of years of data is now great enough that we can experiment with including country fixed effects, as shown in Table 10 of the online statistical appendix. The results are remarkably similar.[16]

The second and third columns of Table 2.1 use the same six variables to estimate equations for national averages of positive and negative affect, where both are based on answers about yesterday's emotional experiences (Box 2.2 explains how these measures are constructed). In general, emotional measures, especially negative ones, are differently and much less fully explained by the six variables than are life evaluations. GDP per capita and healthy life expectancy have significant effects on life evaluations,[17] but not, in these national average data, on positive emotions.[18] However, the social variables do have significant effects on both positive and negative emotions. Bearing in mind that positive and negative emotions are measured on a 0–1 scale, while life evaluations are on a 0–10 scale, having someone to count on can be seen to have similar proportionate effects on positive and negative emotions as it does on life evaluations. Freedom and generosity have even larger associations with positive emotions than with the Cantril Ladder. Negative emotions are significantly reduced by social support, a sense of freedom, and the absence of corruption.

In the fourth column, we re-estimate the life evaluation equation from column 1, adding both positive and negative emotions to partially implement the Aristotelian presumption that sustained positive emotions are important supports for a good life.[19] The results continue to buttress a finding in psychology that the existence of positive emotions matters more than the absence of negative ones when predicting either longevity[20] or resistance to the common cold.[21] Consistent with this evidence, we find that positive affect has a large and highly significant impact in the final equation of Table 2.1, while negative affect has none.

As for the coefficients on the other variables in the fourth column, the changes are substantial only on those variables – especially freedom and generosity – that have the largest impacts on positive affect. Thus, we can infer that positive emotions play a strong role in supporting life evaluations and that much of the impact of freedom and generosity on life evaluations is channelled through their influence on positive emotions. That is, freedom and generosity have large impacts on positive affect, which in turn has a major impact on life evaluations.

Table 2.1: Regressions to explain average happiness across countries

	Dependent Variable			
Independent Variable	**Cantril Ladder**	**Positive Affect**	**Negative Affect**	**Cantril Ladder**
Log GDP per capita	0.328	-0.017	-0.0008	0.366
	(0.062)***	(0.01)*	(0.007)	(0.061)***
Social support	2.686	0.325	-0.343	2.021
	(0.336)***	(0.055)***	(0.043)***	(0.348)***
Healthy life expectancy at birth	0.032	-0.0007	0.003	0.033
	(0.01)***	(0.001)	(0.001)***	(0.01)***
Freedom to make life choices	1.518	0.381	-0.089	0.694
	(0.295)***	(0.045)***	(0.039)**	(0.276)**
Generosity	0.382	0.081	0.026	0.198
	(0.243)	(0.032)**	(0.027)	(0.232)
Perceptions of corruption	-0.669	-0.016	0.095	-0.645
	(0.249)***	(0.027)	(0.021)***	(0.235)***
Positive affect				2.212
				(0.326)***
Negative affect				0.147
				(0.379)
Year fixed effects	Included	Included	Included	Included
Number of countries	155	155	155	155
Number of obs.	2234	2229	2233	2228
Adjusted R-squared	0.761	0.437	0.35	0.785

Note: This is a pooled OLS regression for a tattered panel explaining annual national average Cantril ladder responses from all available surveys from 2005 through 2024. See Box 2.2 for detailed information about each of the predictors. Coefficients are reported with robust standard errors clustered by country (in parentheses). ***, **, and * indicate significance at the 1, 5, and 10 percent levels respectively.

Box 2.2: Explaining the variables in Table 2.1

GDP per capita is in terms of Purchasing Power Parity (PPP) adjusted to constant 2021 international dollars, taken from the World Development Indicators (WDI) by the World Bank (version 34, last updated on 28 October 2024). See the online statistical appendix for more details. GDP data for 2024 are not yet available, so we extend the GDP time series from 2023 to 2024 using country-specific forecasts of real GDP growth from the OECD Economic Outlook No. 116 (Edition 2024/2) or, if missing, from the World Bank's Global Economic Prospects (last updated: 11 June 2024), after adjustment for population growth. The equation uses the natural log of GDP per capita as this form fits the data significantly better than GDP per capita.

The time series for **healthy life expectancy** at birth are constructed based on data from the World Health Organization (WHO) Global Health Observatory data repository, with data available up to 2021 (last updated: 2 August 2024). To match this report's sample period (2005–2024), interpolation (when necessary) and extrapolation are used. See the online statistical appendix for more details.

Social support is the national average of the binary responses (0=no, 1=yes) to the Gallup World Poll (GWP) question "If you were in trouble, do you have relatives or friends you can count on to help you whenever you need them, or not?"

Freedom to make life choices is the national average of binary responses to the GWP question "Are you satisfied or dissatisfied with your freedom to choose what you do with your life?"

Generosity is the residual from regressing the national average of GWP responses to the donation question "Have you donated money to a charity in the past month?" on log GDP per capita.

Perceptions of corruption is the average of binary answers to two GWP questions: "Is corruption widespread throughout the government or not?" and "Is corruption widespread within businesses or not?" Where data for government corruption are missing, the perception of business corruption is used as the overall corruption-perception measure.

Positive affect is defined as the average of previous-day affect measures for laughter, enjoyment, and doing interesting things. The inclusion of doing interesting things (first added for World Happiness Report 2022) gives us three components in each of positive and negative affect, and slightly improves the equation fit in column 4. The general form for the affect questions is: "Did you experience the following feelings during a lot of the day yesterday?" See the online statistical appendix for more details.

Negative affect is defined as the average of previous-day affect measures for worry, sadness, and anger.

The variables we use in our Table 2.1 modelling may be taking credit properly due to other variables or unmeasured factors. There are also likely to be vicious or virtuous circles, with two-way linkages among the variables. For example, there is much evidence that those who have happier lives are likely to live longer,[22] be more trusting and cooperative, and generally better able to meet life's demands.[23] This will double back to improve health, income, generosity, corruption, and a sense of freedom. Collectively, these possibilities suggest that we should interpret the observed relationships with some caution.

Another possible reason for a cautious interpretation of our results is that some of the data come from the same respondents as the life evaluations and are thus possibly determined by common factors. This is less likely when comparing national averages because individual differences in personality and individual life circumstances tend to average out at the national level. To provide even more assurance that our results are not significantly biased because we are using the same respondents to report life evaluations, social support, freedom, generosity, and corruption, we tested the robustness of our procedure by splitting each country's respondents randomly into two groups.[24] We then examined whether the average values of social support, freedom, generosity, and the absence of corruption from one half of the sample explained average life evaluations in the other half of the sample. The coefficients on each of the four variables fell slightly, just as we expected.[25] But the changes were reassuringly small (ranging from 1% to 5%) and were not statistically significant.[26]

Overall, the model explains average life evaluation levels quite well within regions, among regions, and for the world as a whole.[27] On average, the countries of Latin America still have mean life evaluations that are significantly higher than predicted by the model (by about 0.5 on the 0–10 scale). This difference has been attributed to a variety of factors including some unique features of family and social life in Latin American countries.[28] In partial contrast, countries in East Asia have average life evaluations below predictions, although only slightly and insignificantly so in our latest results.[29] This may reflect, at least in part, cultural differences in the way people think about and report on the quality of their lives.[30] It is reassuring that our findings about the relative importance of the six factors are generally unaffected by whether or not we make explicit allowance for these regional differences.[31]

The relative frequency of the other benevolent acts depends on local social and religious norms, as well as the role of private benevolence as a substitute or supplement for institutional social safety nets.

Benevolence ranking

In Table 2.2, we report country rankings for six measures of benevolence. The first three are national average frequencies of people who report engaging once or more in three benevolent acts during the past month – donating, volunteering, and helping a stranger. The donation answers are used, after adjusting for differences in national income, in our Table 2.1 equation in the previous section.[32]

The remaining three indicators are quite different from the first three. Instead of reporting the respondent's *own* benevolent acts, they capture respondents' forecasts of how *others* would behave when facing an opportunity to show benevolence. Specifically, the so-called "wallet questions" ask respondents to say how likely it is that their lost wallet or other valuable object would be returned if found by: (a) a neighbour, (b) a stranger, or (c) a police officer.[33]

The rankings differ among the benevolent acts, and especially between benevolent acts and expected rates of wallet return. The Nordic countries are at the top in the rankings for expected return of wallets, as shown in Table 2.2, and are also much higher than other countries for actual wallet return, an important benevolent act. The relative frequency of the other benevolent acts depends on local social and religious norms, as well as the role of private benevolence as a substitute or supplement for institutional social safety nets. Among countries with incomplete social safety nets, more people fall between the cracks, thereby increasing the need for private benevolence to fill the gap.

Consider, for example, the case of Finland, which has universally available and high-quality health, education, and social support systems. Inequality of wellbeing is low in Finland and our data suggest a correspondingly lower need for private charity. A preference for universal over targeted social

Table 2.2: Country rankings for six measures of benevolence (part 1)

Gallup World Poll (2022–2024), World Risk Poll (2019)

Country	Rankings by:						
					Wallet returned by:		
	Cantril Ladder	Donated	Volunteered	Helped a stranger	Neighbour	Stranger	Police
Finland	1	39	75	96	3	5	2
Denmark	2	25	64	76	8	6	18
Iceland	3	5	77	125			
Sweden	4	15	87	90	4	32	11
Netherlands	5	9	42	134	1	4	6
Costa Rica	6	92	84	36	98	128	104
Norway	7	11	43	101	2	1	1
Israel	8	32	61	84	42	69	55
Luxembourg	9	31	41	135	24	43	8
Mexico	10	102	89	61	126	120	136
Australia	11	20	34	39	7	14	12
New Zealand	12	22	22	65	10	8	5
Switzerland	13	21	56	137	11	19	7
Belgium	14	35	68	118	33	64	35
Ireland	15	7	36	88	9	21	16
Lithuania	16	110	125	131	134	132	56
Austria	17	10	59	111	5	7	4
Canada	18	17	25	47	12	18	14
Slovenia	19	37	47	127	15	31	24
Czechia	20	38	73	66			
United Arab Emirates	21	16	19	67	51	12	13
Germany	22	26	67	82	6	15	3
United Kingdom	23	4	54	108	14	27	15
United States	24	12	15	12	17	52	25
Belize	25	100	37	23			
Poland	26	125	143	146	75	90	43
Taiwan Province of China	27	42	112	86	43	22	27
Uruguay	28	75	95	34	60	74	91
Kosovo	29	23	139	63	72	45	46
Kuwait	30	33	46	27	19	11	
Serbia	31	41	145	140	74	54	53
Saudi Arabia	32	48	92	41	16	25	20
France	33	52	38	143	21	62	22
Singapore	34	14	32	35	25	56	9

Table 2.2: Country rankings for six measures of benevolence (part 2)
Gallup World Poll (2022–2024), World Risk Poll (2019)

Country	Rankings by:						
					Wallet returned by:		
	Cantril Ladder	Donated	Volunteered	Helped a stranger	Neighbour	Stranger	Police
Romania	35	119	140	110	86	87	49
Brazil	36	78	85	58	59	75	84
El Salvador	37	128	70	85	113	129	117
Spain	38	44	97	83	13	41	10
Estonia	39	36	82	102	31	65	31
Italy	40	65	106	121	64	100	32
Panama	41	104	62	75	125	115	106
Argentina	42	105	94	52	88	72	128
Kazakhstan	43	51	122	138	30	42	65
Guatemala	44	98	20	71	118	119	127
Chile	45	68	105	40	85	99	118
Viet Nam	46	136	124	122	78	101	52
Nicaragua	47	89	50	69	136	136	130
Malta	48	6	55	105	54	70	48
Thailand	49	8	81	87	121	138	83
Slovakia	50	88	103	136	57	91	57
Latvia	51	54	126	117	50	88	61
Oman	52	34	83	20			
Uzbekistan	53	29	107	100	18	49	19
Paraguay	54	76	30	48	120	106	131
Japan	55	131	104	147	36	26	34
Bosnia and Herzegovina	56	28	128	113	53	48	58
Philippines	57	106	6	51	95	20	64
Republic of Korea	58	53	93	109	58	17	23
Bahrain	59	27	63	31	23	9	17
Portugal	60	101	120	116	47	84	40
Colombia	61	130	100	80	119	113	121
Ecuador	62	118	90	92	133	134	122
Honduras	63	70	31	43	107	116	125
Malaysia	64	19	28	81	80	122	66
Peru	65	124	88	79	132	121	124
Russian Federation	66	60	91	64	34	37	86
Cyprus	67	56	76	70	26	39	33

Table 2.2: Country rankings for six measures of benevolence (part 3)
Gallup World Poll (2022–2024), World Risk Poll (2019)

Country	Rankings by:						
					Wallet returned by:		
	Cantril Ladder	Donated	Volunteered	Helped a stranger	Neighbour	Stranger	Police
China	68	85	74	123	49	61	21
Hungary	69	58	119	21	32	112	28
Trinidad and Tobago	70	50	40	3			
Montenegro	71	84	136	142	66	47	68
Croatia	72	126	134	144	81	85	45
Jamaica	73	108	9	1	77	81	103
Bolivia	74	117	66	72	137	126	133
Kyrgyzstan	75	40	78	55	40	92	98
Dominican Republic	76	96	21	25	108	127	134
Mongolia	77	30	18	132	99	102	60
Mauritius	78	69	29	119	89	68	85
Libya	79	73	80	30	20	10	72
Republic of Moldova	80	116	127	73	91	94	88
Greece	81	138	114	104	90	108	38
Venezuela	82	111	23	8	111	131	137
Indonesia	83	1	1	59	73	117	54
Algeria	84	114	121	99	56	3	62
Bulgaria	85	94	146	129	63	66	47
North Macedonia	86	46	138	130	122	98	120
Armenia	87	134	131	89	70	29	80
Hong Kong SAR of China	88	43	102	114	76	79	94
Albania	89	72	135	103	128	124	114
Tajikistan	90	122	4	60	48	96	41
Georgia	91	146	86	37	45	60	51
Nepal	92	59	26	120	112	107	92
Lao PDR	93	55	108	141	135	137	100
Türkiye	94	112	132	115	55	77	29
South Africa	95	113	60	54	96	105	95
Mozambique	96	103	24	93	79	55	82
Gabon	97	127	115	22	131	118	119
Côte d'Ivoire	98	95	96	50	93	59	59
Iran	99	18	117	29	28	2	30

Table 2.2: Country rankings for six measures of benevolence (part 4)
Gallup World Poll (2022–2024), World Risk Poll (2019)

Country	Rankings by:						
					Wallet returned by:		
	Cantril Ladder	Donated	Volunteered	Helped a stranger	Neighbour	Stranger	Police
Congo	100	115	58	24	117	58	116
Iraq	101	77	110	17	29	30	81
Guinea	102	61	12	16	67	73	115
Namibia	103	133	57	53	97	78	77
Cameroon	104	99	65	33	103	83	97
Nigeria	105	45	5	7	71	33	126
Azerbaijan	106	90	123	62	37	28	67
Senegal	107	79	35	6	27	35	26
State of Palestine	108	137	130	95	44	16	50
Pakistan	109	71	109	133	83	36	
Niger	110	109	69	13	35	51	79
Ukraine	111	3	45	9	62	40	109
Morocco	112	144	129	32	69	95	78
Tunisia	113	139	118	77	82	76	99
Mauritania	114	86	44	42	61	67	87
Kenya	115	24	3	4	84	93	96
Uganda	116	74	39	19	101	111	89
Gambia	117	13	16	14	46	50	63
India	118	57	10	74	115	86	93
Chad	119	66	17	28	102	109	129
Burkina Faso	120	81	49	56	87	71	36
Benin	121	121	99	124	124	89	73
Somalia	122	47	71	49			
Mali	123	132	53	98	94	24	110
Cambodia	124	64	144	145	138	139	107
Ghana	125	63	14	44	65	34	75
Myanmar	126	2	48	106	105	125	102
Togo	127	123	79	107	114	82	69
Jordan	128	135	142	57	41	46	37
Liberia	129	91	2	2	104	53	111
Madagascar	130	97	8	94	139	135	123
Zambia	131	83	52	10	109	103	101
Ethiopia	132	49	51	91	110	130	108
Sri Lanka	133	62	7	45	68	13	44
Bangladesh	134	80	111	11	123	97	105

Table 2.2: Country rankings for six measures of benevolence (part 5)
Gallup World Poll (2022–2024), World Risk Poll (2019)

Country	Rankings by:						
					Wallet returned by:		
	Cantril Ladder	Donated	Volunteered	Helped a stranger	Neighbour	Stranger	Police
Egypt	135	143	147	38	38	44	76
Tanzania	136	67	116	112	116	63	74
Eswatini	137	129	72	26	129	123	71
Lesotho	138	145	101	68	52	57	39
Comoros	139	93	11	78			
Yemen	140	147	141	128	22	23	112
DR Congo	141	87	27	46			
Botswana	142	141	113	15	92	104	42
Zimbabwe	143	140	98	97	106	110	113
Malawi	144	120	33	18	100	38	90
Lebanon	145	107	137	139	39	80	70
Sierra Leone	146	82	13	5	127	114	132
Afghanistan	147	142	133	126	130	133	135

assistance may also help to explain their relatively low rankings for the three benevolent acts other than the return of lost wallet return. In the Finnish case, the contrast between the two sets of rankings is especially marked between the frequency of helping strangers and that of actual and expected return of lost wallets. All international wallet-dropping experiments have shown Finland and the other Nordic countries to be among the best places to lose your wallet. Given that the return of a lost wallet is a very powerful way of helping strangers, the low ranking for helping strangers may reflect fewer strangers around who need help. A lost wallet exposes an immediate need, and that call is indeed answered readily in the Nordic countries.[34]

There are a few countries where the ranking for helping strangers is very high, while the ranking for donating to charity is very low. Jamaica, Liberia, and Sierra Leone are in the top ranks for helping strangers but have donation rankings that are 80 or more places lower. Nigeria and Kenya, also in the top ten for helping strangers, provide a less extreme contrast, but still rank much higher for helping strangers than for donating. All five countries are near the bottom of the ranking for expected wallet return by police. People generally want to help others and are likely to choose the best means available.[35] Where institutional structures are weak, helping strangers in need probably represents a far more effective channel than donations to charities. In these countries, charities are fewer and less likely to have the credible and efficient structures that characterise effective benevolence, as illustrated in our review of published research below.

Comparing actual and expected benevolence

The wallet questions are of special interest as they are open to experimental testing in two ways. First, they can be used to see whether international differences in expected wallet return (by a stranger) match international differences in the actual return of lost wallets. They do. The first recorded international wallet drop experiment was by Reader's Digest and they repeated it recently in one city in each of 16 countries.[36] Expected wallet return predicts actual return better in 1996 (r=0.71), than in 2018 (r=0.19), with some different cities involved. A much larger experiment covering 40 countries, with over 17,000 wallets handed in at societal institutions rather than randomly dropped,[37] also shows a high correlation (r=0.62) with expected wallet returns by strangers. These results show that cross-country differences in the expected benevolence of others have the power to predict actual benevolent behaviours by others.

The second valuable use of the wallet questions is to see whether people are too optimistic or too pessimistic about the benevolence of others. This was first done by comparing experimentally dropped wallets in Toronto with large samples of answers from Toronto respondents to the Canadian General Social Survey. The expected rate of return was 23% and the actual return was over 80%.[38] In the same vein, the 40-country study showed actual return to be much higher than expected (1.8 times). Finally, two-thirds of 200 wallets dropped in 20 North American cities were returned, far higher than the author expected,[39] and double that expected by US respondents to the wallet stranger question in our 2019 data.[40] Collectively, these data on expected and actual wallet return show that people are far too pessimistic about the benevolence of others.[41] We show later that expected benevolence is a substantial predictor of life satisfaction, meaning that people may be made needlessly unhappy by their unwarranted pessimism.

These data on expected and actual wallet return show that people are far too pessimistic about the benevolence of others.

The discussion above relates to wallets found by strangers since that is the answer most open to testing by experiments. How might we expect the three different wallet measures to relate to one another and to other measures? The expected return by a neighbour is an indicator of the local social context. Thus, individual answers to "having someone to count on" are more highly correlated with the neighbour wallet answers than with the other wallet answers. On the other hand, the police answers are positively related to variables reflecting how highly people rate their public institutions, especially the honesty and efficiency of the judicial system. The answers for return by strangers are positively correlated with the other wallet answers, and, as noted above, to the actual return of dropped wallets. Across nations, actual wallet returns were found to be highly correlated (r=0.65) with survey measures of social trust.[42]

The police answers are most closely related to how people rate the quality of their public institutions. These institutions are often national in scope and differ greatly among countries. Thus, it might be expected that international differences in police answers would explain a larger share of the variations among individuals. This is indeed the case, as the international share of the variance of the police answers is much higher than for the other two.[43]

Benevolence trends

In this section, we consider benevolence across the globe and over time. We offer a brief summary of the relevant research before presenting new data.

A review of previous research

Recent evidence shows that empathy – imagining others' perspectives and feeling compassion for others – has fluctuated over time. For instance, these two empathy types declined in American young adults from 1979–2009.[44] Yet, after the global financial crisis, empathy increased in both

American and Chinese youth, and benevolent values increased in European youth.[45]

Other research has assessed actual helping behaviours (e.g., picking up dropped papers) in 24 US cities over time. Mirroring the empathy decline discussed above, between 1994 (53%) and 2005 (49%), Americans were less likely to help strangers over time.[46] Larger, denser, and lower-income cities had lower rates of helping behaviours. Yet, this trend was not observed consistently everywhere. A behavioural study found a 10 percent decline in helping behaviour (mailing lost letters) between 2001 and 2011 in the United States, but not in Canada.[47] Yet, long-term behavioural studies have found increased cooperation among Americans between 1956 to 2017.[48]

As for self-reported benevolent behaviours, there were increases in helping strangers between 2010/2011 and 2016/2017 globally, but not giving time or money to charities.[49] Yet, after this, *World Happiness Report 2024* found increases in helping strangers, volunteering, and donating, from 2017–2019 (pre-COVID) to 2020–2023 (during COVID).[50] Here, we go beyond previous research by examining a longer time frame.

There have been few studies examining cross-cultural differences in benevolent traits and acts, with sometimes differing results.[51] For example, one study examined cross-cultural variations in empathy across 63 countries, finding higher empathy scores in more collectivistic countries.[52] However, researchers examining self-reported benevolent behaviours in 66 countries found higher rates of charitable giving in more individualistic countries.[53]

The 40-country wallet-dropping study discussed above found that wallets were returned 40% of the time when they had no money, and over half (51%) the time when they contained money.[54] That return rates were higher for the wallets including money provides a likely example of the 'clear positive impact' effect we describe below. Returns were more likely in countries with more universalist values (which are related to individualism). These results are supported by another study with 21 countries, which found that people were less likely to help strangers in cultures with more collectivist values.[55]

Taken together, empathy is higher in collectivist contexts, but helping strangers (both self-reported and actual) appears to be lower in these contexts. This apparent discrepancy may, at least in part, stem from who respondents have in mind as a target when answering questions about empathy as well as how people compare themselves to their peers.[56] In particular, it is possible that empathy measures capture feelings toward people you know well, which is very different from more impersonal forms of helping (like helping strangers and returning lost wallets). Indeed, new research suggests that providing assistance to family and strangers predicts wellbeing through feelings of autonomy while supporting friends increases feelings of closeness.[57]

How is global benevolence changing?

In Figure 2.2, we show the global trends in three types of benevolent acts: donating, volunteering, and helping strangers. We also include a variable called 'prosocial', which is equal to 1 for any respondent who has done any of the three benevolent acts during the past month. We have noted the post-COVID increases in benevolent acts in each of our past three reports, being struck by the longevity of the increases appearing first in 2020. In the 2024 data, we see a significant decline in the frequency of benevolent acts since 2023.[58] However, the size and persistence of the post-COVID increases in benevolent acts are such that even in 2024, four years after the onset of COVID, all three benevolent acts were still more than 10% above their pre-pandemic levels in 2017–2019.[59]

Even in 2024, four years after the onset of COVID, all three benevolent acts were still more than 10% above their pre-pandemic levels in 2017–2019.

Figure 2.2: Global trends in benevolence
Gallup World Poll (2005-2024)

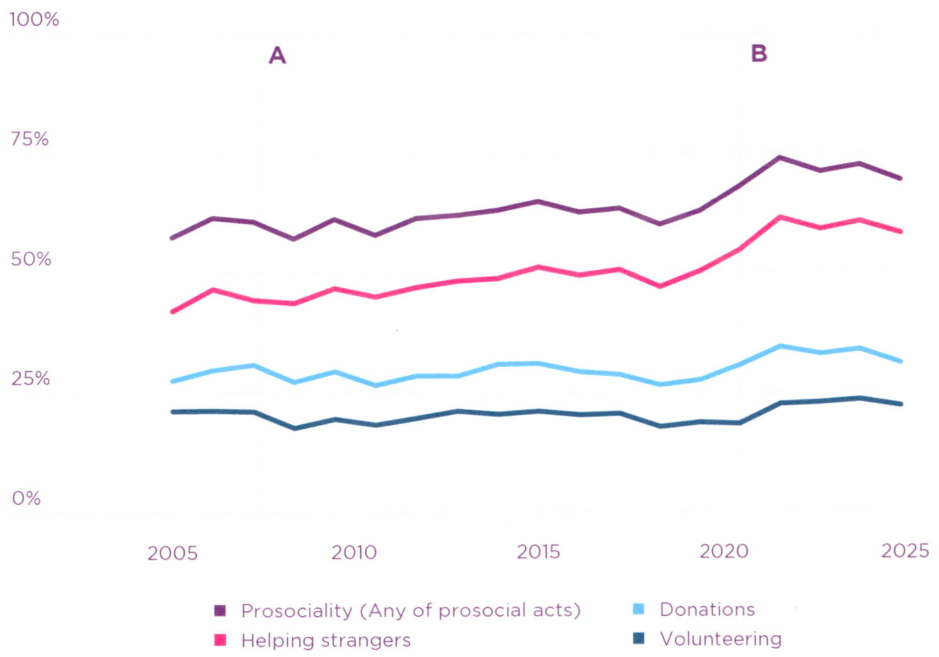

Note: The 95% confidence regions fall within the width of the lines as plotted.
The vertical dotted lines indicate (A) 2008 — the economic recession and (B) 2020 — COVID-19.

Figure 2.3 illustrates benevolence trends for 10 global regions. In almost all regions, helping strangers is the most common benevolent act, while volunteering is generally the least common.

There are notable regional differences in the modes of benevolence. For instance, donations have been among the most common benevolent acts in three regions: Southeast Asia; Western Europe; and North America, Australia, and New Zealand (NANZ) - although in recent years, helping strangers has become the most prevalent. In contrast, donations are the least common form of benevolence in Sub-Saharan Africa and Latin America and the Caribbean, where helping strangers is the primary benevolent action. Volunteering was more commonly practised in the Commonwealth of Independent States until 2014, at which point donations became more frequent.

Regional differences may represent cultural variations that shape norms for caring for others.[60] In regions such as Central and Eastern Europe, Latin America and Caribbean, and Middle East and North Africa, an informal benevolent act, such as helping strangers, appears to be more common than other formal acts of benevolence such as donating and volunteering. On the contrary, regions including Southeast Asia, Western Europe, and NANZ tend to engage more in formal helping behaviours, particularly through sharing material resources. This may also be influenced by institutions and policies in each

Figure 2.3: Regional trends in benevolence
Gallup World Poll (2005–2024)

Note: The 95% confidence regions fall within the width of the lines as plotted.
The vertical dotted lines indicate (A) 2008 — the economic recession and (B) 2020 — COVID-19.

country that encourage donations by monetary rewards, such as tax credits.[61]

We also noticed regional differences in benevolence trajectories. The post-COVID increase in benevolent acts was more prominent in regions such as NANZ and Western Europe, and less pronounced in regions like Sub-Saharan Africa, the Middle East and North Africa, and Southeast Asia. Meanwhile, some regions, such as Central and Eastern Europe and Latin America and the Caribbean, showed an immediate surge in benevolence after COVID-19, but this increase declined soon after. We observe an overall decline in benevolence between 2023 and 2024 across most regions, except for Southeast Asia, South Asia, and Sub-Saharan Africa, where levels remained stable. Nevertheless, the post-COVID increase in benevolent acts (relative to the 2017–2019 average) remains almost universally intact. NANZ is the only region where the prosocial variable was not significantly higher[62] in 2024 than in the pre-COVID years (2017–2019), but even in that region, the helping of strangers is still significantly higher than before the pandemic.

Benevolence and happiness

In this section, we review the burgeoning research literature suggesting that the wellbeing benefits of benevolent acts depend fundamentally on their motivations, the ways in which they are designed and delivered, and the extent of cooperation and collaboration among givers and receivers. In short, we find that benevolent actions deliver greater wellbeing benefits when they involve three Cs: caring connections, choice, and a clear positive impact.[63]

Caring connections

Much research supports this chapter's finding that generous behaviours are associated with — and often directly cause — increased wellbeing among givers across the world.[64] Even children as young as two years old feel pleasure from giving to others.[65] Previous research examines various types of benevolent acts or ways in which people may help one another that include and go beyond the Gallup World Poll questions examined in this chapter. These range from the more formal or institutional giving in nonprofit organisations (e.g., donating money, volunteering time) and workplaces (e.g., helping coworkers) to more informal everyday forms of giving like helping friends, neighbours, family, and strangers.

Yet, there is evidence to suggest that the wellbeing benefits of benevolent acts are often stronger for both helpers and beneficiaries in *caring communities*. Thus, benevolent actions may deliver greater wellbeing benefits to the extent to which they involve more social connectedness and caring motivations.

There is some accuracy to the saying, "*charity begins at home*". Most often, people care for and share their resources with others they regularly see, whether family,[66] friends, coworkers, or acquaintances. Giving behaviours often occur within social contexts and simply witnessing someone else giving has positive effects on wellbeing.[67] Such behaviours can also spread through people's social networks,[68] thus creating and reinforcing caring communities.

When social or relational aspects of giving are stronger, this can amplify wellbeing among givers. Several social and relational factors demonstrate this. For example, having *direct social engagement* can amplify givers' wellbeing. When volunteers engage directly with others, they experience higher wellbeing compared to when they have more indirect roles.[69] In addition, *being closer to the recipient* can also amplify wellbeing effects. For example, research finds that spending money on stronger social ties increases happiness more than spending on weaker social ties.[70] When directly compared, giving to loved ones increases reward centre activation in the brain more than donating to charity.[71]

Recipients also benefit more from closeness. Although receiving gifts can sometimes lead to feelings of indebtedness, receiving gifts can also lead to happiness, especially when gifts come from stronger (as opposed to weaker) social ties.[72] Even when not directly interacting with people, loved ones can inspire us to give and help. For example, simply *being socially motivated*

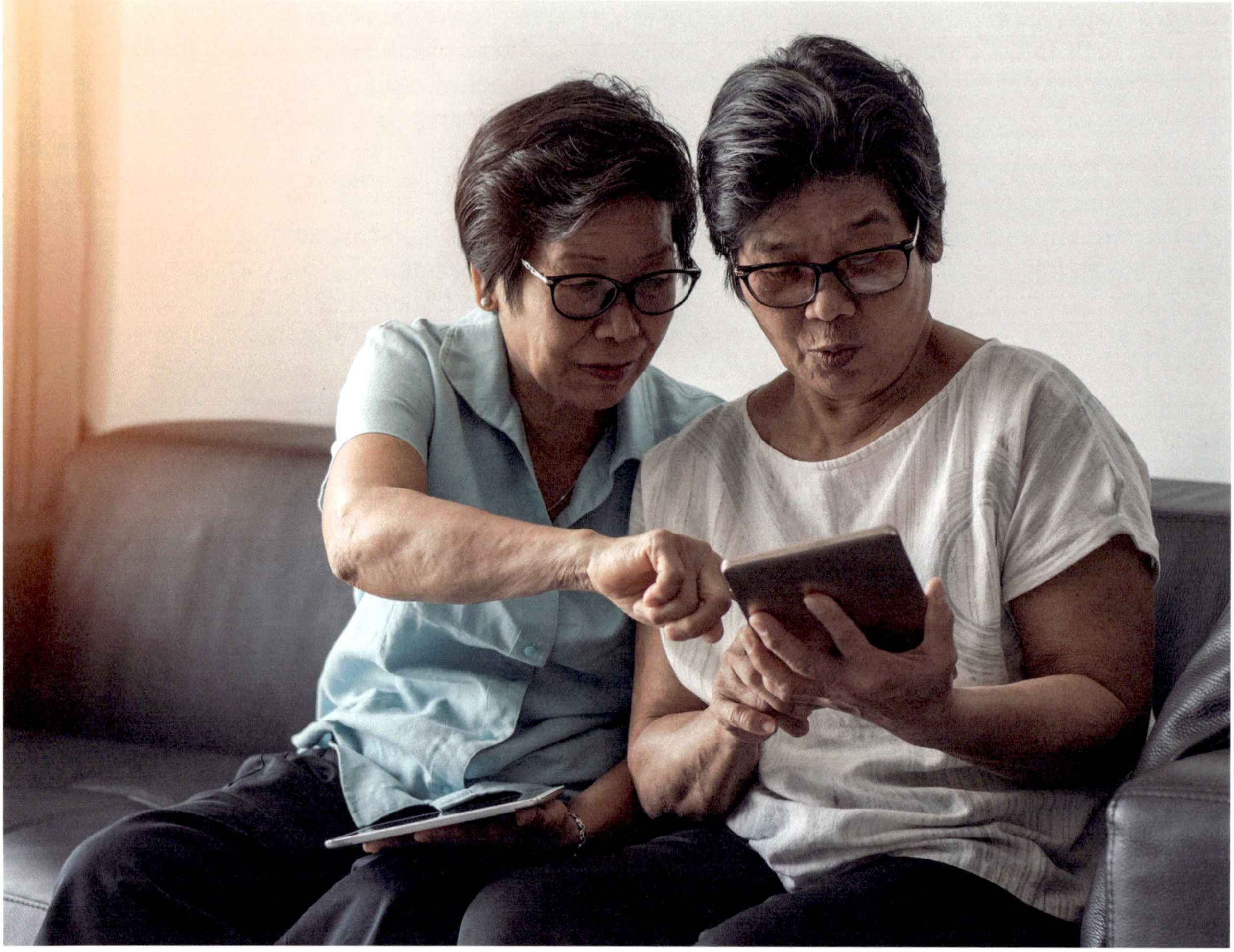

to give can amplify benefits to givers, as demonstrated by research finding that those who volunteer for social reasons – because it's important to their loved ones – have higher self-esteem, wellbeing, self-efficacy, and connectedness,[73] and a lower risk of mortality among older adults.[74]

In caring communities, the lines between "giver" and "recipient" blur as everyone regularly occupies both roles through interactions over time.[75] Research on adolescents[76] and older adults[77] has found that social support interactions that are more balanced – with similar levels of giving and receiving – are associated with higher wellbeing, fewer depressive symptoms, and even a lower risk of dying. However, other research suggests that low levels of both giving and receiving, even if balanced, are associated with lower wellbeing.[78] Thus, increased wellbeing is more likely when people are embedded in mutual and frequent caring interactions.

Although giving within caring communities can support increased meaning and joy, people often give beyond their own immediate groups, to strangers both near and far, and, as reviewed below, this still feels good. We review key psychological pathways that can extend one's circle of compassion more broadly beyond the people we know and love, creating a sense of caring community across our cities, countries, and even globally, to people we might not ever directly interact with.

One such pathway is *caring personality traits*, like compassion or altruism. People with caring personality traits tend to prioritise others' needs and wellbeing, and they are often kind to themselves as well.[79] Notably, they also tend to have more balanced social support interactions in their relationships,[80] which has implications for their wellbeing, as discussed above. But their care extends beyond people they directly interact with, often giving to and helping strangers in need.[81] Much research supports the idea that more compassionate people have higher wellbeing and experience fewer mental health symptoms.[82] Even in high-stress jobs, those who are more compassionate report lower stress and burnout.[83] These wellbeing effects run deep, translating into lower stress hormones during stressful situations[84] and even a lower risk of dying among older adults.[85]

A related pathway that can extend one's circle of compassion is experiencing *caring emotions*, which increase altruistic motivations, making people more likely to help others.[86] For example, to the extent that people experience increased feelings of benevolence after giving to distant others (compared to close others), these feelings increase their happiness.[87] Thus, both compassionate traits and feelings can feel rewarding to people, even when directed toward strangers.

Another such pathway is *caring motivations*. People who have such other-focused motivations experience more positive emotions and higher self-esteem.[88] When following such people over time, researchers find that having caring motivations at one time point predicts later increased positive emotions and self-esteem, fewer anxiety and depression symptoms, and decreased feelings of loneliness, isolation, and other negative feelings.[89]

People can have caring motivations in general, but such motivations are often applied to specific prosocial behaviours. Although prosocial behaviours may appear to be other-focused, there are many reasons that people engage in them, such as self-enhancement, indicating that they can be motivated by self- and/or other-oriented concerns.[90] Research consistently finds that not all acts of kindness offer equivalent wellbeing benefits and that self-focused motives may undermine the emotional rewards that typically follow other-focused prosocial behaviour. For example, among volunteer healthcare workers, other-oriented motives for helping predict higher life satisfaction, but self-oriented motives do not.[91] Other research confirms and extends this, finding that volunteers with altruistic motivations have higher self-esteem, wellbeing, self-efficacy, and connectedness,[92] and volunteers who value others report decreased distress.[93] Even simply recalling an instance of other-focused helping can increase positive emotions compared to recalling an instance of self-focused helping.[94]

Such other-oriented motivations also benefit nonprofit organisations, in that volunteers with altruistic motivations also have increased satisfaction with their volunteer role and more intentions to continue volunteering in the future.[95] Remarkably, other-oriented older adult volunteers have a lower risk of dying when tracked over time compared to those with more self-focused motivations.[96]

Research on blood donations displays a similar pattern on physical pain measures. In one recent study, people reported less intense pain from a vein puncture when blood was drawn for the purpose of post-earthquake medical use than when blood was drawn for personal medical tests.[97] Taken together, research on caring traits, feelings, and motivations suggest that the reason that people give is as important as the actual giving behaviours in terms of promoting wellbeing.

Overall, giving in communities of care is more likely to lead to emotional rewards than giving in disconnected or less caring contexts. These communities can consist of people we interact with in our daily lives,[98] and can also imaginatively extend beyond known others, through caring traits, feelings, and motivations. Next, we discuss two other features that may be prevalent in such caring communities: freely choosing one's benevolent behaviours, and caring about the impact of them.

Choice

Caring communities provide people with a variety of ways to help one another as well as the choice of how to do so. For example, a teenager living in a caring community can choose to volunteer at the local homeless shelter, mentor younger children in a sport they love, or assist an elderly neighbour with grocery shopping.

The opportunity to choose how to help others is important because people are more likely to feel happy after giving when they have a sense of autonomy or personal choice in how they help.[99] Evidence supporting this claim comes from studies exploring different kinds of prosocial behaviour and research methods. For instance, in the realm of volunteering, people can choose to help others for reasons that they find personally valuable (autonomous motivations) or reasons that are forced upon them (controlled motivations), and the reasons for volunteering matter. A large survey conducted with over 700 college students in China who volunteered during COVID found that volunteers who endorsed more autonomous motivations for their actions experienced greater satisfaction, meaning, and happiness.[100] Meanwhile, volunteers who expressed higher levels of controlled motivations reported lower satisfaction, meaning, and happiness. Similar findings have been observed using a large data set from the National Longitudinal Study of Adolescent and Adult Health in the United States. Youth volunteers who provide assistance voluntarily also reported lower levels of depression, while volunteers who say their volunteering is "required by others" do not.[101]

Similar findings have been observed when people give or transfer money. For instance, in one study, 80 students were provided with a few dollars to distribute between themselves and another person. Afterwards, each student reported their wellbeing. Importantly, half of the students were allowed to choose how much money, if any, they wanted to give to the other person (high-choice). Meanwhile, the other half of students were not given a choice in how the money was distributed, they were simply told to distribute the money between themselves and the other person (low-choice). Participants reported experiencing higher levels of happiness when they gave more money to the other person, but only when the participant had a choice over how much was given.[102] Harbaugh and colleagues (2007) arrived at a similar conclusion when looking at brain activation as people donated to a local charity while in an fMRI scanner. People showed activation in parts of the brain that are commonly associated with pleasurable tasks or enjoyable activities while giving, and this pattern of activity was greatest when people made voluntary donations.

The importance of feeling as if one has exercised choice when helping others can even be detected upon reflection. In one experiment, individuals were asked to write about a time they chose to help another person and a time in which they helped another person but had little choice. After writing each description, participants reported their current happiness. Consistent with the evidence reviewed above, people reported greater momentary wellbeing after recalling a time they chose to help someone else as opposed to a time they did not have a choice about how to help.[103]

While not all situations allow people total freedom when deciding how to engage in prosocial action, past research suggests it can be helpful to provide even a small degree of choice. For instance, when people are required to complete a prosocial act, such as contribute to their child's school or religious community, it could be helpful to allow people to decide when (which date and time) and how (volunteer at a fundraiser, donate from home) they would like to help. This flexibility may protect one's sense of choice and, in turn, bolster the joy of giving. If this possibility is not viable, some research suggests that reminding people they have some freedom to choose whether or not to help can be beneficial. For example, in one study, 104 students were instructed to help on a task.[104] Half the students were told it was "entirely their choice whether to help or not" while the other half of students were told that they "should help out." Afterwards, students reported their wellbeing. Students who were reminded of their choice to help reported greater happiness.

There are a few potential reasons why voluntary (as opposed to involuntary) giving may lead to greater happiness. One possibility is that freely chosen giving may fulfil a basic human urge to act in a way that one chooses. Whether it be deciding what to eat for dinner or how to assist others, having the freedom to make your own choices is a strong predictor of happiness.[105]

Another reason that voluntary giving may lead to greater happiness is that it allows people to show and act upon their personal preferences when it comes to giving. If someone has a particular affinity for the environment or caring for children, autonomous helping provides a chance to help in this way. Finally, unforced giving may be especially likely to bolster happiness because it provides people with evidence that they are a kind and generous person, revealing their caring traits. Unforced giving is particularly powerful proof because one's kind act was self-chosen and not required by external pressures, such as government legislation or educational requirements.

Clear positive impact

Caring communities may also encourage generosity by providing clearer opportunities to see how one's actions have made a positive impact on others. Indeed, caring communities may be more likely to foster a clear dialogue or exchange of information that allows recipients to relay what they need, and for helpers to appreciate how their assistance has been effective.

Classic research in psychology has shown that recognising how one's actions are effective is a key predictor of whether one helps. For instance, research on what is called "the identifiable victim effect" has documented that people are more likely to help one clear identifiable target in need over a larger number of unknown targets.[106] While there are various reasons for why people respond in this manner, one contributing factor is that people can more clearly see or imagine how their assistance will be effective when helping a singular, detailed target over a larger number.[107]

Beyond encouraging people to act, recognising how one's actions positively impact others amplifies the joy of giving. For instance, in one experiment, 120 students were given 10 Canadian dollars and were then invited to donate some, none, or all to a charity before reporting their happiness.[108] Critically, half of the students were asked to donate to a charity that made the positive impact of their work incredibly clear by stating that every $10 purchases a bed net for a child in need to stop the spread of malaria. Meanwhile, the other half of students were asked to donate to another charity that also helped children in the same region but did not clearly explain how the donation would be used. When the impact of the donation was clear, people who donated more money to charity reported higher happiness. Yet, when the impact of giving was not clear, people who gave more money did not report greater happiness. These findings suggest that creating or fostering opportunities for people to appreciate how they have helped others can increase the wellbeing of the helper.

Indeed, similar findings have been observed among living kidney donors who tend to report the greatest positive reactions about their actions when they feel like their donation was impactful in helping the recipient survive.[109] Thus, creating an opportunity for recipients to meet the donor and express their appreciation may heighten feelings of impact and the joy of giving. This may be why many nonprofits go to great lengths to facilitate these connections and why some organisations, like blood donor clinics, now allow donors to receive messages when their contributions have been directed to a recipient. Finally, feelings of efficacy may help to explain why activism behaviour is not associated with wellbeing while other forms of helping often are.[110] The positive effects of one's activism may be harder to see given the nature of many ongoing social problems.

Our new evidence

How are individual benevolent actions and perceptions of social benevolence linked to life evaluations? To answer this question, we use data from the 2019 Gallup World Poll which includes individual data for the wallet questions discussed earlier.[111] This allows us to see how personally engaging in prosocial behaviour - versus believing that others will act in a benevolent way - is linked to wellbeing.

Figure 2.4 shows that people's prosocial actions and their beliefs about the benevolence of others are both important predictors of individual life evaluations when assessed within the same equation.[112]

Figure 2.4: How benevolent acts and expected wallet return affect life evaluations

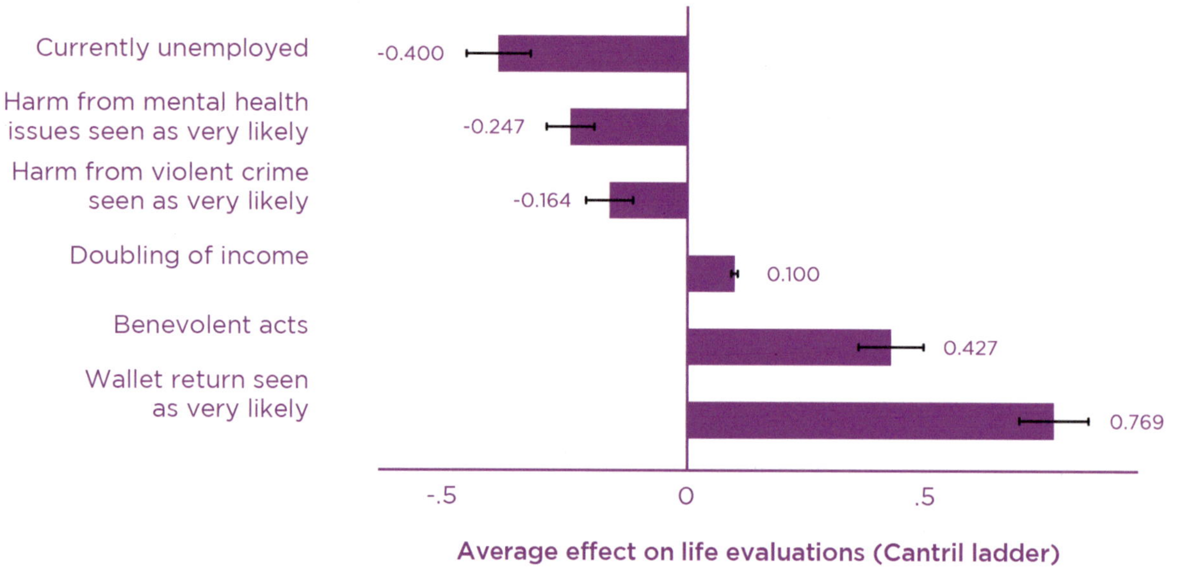

The 'benevolent acts' variable is the average frequency for the three benevolent acts, with a value of 1 for individuals who performed each of the three acts in the previous month. If the acts are included as separate variables, each of them has a significant effect, about twice as large for donations as for volunteering or helping strangers.[113]

Expected wallet return also has a large positive effect, almost twice that for benevolent acts. Believing that others would return a wallet predicts a larger boost to life satisfaction than a doubling of income. Believing that your lost wallet would very likely be returned is accompanied by life satisfaction that is higher by more than three-quarters of a point on the 0–10 scale.[114] This effect is almost twice as large as being unemployed. It is also higher than the negative effects of comparably measured expected harms from mental health issues or violent crime.[115]

Benevolence and inequality

The distribution of wellbeing among individuals and population groups is as important as its average. In several recent editions of the World Happiness Report, we have documented sharp increases in the inequality of wellbeing globally[116] and in most global regions[117] as well as growing gaps between the top and bottom halves of the population.[118]

Our latest analysis, shown in Figure 2.5, separates the trends in wellbeing inequality within countries and between countries. The purple line shows that the average within-country inequality of wellbeing has increased by about one-quarter over the past two decades.[119] As shown by the pink line, the between-country inequality of wellbeing has remained essentially unchanged at 0.2.[120] The international share of total individual variance in life evaluations, as shown by the blue line,[121] has thus dropped, from 0.236 to 0.187.

Past reports have shown that wellbeing inequality itself lowers national average happiness, while

Figure 2.5: Wellbeing inequality is growing within countries, but not across countries
Gallup World Poll (2005–2024)

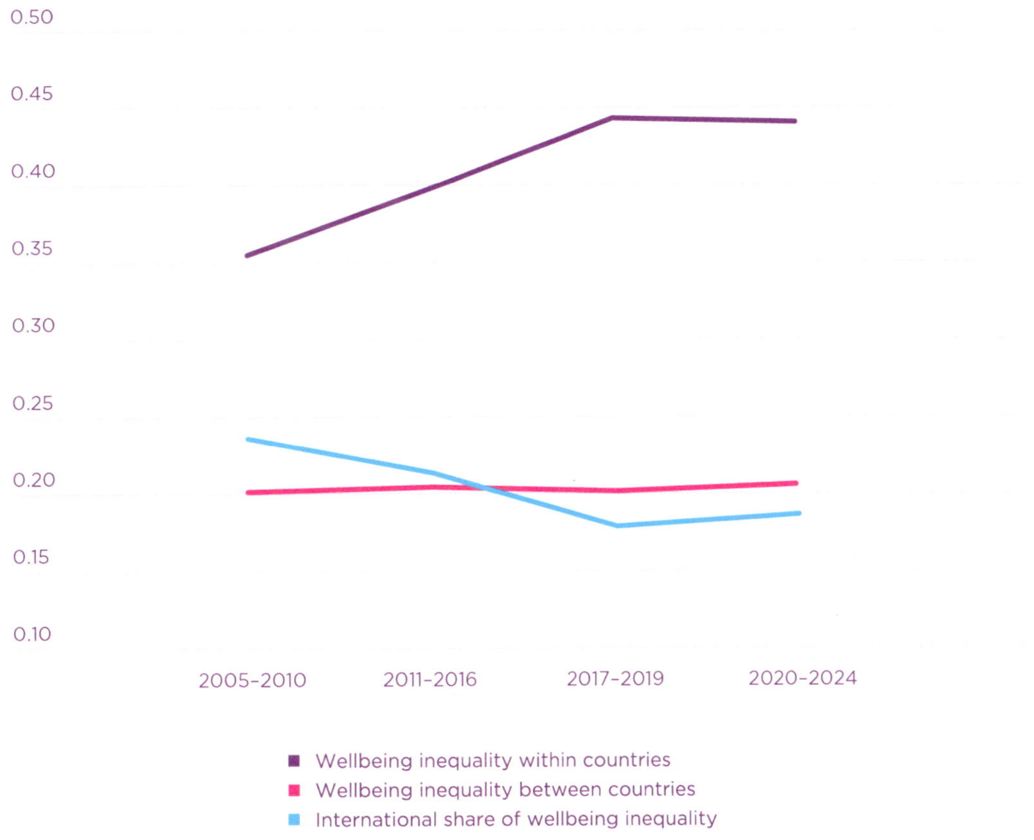

World Happiness Report 2020 broke new ground in showing that living in countries with high social and institutional trust is of significantly greater value for those afflicted by ill health, unemployment, unsafe streets, and difficult family circumstances. These afflictions are all more common among those reporting lower life satisfaction. Thus, countries with higher trust have less inequality of wellbeing, as revealed by using interaction terms between life circumstances and measures of trust.[122]

Furthermore, caring and sharing reduce wellbeing inequality by being more valuable to those facing less fortunate life circumstances. Data from the European Social Survey (2002–2022) show that people who judge most people to be both fair and kind suffer materially less from being subject to unemployment, ill health, discrimination, or unsafe streets.[123] The results are shown in Figure 2.6. These are large effects, so the scale is different than in Figure 2.4.

Caring and sharing reduce wellbeing inequality by being more valuable to those facing less fortunate life circumstances.

Figure 2.6: The protective effects of positive perceptions
European Social Survey (2002–2022)

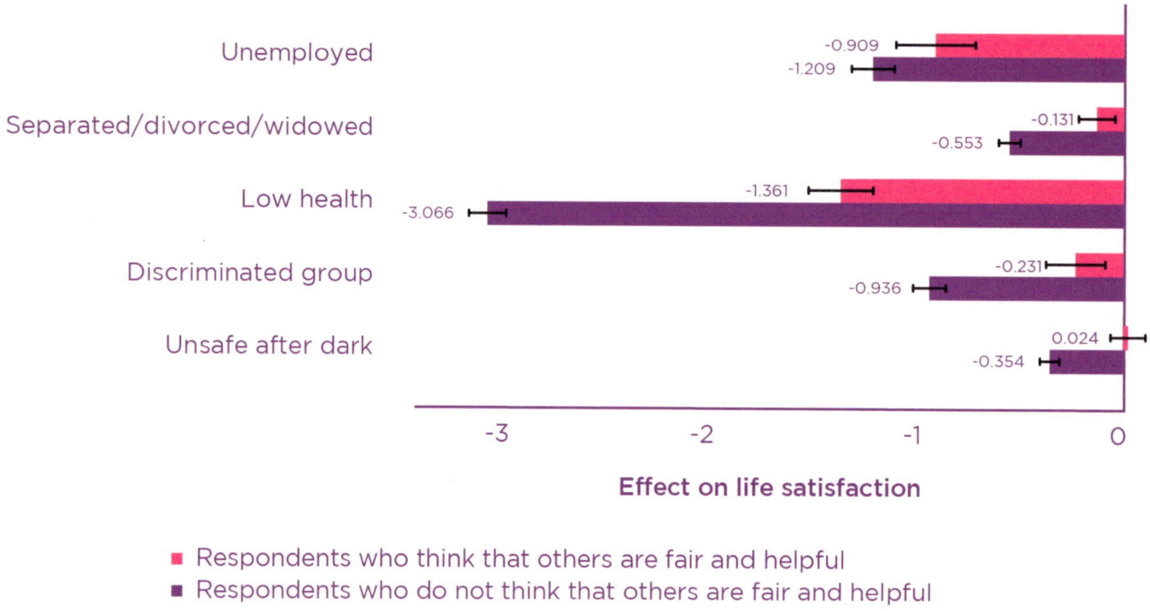

These interactions come from an equation using European Social Survey data from 2002–2023 that also includes a large main effect, four-fifths of a point on the 0–10 scale (as shown in Figure 2.7) for anyone living in a country where they judge others to be both fair and helpful. Figure 2.7 also shows substantial positive effects from frequent social meetings,[124] high social trust,[125] and trust in the police.[126] The scale in Figure 2.7 is the same as in Figure 2.6.

Benevolence between countries

Official Development Assistance (ODA) refers to government-provided aid aimed at fostering economic development and improving the welfare of low- and middle-income countries. It can be considered a form of prosocial behaviour when viewed from the perspective of international relations and collective welfare. However, ODA is also associated with motivations such as strategic interests, political alliances, or economic benefits. This duality of altruistic versus self-interested motives makes ODA a nuanced example of prosocial behaviour.

To better capture the prosocial component, we focus on untied ODA i.e., aid that is freely and fully available to finance procurement from all countries. It differs from tied or partially tied aid which restricts the procurement of goods or services to the donor country or a specific group of countries. Tied aid can negatively affect aid effectiveness by increasing the costs of goods and services, and possibly deflecting policy priorities. Moreover, untied aid represents the majority of total ODA commitment by many donors, thus we focus on untied ODA in our analysis.[127]

Figure 2.7: The effects of trust, social meetings, and positive perceptions on life satisfaction
European Social Survey (2002–2022)

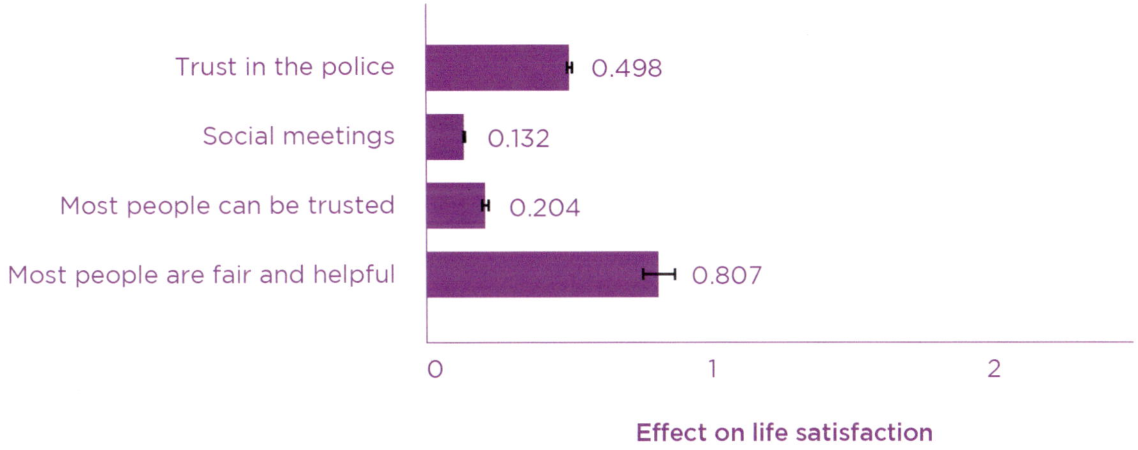

Another potential form of international prosocial behaviour is hosting refugees, as it reflects actions taken by countries to assist vulnerable populations in need, often at considerable cost and with little or no immediate benefit to the host country. The number of forcibly displaced people worldwide has risen significantly in recent years. According to UNHCR's 2024 Mid-Year Trends report, 122.6 million people worldwide were forcibly displaced in the first half of 2024 due to persecution, violence, conflict, human rights abuses, or events disrupting public order. Of these, 32 million are refugees under UNHCR's mandate. While hosting refugees often aligns with prosocial and altruistic values, it does not always stem purely from voluntary goodwill. Geopolitical pressures, international legal obligations, geographical location, and domestic political considerations frequently play substantial roles. For instance, low- and middle-income countries host 71% of the world's refugees and other people in need of protection, and 69% of refugees and others in need of international protection reside in neighbouring countries. These figures illustrate that hosting refugees is not always a purely prosocial activity. Given that current data do not distinguish between voluntarily and involuntarily hosted refugees, interpreting the estimated relationship between the refugee ratio and happiness requires caution.

While numerous empirical studies at the individual level demonstrate that prosocial spending enhances the giver's happiness, research on how prosocial behaviours by nations affect donor happiness remains scarce. Most individual-level studies focus on the psychological mechanisms behind giving, such as the warm glow effect or a sense of altruistic satisfaction, which contribute to increased subjective wellbeing. However, at the national level, prosocial behaviours like ODA and refugee hosting involve complex dynamics that extend beyond individual motivations. These behaviours are influenced by political, economic, and social factors which can shape the perceptions and experiences of citizens in donor countries. This section addresses this gap by analysing the relationship between national-level prosocial actions and donor happiness on a global scale. By examining data on ODA and refugee hosting alongside measures of national wellbeing, this analysis explores whether collective generosity translates into increased happiness for donor populations.

We present the data for ODA and refugees first and then proceed to empirical analysis. Table 2.3 shows the top 20 countries ranked by untied ODA commitments as a share of GDP in 2022. It refers to untied bilateral commitments that consist of grants and grant-like contributions as well as loans in current dollars.[128] We calculate the untied ODA as a share of PPP-adjusted GDP and per capita value ($US). The GDP and population data are retrieved from the World Bank's World Development Indicators (WDI). The table reveals that Western and Northern European countries generally exhibit higher per capita values among the donor countries. In 2022, the top five countries are Norway, Switzerland, Sweden, Germany, and Denmark.

Table 2.3: Untied Official Development Assistance (ODA)
OECD (2022), World Bank (2022)

Rank	Donor	Ratio of untied ODA to GDP (%)	Untied ODA per capita ($)	Untied ODA (million $)
1	Norway	0.54	673.63	3,676.07
2	Switzerland	0.36	324.46	2,847.77
3	Sweden	0.33	218.82	2,294.77
4	Germany	0.31	211.86	17,753.10
5	Denmark	0.28	218.05	1,287.15
6	France	0.26	148.66	10,118.30
7	Luxembourg	0.24	351.80	229.76
8	Canada	0.22	137.72	5,362.77
9	Netherlands	0.21	164.73	2,915.89
10	Japan	0.19	89.75	11,230.20
11	Iceland	0.17	123.98	47.36
12	Australia	0.11	69.04	1,796.07
13	United States	0.10	81.83	27,270.10
14	New Zealand	0.10	51.25	262.26
15	Republic of Korea	0.10	50.44	2,606.33
16	United Kingdom	0.09	54.13	3,669.26
17	Belgium	0.09	59.27	692.33
18	Ireland	0.08	106.85	551.96
19	Austria	0.06	41.01	370.85
20	Spain	0.05	25.32	1,209.39

Notes: ODA share of GDP and per capita ODA are calculated by authors using ODA data from OECD as well as GDP and population data from WDI. ODA in this table refers to total untied bilateral commitments that consist of grants and grant-like contributions as well as loans in current dollars.

Table 2.4 presents the top 20 countries ranked by the ratio of resident refugees to population in 2022. The total numbers of refugees in each country are also reported. Refugee statistics are sourced from UNHCR, while population data is retrieved from the World Bank's WDI to calculate refugees as a percentage of the population. The table shows that population refugee shares are highest in countries in the Middle East, Africa, and Europe. Specifically, Lebanon, Jordan, Montenegro, Türkiye, and Czechia were the countries with the highest refugee ratios in 2022.

To assess the possible links to happiness, we added 'ODA as % of GDP' and 'refugees as % of the population' to our Table 2.1 equations for life evaluations and emotions, as reported in the online ODA appendix. The ODA share is positively linked to life evaluations, while the refugee share is negatively linked to both life evaluations and

Table 2.4: Resident refugees in 2022
UNHCR (2022), World Bank (2022)

Rank	Country	Number of refugees	Refugees as % of population
1	Lebanon	818,861	14.9
2	Jordan	697,761	6.2
3	Montenegro	32,438	5.3
4	Türkiye	3,568,259	4.2
5	Czechia	435,212	4.1
6	Republic of Moldova	105,374	4.1
7	Iran	3,425,091	3.9
8	Chad	592,764	3.3
9	Uganda	1,463,523	3.1
10	Estonia	40,806	3.0
11	Austria	258,613	2.9
12	South Sudan	308,369	2.8
13	Bulgaria	176,297	2.7
14	Sweden	277,726	2.6
15	Poland	971,129	2.6
16	Germany	2,075,445	2.5
17	Lithuania	67,638	2.4
18	Sudan	1,097,128	2.3
19	Cyprus	29,280	2.3
20	Mauritania	100,981	2.1

positive emotions. Neither ODA nor refugee shares show significant links to negative emotions. The life evaluation links for ODA and refugee shares are similar in magnitude after allowing for the different cross-country variability of ODA and refugee shares. An increase in ODA by one standard deviation is associated with a life evaluation higher by .07 points, while an increase of the same relative size in the refugee share is associated with an average life evaluation that is lower by .06 points. The data in Table 2.4 show that refugee shares are highest in countries that share borders with countries in turmoil or on escape routes from those countries. The plausible positive effects of offering a safe haven for refugees are being swamped by the pressures that fleeing populations place on neighbouring countries that are often themselves short of the basics of life.

The results in this section echo the general pattern of results surveyed earlier in the chapter: benevolent acts are more likely to support higher levels of happiness if they embody caring connections, choice, and a clear positive impact. These features are less typical for refugee flows, which are frequently driven by circumstances rather than caring connections or choice, and impose costs on both refugees and hosts. ODA is more likely to involve all of the three C's. Since these are national decisions, however, they may reflect national interests not fully shared by the individuals whose happiness is being assessed. Furthermore, both the decisions to donate and the possible benefits in the receiving countries are far removed from the daily lives of individuals in the donor countries, whose immediate surroundings and social connections are likely to be of much greater importance. As for future trends in ODA, it is apparently being cut entirely in the United States, while being cut back in a number of other donor countries.[129]

National decisions are likely to have more impacts on the wellbeing of individuals in other countries to the extent that the resulting actions create or destroy peace. Overall rankings of individual life evaluations are especially low in countries wracked by violence, notably Afghanistan and Lebanon, and other countries not even in the rankings, such as Sudan and Syria, because conditions have been too unsafe to permit surveys.

International caring and sharing are likely to be of most benefit to global life evaluations, especially in conflict-ridden countries, to the extent that they can create peace and heal the wounds of past conflicts. Forgiveness is associated with higher life satisfaction in many studies,[130] especially where supported by favourable attitudes and beliefs about others.[131] These supporting attitudes are hardest to establish in international settings and where there are long-standing animosities and grievances. This makes it especially challenging to achieve forgiveness and reconciliation in conflict and post-conflict settings.[132] Our findings in this chapter expose the need for the kind of caring and sharing that delivers peace, forgiveness and reconciliation. Building a broader international network of caring connections seems to us a first necessary step, using such efforts to supplant the force of arms.

Endnotes

1 This was reported for the seven deep-dive countries in *World Happiness Report 2023* and for the global sample of Gallup World Poll countries in *World Happiness Report 2024*.

2 Donations are the only one of the three benevolent acts to have a statistically significant relation to average life evaluations at the national level. In our results using individual-level data later in the chapter we find significant positive linkages for all three benevolent acts. In our Table 2.1 results the donation variable removes the influence of log GDP, leaving the coefficient on income to include those effects that flow through their support for larger donations.

3 A country's average answer to the Cantril Ladder question is exactly equivalent to a notion of average underlying satisfaction with life under an assumption of 'cardinality': the idea that the difference between a 4 and a 3 should count the same as the difference between a 3 and a 2, and be comparable across individuals. Some social scientists argue that too little is known about how people choose their answer to the Cantril Ladder question to make this assumption and that if it is wrong enough, then rankings based on average survey responses could differ from rankings based on underlying satisfaction with life (Bond & Lang, 2019). Other researchers have concluded that answers to the Cantril Ladder question are indeed approximately cardinal (Bloem & Oswald, 2022; Ferrer-i-Carbonell & Frijters, 2004; Kaiser & Oswald, 2022; Krueger & Schkade, 2008).

4 For any pair of countries, the confidence intervals for the *means* (depicted in Figure 2.1 as whiskers) can be used to gauge which country's mean is higher than the other, accounting for statistical uncertainty in the measurement of each. The confidence interval for a country's *rank* (given in Figure 2.1 as text, in the form (4-8) represents a range of possible values for the ranking of their mean among all countries, accounting for uncertainty in the measurement of all of the means (following Mogstad et al. 2024). The ranges are constructed so that the chance that the range does not contain the country's true rank is no more than 5%.

5 Not every country has a survey every year. The total sample sizes are reported in the online Statistical Appendix and are reflected in Figure 2.1 by the size of the 95% confidence intervals for the mean, indicated by horizontal lines. The confidence intervals are naturally tighter for countries with larger samples.

6 See Helliwell et al. (2020) for a detailed analysis of the life satisfaction of immigrants to the United Kingdom and Canada from many source countries.

7 In 2013 there was no significant difference between the male and female responses while in this report (with surveys from 2022 and 2023) the male life evaluations are one-third higher than for females (1.57/1.16). This is the largest gender gap ever seen in our reports, where at the global level there is a slight advantage favouring females.

8 Going from the larger to the smaller gains, these were Portugal, Iceland, Germany, and Finland.

9 All with increases of 0.2 points or more on the 0-10 scale.

10 All with drops of 0.17 or more on the 0-10 scale.

11 The online statistical appendix contains alternative forms without year effects (Table A9) and a repeat version of the Table 2.1 equation showing the estimated year effects (Table A8). These results continue to confirm that inclusion of year effects makes no significant difference to any of the coefficients. In these aggregate equations, adding country fixed effects (as in Table A10) lowers the coefficients on relatively slowly moving variables where most of the variance is across countries rather than over time, such as healthy life expectancy. Our equations based on individual observations (e.g., Figure 2.4), where income and health are measured by individual-level variables, include both year and country fixed effects, with coefficients very similar to those estimated without fixed effects.

12 The definitions of the variables are shown in Box 2.2, with additional detail in the online statistical appendix.

13 The model's predictive power is little changed if the year fixed effects in the model are removed, with adjusted R-squared falling only from 0.761 to 0.756.

14 The data and rankings for the 2022-2024 averages for the six variables are to be found in Figures 48-65 of the online statistical appendix. The rankings for positive affect are in Figures 66-68, and for negative affect in Figures 69-71. The underlying annual data used in estimating the equations shown in Table 2.1 are currently not available on our website. We can at this time provide the data to researchers approved by Gallup.

15 For example, unemployment responses at the individual level are available in most waves of the Gallup World Poll. While they show an effect size similar to that found in other research, the coefficient has never been significant in the country-level equation and their inclusion does not influence the size of the other coefficients.

16 The main differences are a larger income effect, presumably flowing from the cyclical variations, and the insignificance of the healthy life expectancy effect, probably due to its trend-like variation in most countries.

17 Below, we use the term 'effect' when describing the coefficients in these regressions; some caveats to this interpretation are discussed later in this section.

18 In the equation for negative affect, healthy life expectancy takes a significant positive coefficient, despite its positive simple correlation with life evaluations in this aggregate dataset. This may be due to the fact that in the global sample there is a positive correlation between age and the frequency of reports of negative emotions. Countries with higher healthy life expectancies have respondents who are older on average, since the sample data are weighted to replicate the actual age shares of the population.

19 This influence may be direct, as many have found, e.g., De Neve et al. (2013). It may also embody the idea, as made explicit in Fredrickson's broaden-and-build theory (Fredrickson, 2001), that good moods help to induce the sorts of positive connections that eventually provide the basis for better life evaluations.

20 See, for example, the well-known study of the longevity of nuns, Danner et al. (2001).

21 See Cohen et al. (2003) and Doyle et al. (2006).

22 The meta-analysis by Chida and Steptoe (2008) shows significant linkages from positive affect to health, independent of the effects of negative affect. For a recent survey of the links running from positive emotions and life evaluations to subsequent morbidity and mortality, see Pressman et al. (2019).

23 The prevalence of these feedbacks was documented in Chapter 4 of *World Happiness Report 2013*, De Neve et al. (2013).

24 For more detail, see Table 10 of the online Statistical Appendix for *World Happiness Report 2018*.

25 We expected the coefficients on these variables (but not on the variables based on non-survey sources) to be reduced to the extent that idiosyncratic differences among respondents tend to produce a positive correlation between the four survey-based factors and the life evaluations given by the same respondents. This line of possible influence is cut when the life evaluations are coming from an entirely different set of respondents than are the four social variables. The fact that the coefficients are reduced, but only very slightly, suggests that the common-source link is real but very limited in its impact.

26 The coefficients on GDP per capita and healthy life expectancy were affected even less than were the coefficients on the survey variables, and in the opposite direction in the case of the income measure, being increased rather than reduced, once again just as expected. The changes were very small because the data come from other sources and are unaffected by our experiment. However, the income coefficient does increase slightly since income is positively correlated with the other four variables being tested, so that income is now able to pick up a fraction of the drop in influence from the other four variables. We also performed an alternative robustness test using the previous year's values for the four survey-based variables. Because each year's respondents are from a different random sampling of the national populations, using the previous year's average data also avoids using the same respondent's answers on both sides of the equation. This alternative test produced similarly reassuring results as shown in Table 13 of the online Statistical Appendix in *World Happiness Report 2018*. The Table 13 results are very similar to the split-sample results shown in Tables 11 and 12, and all three tables give effect sizes very similar to those in Table 2.1 in the main text. Because the samples change only slightly from year to year, there was no need to repeat these tests with this year's sample.

27 Actual and predicted national and regional average 2022–2024 life evaluations are plotted in Figure 72 of the online statistical appendix. The 45-degree line in each part of the figure shows a situation where the actual and predicted values are equal. A predominance of country dots below the 45-degree line shows a region where actual values are below those predicted by the model, and vice versa. Southeast Asia provides the largest current example of the former case, and Latin America of the latter.

28 See Rojas (2018) and Chapter 4 of this report.

29 If special variables for Latin America and Southeast Asia are added to the equation in column 1 of Table 2.1, the Latin American coefficient is +0.49 (t=5.3) while that for Southeast Asia is -0.3 (t=2.1). Special variables for East Asia and South Asia are not significant.

30 See Chen et al. (1995) for differences in response style, and Chapter 6 of *World Happiness Report 2022* for data on regional differences in variables thought to be of special importance in Asian cultures.

31 One slight exception is that the negative effect of corruption is estimated to be slightly larger (0.87 rather than 0.73), although not significantly so, if we include a separate regional variable for Latin America. This is because perceived corruption is worse than average in Latin America and its happiness effects there are offset by stronger close-knit social networks, as described in Rojas (2018). The inclusion of a special Latin American variable thereby permits the corruption coefficient to take a higher value.

32 The donation variable on its own adds more to the explanatory power of the Table 2.1 equation than does an average of the three variables. In a later section of this chapter we use individual-level data to show that each of the three benevolent acts is associated with reported life evaluations, with the donation effect significantly larger than the others.

33 The data are drawn from the Lloyd's Register Foundation World Risk Poll, included as part of the 2019 round of the Gallup World Poll. The answers were on a three-point scale anchored by 'not at all likely' at 0 and very likely at 1. The middle option, 'somewhat likely' was coded as 0.5. The same question form and scale were also used to assess negative rather than positive risks, including the risks from violent crime and from mental health problems. This made it possible to compare the subjective wellbeing effects of positive and negative risks evaluated in comparable terms, as will be reported later in the chapter.

34 If we compare the data from countries in the 40-country study (Cohn et al., 2014) we find *actual* wallet return in the three Nordic countries in the study (Sweden, Denmark and Norway) to be 81%, compared to 47% in the other countries in their global sample. The *expected* wallet return by strangers (as ranked in Table 2.2) is also much more likely in the Nordic countries, at 47%, than in the other countries, at 28%. Hence, perceptions match reality in that wallet return is higher in the Nordic countries than in other places. Yet, in the Nordic countries and elsewhere, people underestimate the kindness of others, as will be shown later in this chapter. These results are all shown graphically in the first of the online wallet figures.

35 Kenya and Liberia are both near the top of the ranking for volunteering.

36 Reader's Digest (2023).

37 Cohn et al. (2019). Employees of societal institutions were approached and asked to take the lost object to the next stage on its return to the imaginary owner.

38 See Helliwell and Wang (2011) for the fullest description, and also Helliwell and Aknin (2018) and Helliwell et al. (2019).

39 Rober (2019). He also did sub-experiments to show that the unexpectedly high wallet return rate applied equally to rich and poor, and equally to males and females.

40 Of the 20 cities, 18 were in the United States and 2 in Canada. The expected rate of return for wallets found by strangers is 34% in the US and 39% in Canada.

41 These results are also shown in the second of the online wallet figures.

42 Knack (2001). His analysis was based on the first Reader's Digest wallet drops, in the late 1990s.

43 The shares of explained variance are given by the r^2 on the full set of country fixed effects. These are .206 for police, .125 for neighbours, and .080 for strangers.

44 See Konrath et al. (2011).

45 See Konrath et al. (2023); Sortheix et al. (2019); Yan et al. (2017).

46 Levine et al. (2008).

47 See Hampton (2016) in which the result is based on modelling that controls for differences in average neighbourhood characteristics. It shows a 10% drop in the United States, but no significant change in Canada. In 2001 there was no variation in altruistic behavior based on neighborhood diversity. However, areas of the United States where the proportion of non-citizens increased since 2001 experienced reduced helping; the opposite was found in Canada.

48 Yuan et al. (2022).

49 Smith (2019).

50 Helliwell et al. (2024).

51 See Aydinli et al. (2013) for a more detailed summary.

52 See Chopik et al. (2017).

53 Luria et al. (2015).

54 Our analysis of actual and expected returns uses actual returns from the wallet+money experiments of Cohn et al. (2019) and expected returns from the 2019 Lloyd's Register Foundation World Risk Poll. Actual return averages 1.83 times expected return.

55 See Knafo et al. (2009). See also Rhoads et al. (2021) for an evaluation of seven kinds of altruism across countries and variation showing higher levels in individualistic countries.

56 See Heine et al., (2002) for challenges in cross-cultural comparisons of subjective Likert scales.

57 See Ju et al. (2025).

58 These declines in the average frequency appear significantly (as indicated by non-overlapping 95% confidence regions) for each of the three benevolent acts as well as the prosocial umbrella variable. Donation frequency was down from 34.6% in 2023 to 31.7% in 2024, while helping strangers fell from 61.4% to 58.8%, and volunteering from 24.2% to 22.9%. The prosocial frequency fell from 73.1% to 70.0%.

59 For prosocial, 70% in 2022–2024 vs 63% 2017–2019. For donations, 31.7% vs 28%. For helping strangers 58.8% vs 50%, and for volunteering 22.9% vs 20%. These differences in means are all highly significant, $p<.001$.

60 Joshanloo and Bond (2023) use Gallup World Poll data from 2019 to investigate the extent to which religiosity, trust, income and individualism, which vary among countries and cultures, serve to moderate the links between volunteering and life satisfaction. When all moderations considered together, only trust and religiosity appear, both with negative moderating effects (their Model 5).

61 https://globalindices.indianapolis.iu.edu/environment-index/index.html

62 We speak of a significant difference for the regional data if there is no overlap in the 95% confidence levels for the estimate of the 2017–2019 and 2024 means.

63 Aknin and Whillans (2021); Dunn et al. (2014); Ryan and Deci (2000).

64 Aknin et al. (2013); Aknin et al. (2022); Curry et al. (2018); Hui et al. (2020); Joshanloo and Bond (2023); Kushlev et al. (2022). In addition, two chapters in previous World Happiness Reports have considered the joy of giving (Aknin et al., 2019; Rhoads & Marsh, 2023a).

65 Aknin et al. (2012); Song et al. (2020); Yang (2024).

66 See Chapter 4 of this report for more on caring and connection in family settings.

67 Algoe and Haidt (2009); Chancellor et al. (2018).

68 Fowler and Christakis (2010); Jordan et al. (2013); Tsvetkova and Macy (2014).

69 Wheeler et al. (1998).

70 Aknin et al. (2011).

71 Inagaki and Ross (2018).

72 Zhang et al. (2021).

73 Stukas et al. (2016).

74 Konrath et al. (2012).

75 In the psychology literature, relationships in which the primary concern is the person's welfare and people assist one another without monitoring contributions and/or expectation of reciprocity is called a "communal relationship" (see Clark & Mills, 2012).

76 Gallagher et al. (2022).

77 Chen et al. (2021); Xia et al. (2024).

78 Tham et al. (2024).

79 García-Campayo et al. (2024). Other research also finds that extreme altruists (i.e., individuals who donate organs to strangers) are higher on honesty-humility and general unselfishness (Rhoads et al., 2023; Rhoads & Marsh, 2023b).

80 Crocker and Canevello (2008); Huo et al. (2019).

81 Yin and Wang (2023).

82 Au et al. (2011); Depow et al. (2021); Ironson et al. (2002); Morelli et al. (2015); Peng et al. (2023); Steffen and Masters (2005).

8 Burtson and Stichler (2010); Von Harscher et al. (2018); Waddimba et al. (2021).

84 Ho et al. (2014).

85 Qu et al. (2020).

86 Batson (2011).

87 Das et al. (2023).

88 Le et al. (2013).

89 Canevello and Crocker (2011; 2017); Crocker and Canevello (2008); Crocker et al. (2010).

90 For volunteering: Clary and Snyder (1999); for charitable giving: Konrath and Handy (2018).

91 Veerasamy et al. (2015).

92 Stukas et al. (2016).

93 Poulin (2014).

94 Wiwad and Aknin (2017).

95 Stukas et al. (2016).

96 Konrath et al. (2012); Poulin (2014).

97 Wang et al. (2020). See also Brethel-Haurwitz et al. (2018) for research showing that altruistic kidney donors have a higher pain tolerance than controls.

98 Including those we share meals with, see Chapter 3 in this year's report.

99 Aknin and Whillans (2021); Dunn et al. (2014); Weinstein and Ryan (2010).

100 Gebauer et al. (2008); Qu et al. (2024).

101 Kim and Morgül (2017); see also work by Rinner et al. (2017) which shows that providing help that is freely chosen predicts greater personal wellbeing, but providing help that is forced predicts lower personal wellbeing.

102 Weinstein and Ryan (2010).

103 Lok and Dunn (2020).

104 Weinstein and Ryan (2010).

105 Ryan and Deci (2000).

106 Jenni and Lowenstein (1997).

107 Lee and Feely (2016); These considerations may also shape how people give to charity; see Chapter 8 for a discussion of how some charities may have a larger positive impact on recipients' wellbeing.

108 Aknin et al. (2013); see also Martela and Ryan (2016).

109 Switzer et al. (1996).

110 Ballard et al. (2019).

111 The data for the likelihood of wallet return as well as comparably measured negative risks posed by violent crime and poor mental health, are drawn from the Lloyd's Register Foundation World Risk Poll for 2019.

112 The estimated equation for the Cantril Ladder includes country fixed effects for which the coefficients are not shown. Control variables included in the equation but not shown in the figure are gender, age and age squared, married or living as a couple, separated divorced or widowed, having a health problem, and having a college education. The sample includes 123,050 observations from the 2019 Gallup World Poll supplemented by additional questions asked for the Lloyd's Register Foundation World Risk Poll.

113 The coefficients are .246 for donations, .100 for helping strangers and .089 for volunteering.

114 To guard against the possibility that this effect would be inflated by the use of individual data for expected wallet return, we have used the data for actual return of wallets in the 40 country sample of Cohn et al. (2019), and find that it adds significantly to the explanation of life evaluations in those countries. The estimated coefficient on the Cohn et al. (2019) return rate for wallets with money is 1.87 (t=4.4).

115 Standardised betas provide another way of comparing the explanatory power of the key variables in the equation. The standardised betas are +0.078 for expected wallet return (t=20.5), +0.048 (t=12.2) for performing all three benevolent acts in the previous month, -0.024 (t=6.5) for believing harm from violent crime very likely, -0.035 (t=9.3) for believing mental health problems very likely, and -0.040 (t=11.5) for being currently unemployed. The wallet return effect explains almost as large a share of total variance as the log of household income at 0.108 (t=23.7) or the five-point scale for self-assessed health, -0.097 (t=26.8).

116 The lower right panel of Figure 2.2 of *World Happiness Report 2022* shows that global inequality of wellbeing, not population-weighted, but treating each country with equal weight, increased by about 20% from 2007 to 2019.

117 Figure 2.5 of *World Happiness Report 2022* shows inequality converging between the two parts of Europe and generally rising in all other regions.

118 This is shown in Figures 2.3 and 2.4 of *World Happiness Report 2023*.

119 As measured by the within-country coefficient of variation (COV), within-country inequality has grown from .35 to .44 from 2005–2010 to 2020–2024.

120 As measured by the coefficient of variation (COV) of country level average life evaluations in each of the four time periods.

121 As measured by the r2 of a regression of individual ladder answers on a set of country dummies.

122 See Chapter 7 of this report for a discussion of how trust and life satisfaction predict voting behaviour.

123 Wu and Nugent (2024) do a parallel analysis using Gallup World Poll data, with broadly similar results to those shown here for ESS data.

124 Represented using a seven part Likert scale where 0 is none and 1 is every day.

125 These are the answers, converted to a 0 to 1 scale, to a question about how much others can be trusted, with a zero to 10 scale ranging from not at all to completely.

126 This variable is like that for general social trust, but in this case referring to trust in police.

127 As noted in the online ODA appendix, the two series are highly correlated (0.971) so that the regression results are very similar using the alternative definitions.

128 The untied ODA data are obtained from "DAC7B: Aid (ODA) tying status" in the OECD Data Explorer.

129 Gulrajani and Pudussery (2025).

130 For a meta-analysis, see Gao et al. (2022). For an earlier study with more equivocal findings, see Muñoz Sastre et al. (2003).

131 Toussaint and Friedman (2009).

132 See Mullet et al. (2021).

References

Aknin, L. B., Barrington-Leigh, C. P., Dunn, E. W., Helliwell, J. F., Burns, J., Biswas-Diener, R., Kemeza, I., Nyende, P., Ashton-James, C. E., & Norton, M. I. (2013). Prosocial spending and well-being: Cross-cultural evidence for a psychological universal. *Journal of Personality and Social Psychology, 104*(4), 635–652.

Aknin, L. B., Dunn, E. W., Whillans, A. V., Grant, A. M., & Norton, M. I. (2013). Making a difference matters: Impact unlocks the emotional benefits of prosocial spending. *Journal of Economic Behavior & Organization, 88*, 90-95.

Aknin, L. B., Dunn, E. W., & Whillans, A. V. (2022). The emotional rewards of prosocial spending are robust and replicable in large samples. *Current Directions in Psychological Science, 31*(6), 536-545.

Aknin, L. B., Hamlin, J. K., & Dunn, E. W. (2012). Giving leads to happiness in young children. *PLoS one, 7*(6), e39211.

Aknin, L. B., Sandstrom, G. M., Dunn, E. W., & Norton, M. I. (2011). It's the recipient that counts: Spending money on strong social ties leads to greater happiness than spending on weak social ties. *PLoS one, 6*(2), e17018.

Aknin, L. B., & Whillans, A. V. (2021). Helping and happiness: A review and guide for public policy. *Social Issues and Policy Review, 15*(1), 3-34.

Aknin, L. B., Whillans, A. V., Norton, M. I., & Dunn, E. W. (2019). Happiness and prosocial behavior: An evaluation of the evidence. *World Happiness Report*, 67-86.

Algoe, S. B., & Haidt, J. (2009). Witnessing excellence in action: The 'other-praising' emotions of elevation, gratitude, and admiration. *The Journal of Positive Psychology, 4*(2), 105-127.

Au, A., Wong, A., Lai, M. & Chan, C. (2011). Empathy, coping, social support, and mental health in local and migrant adolescents in Beijing. *International Journal on Disability and Human Development, 10*(3), 173-178.

Aydinli, A., Bender, M., & Chasiotis, A. (2013). Helping and volunteering across cultures: Determinants of prosocial behavior. *Online Readings in Psychology and Culture, 5*(3), 1-27.

Ballard, P. J., Hoyt, L. T., & Pachucki, M. C. (2019). Impacts of adolescent and young adult civic engagement on health and socioeconomic status in adulthood. *Child development, 90*(4), 1138-1154.

Batson, C. D. (2011). Altruism in humans. Oxford University Press.

Bloem, J. R., & Oswald, A. J. (2022). The analysis of human feelings: a practical suggestion for a robustness test. *Review of Income and Wealth*, 68(3), 689-710.

Brethel-Haurwitz, K. M., Cardinale, E. M., Vekaria, K. M., Robertson, E. L., Walitt, B., VanMeter, J. W., & Marsh, A. A. (2018). Extraordinary Altruists Exhibit Enhanced Self–Other Overlap in Neural Responses to Distress. *Psychological Science, 29*(10), 1631-1641.

Bond, T. N., & Lang, K. (2019). The sad truth about happiness scales. *Journal of Political Economy*, 127(4), 1629–1640.

Burtson, P. L., & Stichler, J. F. (2010). Nursing work environment and nurse caring: relationship among motivational factors. *Journal of Advanced Nursing, 66*(8), 1819-1831.

Canevello, A., & Crocker, J. (2011). Interpersonal goals, others' regard for the self, and self-esteem: The paradoxical consequences of self-image and compassionate goals. *European Journal of Social Psychology, 41*(4), 422-434.

Canevello, A., & Crocker, J. (2017). Compassionate goals and affect in social situations. *Motivation and Emotion, 41*, 158-179.

Chancellor, J., Margolis, S., Jacobs Bao, K., & Lyubomirsky, S. (2018). Everyday prosociality in the workplace: The reinforcing benefits of giving, getting, and glimpsing. *Emotion, 18*(4), 507-517.

Chen, E., Lam, P. H., Finegood, E. D., Turiano, N. A., Mroczek, D. K., & Miller, G. E. (2021). The balance of giving versus receiving social support and all-cause mortality in a US national sample. *Proceedings of the National Academy of Sciences, 118*(24), e2024770118

Chen, C., Lee, S. Y., & Stevenson, H. W. (1995). Response style and cross-cultural comparisons of rating scales among East Asian and North American students. *Psychological Science*, 6(3), 170–175.

Chida, Y., & Steptoe, A. (2008). Positive psychological well-being and mortality: a quantitative review of prospective observational studies. *Psychosomatic Medicine*, 70(7), 741-756.

Chopik, W. J., O'Brien, E., & Konrath, S. H. (2017). Differences in empathic concern and perspective taking across 63 countries. *Journal of Cross-Cultural Psychology, 48*(1), 23-38.

Clark, M. S., & Mills, J. R. (2012). A theory of communal (and exchange) relationships. In P. A. M. Van Lange, A. W. Kruglanski, & E. T. Higgins (Eds.), *Handbook of theories of social psychology* (pp. 232–250). Sage Publications Ltd. https://doi.org/10.4135/9781446249222.n38

Clary, E. G., & Snyder, M. (1999). The motivations to volunteer: Theoretical and practical considerations. *Current Directions in Psychological Science*, 8(5), 156-159.

Cohn, A., Maréchal, M. A., Tannenbaum, D., & Zünd, C. L. (2019). Civic honesty around the globe. *Science*, 365(6448), 70-73.

Cohen, S., Doyle, W. J., Turner, R. B., Alper, C. M., & Skoner, D. P. (2003). Emotional style and susceptibility to the common cold. *Psychosomatic Medicine, 65*(4), 652-657.

Crocker, J., & Canevello, A. (2008). Creating and undermining social support in communal relationships: The role of compassionate and self-image goals. *Journal of Personality and Social Psychology, 95*(3), 555-575.

Crocker, J., Canevello, A., Breines, J. G., & Flynn, H. (2010). Interpersonal goals and change in anxiety and dysphoria in first-semester college students. *Journal of Personality and Social Psychology, 98*(6), 1009-1024.

Curry, O. S., Rowland, L. A., Van Lissa, C. J., Zlotowitz, S., McAlaney, J., & Whitehouse, H. (2018). Happy to help? A systematic review and meta-analysis of the effects of performing acts of kindness on the well-being of the actor. *Journal of Experimental Social Psychology*, 76, 320-329.

Danner, D. D., Snowdon, D. A., & Friesen, W. V. (2001). Positive emotions in early life and longevity: Findings from the nun study. *Journal of Personality and Social Psychology, 80*(5), 804–813.

Das, G., van Esch, P., Jain, S. P., & Cui, Y. G. (2023). Donor happiness comes from afar: The role of donation beneficiary social distance and benevolence. *International Journal of Research in Marketing, 40*(4), 865-880.

De Neve, J. E., Diener, E., Tay, L., & Xuereb, C. (2013). The objective benefits of subjective well-being. In J. F. Helliwell, R. Layard, & J. Sachs (Eds.), *World Happiness Report 2013* (pp. 54–79). New York: SDSN.

Depow, G. J., Francis, Z., & Inzlicht, M. (2021). The experience of empathy in everyday life. *Psychological Science, 32*(8), 1198-1213.

Doyle, W. J., Gentile, D. A., & Cohen, S. (2006). Emotional style, nasal cytokines, and illness expression after experimental rhinovirus exposure. *Brain, Behavior, and Immunity*, 20(2), 175-181.

Dunn, E. W., Aknin, L. B., & Norton, M. I. (2014). Prosocial spending and happiness: Using money to benefit others pays off. *Current Directions in Psychological Science, 23*(1), 41-47.

Ferrer-i-Carbonell, A., & Frijters, P. (2004). How important is methodology for the estimates of the determinants of happiness?. *The Economic Journal, 114*(497), 641–659.

Fowler, J. H., & Christakis, N. A. (2010). Cooperative behavior cascades in human social networks. *Proceedings of the National Academy of Sciences, 107*(12), 5334-5338.

Fredrickson, B. L. (2001). The role of positive emotions in positive psychology: The broaden-and-build theory of positive emotions. *American Psychologist, 56*(3), 218–226.

Gallagher, S., Haugh, C., Castro Solano, A., de la Iglesia, G., & McMahon, J. (2022). Social support imbalance and depressive symptoms in young adolescents: The negative effect of giving but not receiving. *International Journal of Adolescence and Youth, 27*(1), 528-540.

Gao, F., Li, Y., & Bai, X. (2022). Forgiveness and subjective well-being: A meta-analysis review. *Personality and Individual Differences*, 186, 111350.

García-Campayo, J., Barcelo-Soler, A., Martínez-Rubio, D., Navarrete, J., Perez-Aranda, A., Feliu-Soler, A., ... & Montero-Marin, J. (2024). Exploring the relationship between self-compassion and compassion for others: The role of psychological distress and wellbeing. *Assessment, 31*(5), 1038-1051.

Gulrajani, N., & Pudussery, J. (2025). *With the knives out on development spending, have we reached "Peak aid"?* . The Guardian. https://www.theguardian.com/global-development/2025/jan/23/global-development-economics-donor-spending-refugee-oecd-world-bank-peak-aid

Hampton, K. N. (2016). Why is helping behavior declining in the United States but not in Canada?: Ethnic diversity, new technologies, and other explanations. *City & Community, 15*(4), 380-399.

Heine, S. J., Lehman, D. R., Peng, K., & Greenholtz, J. (2002). What's wrong with cross-cultural comparisons of subjective Likert scales?: The reference-group effect. *Journal of Personality and Social Psychology, 82*(6), 903-918.

Helliwell, J. F., & Aknin, L. B. (2018). Expanding the social science of happiness. *Nature Human Behaviour*, 2(4), 248-252.

Helliwell, J. F., Huang, H., & Wang, S. (2018). New Evidence on Trust and Well-Being. *The Oxford Handbook of Social and Political Trust*, 409.

Helliwell, J. F., Layard, R., Sachs, J. D., De Neve, J.-E., Aknin, L. B., & Wang, S. (Eds.). (2024). World Happiness Report 2024. University of Oxford: Wellbeing Research Centre.

Helliwell, J. F., Shiplett, H., & Bonikowska, A. (2020). Migration as a test of the happiness set-point hypothesis: Evidence from immigration to Canada and the United Kingdom. *Canadian Journal of Economics/Revue canadienne d'économique, 53*(4), 1618–1641.

Helliwell, J. F., & Wang, S. (2011). Trust and wellbeing. *International Journal of Wellbeing, 1*, 1, 42-78.

Ho, S. S., Konrath, S., Brown, S., & Swain, J. E. (2014). Empathy and stress related neural responses in maternal decision making. *Frontiers in Neuroscience, 8*, 152.

Hui, B. P., Ng, J. C., Berzaghi, E., Cunningham-Amos, L. A., & Kogan, A. (2020). Rewards of kindness? A meta-analysis of the link between prosociality and well-being. *Psychological Bulletin, 146*(12), 1084.

Huo, M., Fuentecilla, J. L., Birditt, K. S., & Fingerman, K. L. (2019). Older adults' empathy and daily support exchanges. *Journal of Social and Personal Relationships, 36*(11-12), 3814-3834

Inagaki, T. K., & Ross, L. P. (2018). Neural correlates of giving social support: Differences between giving targeted versus untargeted support. *Psychosomatic Medicine, 80*(8), 724-732.

Ironson, G., Solomon, G. F., Balbin, E. G., O'cleirigh, C., George, A., Kumar, M., ... & Woods, T. E. (2002). The Ironson-Woods Spirituality/Religiousness Index is associated with long survival, health behaviors, less distress, and low cortisol in people with HIV/AIDS. *Annals of Behavioral Medicine, 24*(1), 34-48.

Jordan, J. J., Rand, D. G., Arbesman, S., Fowler, J. H., & Christakis, N. A. (2013). Contagion of cooperation in static and fluid social networks. *PLoS ONE, 8*(6), e66199.

Joshanloo, M., & Bond, M. H. (2023). National wealth, individualism, generalised trust, and religiosity as moderators of the relationship between helping strangers and life satisfaction in 137 societies. *International Journal of Psychology*, 58(2), 178-186.

Ju, E., Qi, H., Zhao, L., Luo, Y., Li, Y., & You, X. (2025). The Longitudinal Relationship between Adolescents' Prosocial Behavior and Well-Being: A Cross-Lagged Panel Network Analysis. *Journal of Youth and Adolescence*, 1-14.

Kaiser, C., & Oswald, A. J. (2022). The scientific value of numerical measures of human feelings. *Proceedings of the National Academy of Sciences, 119*(42), e2210412119.

Kim, J., & Morgül, K. (2017). Long-term consequences of youth volunteering: Voluntary versus involuntary service. *Social Science Research, 67*, 160-175.

Konrath, S., Fuhrel-Forbis, A., Lou, A., & Brown, S. (2012). Motives for volunteering are associated with mortality risk in older adults. *Health Psychology, 31*(1), 87-96.

Konrath, S., & Handy, F. (2018). The development and validation of the motives to donate scale. *Nonprofit and Voluntary Sector Quarterly*, *47*(2), 347-375.

Konrath, S., Martingano, A. J., Davis, M., & Breithaupt, F. (2023). Empathy trends in American youth between 1979 and 2018: an update. *Social Psychological and Personality Science*, 19485506231218360.

Konrath, S. H., O'Brien, E. H., & Hsing, C. (2011). Changes in dispositional empathy in American college students over time: A meta-analysis. Personality and Social Psychology Review, 15(2), 180-198.

Knack, S. (2001) Trust, associational life, and economic performance. In John F. Helliwell and Aneta Bonikowska, eds. *The Contribution of Human and Social Capital to Sustained Economic Growth and Well-Being*, 172-202, Ottawa: HRDC and OECD.

Knafo, A., Schwartz, S. H., & Levine, R. V. (2009). Helping strangers is lower in embedded cultures. *Journal of Cross-Cultural Psychology*, *40*(5), 875-879.

Krueger, A. B., & Schkade, D. A. (2008). The reliability of subjective well-being measures. *Journal of Public Economics*, *92*(8-9), 1833–1845.

Kushlev, K., Radosic, N., & Diener, E. (2022). Subjective well-being and prosociality around the globe: Happy people give more of their time and money to others. *Social Psychological and Personality Science*, *13*(4), 849-861.

Le, B. M., Impett, E. A., Kogan, A., Webster, G. D., & Cheng, C. (2013). The personal and interpersonal rewards of communal orientation. *Journal of Social and Personal Relationships*, *30*(6), 694-710.

Levine, R. V., Reysen, S., & Ganz, E. (2008). The kindness of strangers revisited: A comparison of 24 US cities. *Social Indicators Research*, *85*, 461-481.

Luria, G., Cnaan, R. A., & Boehm, A. (2015). National culture and prosocial behaviors: Results from 66 countries. *Nonprofit and Voluntary Sector Quarterly*, *44*(5), 1041-1065.

Martela, F., & Ryan, R. M. (2016). The benefits of benevolence: Basic psychological needs, beneficence, and the enhancement of well-being. *Journal of Personality*, *84*(6), 750-764.

Mogstad, M., Romano J.P., Shaikh, A. & Wilhelm D. (2024) Inference for Ranks with Applications to Mobility across Neighbourhoods and Academic Achievement across Countries, *The Review of Economic Studies*, *91*, 1, 476–518.

Morelli, S. A., Lee, I. A., Arnn, M. E., & Zaki, J. (2015). Emotional and instrumental support provision interact to predict well-being. *Emotion*, *15*(4), 484-493.

Mullet, E., López, López, W., & Pineda Marín, C. (2021). Forgiveness and reconciliation in post-conflict settings. Transitioning to peace: Promoting global social justice and non-violence, 67-90.

Muñoz Sastre, M. T., Vinsonneau, G., Neto, F., Girard, M., & Mullet, E. (2003). Forgivingness and satisfaction with life. *Journal of Happiness Studies*, *4*(3), 323-335.

Peng, A., Patterson, M. M., & Wang, H. (2023). Attachment, empathy, emotion regulation, and subjective well-being in young women. *Journal of Applied Developmental Psychology*, *84*, 101497.

Poulin, M. J. (2014). Volunteering predicts health among those who value others: Two national studies. *Health Psychology*, *33*(2), 120-129.

Pressman, S. D., Jenkins, B. N., & Moskowitz, J. T. (2019). Positive affect and health: What do we know and where next should we go?. *Annual Review of Psychology*, 70, 627-650.

Qu, G., Ju, E., Qin, G., Chen, X., & Luo, Y. (2024). Profiles of volunteers' motivations and positive experiences relate to their sustained volunteering during the COVID-19 pandemic. *Journal of Community & Applied Social Psychology*, *34*(1), e2748.

Qu, H., Konrath, S., & Poulin, M. (2020). Which types of giving are associated with reduced mortality risk among older adults?. *Personality and Individual Differences*, *154*, 109668.

Reader's Digest (2023) Most honest cities: the Reader's Digest 'lost wallet' test. https://www.rd.com/list/most-honest-cities-lost-wallet-test/

Rhoads, S. A., Gunter, D., Ryan, R. M., & Marsh, A. A. (2021). Global variation in subjective well-being predicts seven forms of altruism. *Psychological Science*, *32*(8), 1247-1261.

Rhoads, S. A., & Marsh, A. A. (2023a). Doing Good and Feeling Good: Relationships Between Altruism and Well-being for Altruists, Beneficiaries, and Observers. In *World Happiness Report 2023* (11th ed., Chapter 4). Sustainable Development Solutions Network.

Rhoads, S. A., & Marsh, A. A. (2023b). Traits of Extraordinary Altruists. In *Encyclopedia of Heroism Studies* (pp. 1-4). Cham: Springer International Publishing.

Rhoads, S. A., Vekaria, K. M., O'Connell, K., Elizabeth, H. S., Rand, D. G., Kozak Williams, M. N., & Marsh, A. A. (2023). Unselfish traits and social decision-making patterns characterize six populations of real-world extraordinary altruists. *Nature Communications*, *14*(1), 1807.

Rinner, M. T., Haller, E., Meyer, A. H., & Gloster, A. T. (2022). Is giving receiving? The influence of autonomy on the association between prosocial behavior and well-being. *Journal of Contextual Behavioral Science*, *24*, 120-125.

Rober, M. (2019) 200 dropped wallets: the 20 most and least honest cities. https://www.youtube.com/watch?v=jnL7sJYblGY

Rojas, M. (2018). Happiness in Latin America has social foundations. In J. F. Helliwell, R. Layard, & J. Sachs (Eds.), *World happiness report 2018* (pp. 114–145). New York: SDSN.

Ryan, R. M., & Deci, E. L. (2000). Self-determination theory and the facilitation of intrinsic motivation, social development, and well-being. *American Psychologist*, *55*(1), 68-78.

Smith, P. B. (2019). Changes in reported nation-level pro-social behavior frequencies over 6 years: a test of alternative predictors. *Social Indicators Research*, *144*(3), 1195-1208.

Song, Y., Broekhuizen, M. L., & Dubas, J. S. (2020). Happy little benefactor: Prosocial behaviors promote happiness in young children from two cultures. *Frontiers in Psychology*, *11*, 1398.

Sortheix, F. M., Parker, P. D., Lechner, C. M., & Schwartz, S. H. (2019). Changes in young Europeans' values during the global financial crisis. *Social Psychological and Personality Science, 10*(1), 15-25.

Steffen, P. R., & Masters, K. S. (2005). Does compassion mediate the intrinsic religion-health relationship?. *Annals of Behavioral Medicine*, *30*(3), 217-224.

Stukas, A. A., Hoye, R., Nicholson, M., Brown, K. M., & Aisbett, L. (2016). Motivations to volunteer and their associations with volunteers' well-being. *Nonprofit and Voluntary Sector Quarterly*, *45*(1), 112-132.

Tham, Y. J., Okamoto, S., & Kobayashi, E. (2024). The importance of examining both the amount and balance of social support: A study on the relationship between social support and subjective well-being of older Japanese adults. *Journal of Community & Applied Social Psychology*, *34*(2), e2783.

Toussaint, L., & Friedman, P. (2009). Forgiveness, gratitude, and well-being: The mediating role of affect and beliefs. *Journal of Happiness Studies*, 10, 635-654.

Tsvetkova, M., & Macy, M. W. (2014). The Social Contagion of Generosity. *PLoS ONE*, *9*(2), e87275.

Veerasamy, C., Sambasivan, M., & Kumar, N. (2015). Life satisfaction among healthcare volunteers in Malaysia: Role of personality factors, volunteering motives, and spiritual capital. VOLUNTAS: International *Journal of Voluntary and Nonprofit Organizations*, *26*, 531-552.

Von Harscher, H., Desmarais, N., Dollinger, R., Grossman, S., & Aldana, S. (2018). The impact of empathy on burnout in medical students: new findings. *Psychology, Health & Medicine*, *23*(3), 295-303.

Waddimba, A. C., Bennett, M. M., Fresnedo, M., Ledbetter, T. G., & Warren, A. M. (2021). Resilience, well-being, and empathy among private practice physicians and advanced practice providers in Texas: A structural equation model study. *Mayo Clinic Proceedings: Innovations, Quality & Outcomes*, *5*(5), 928-945.

Wang, Y., Ge, J., Zhang, H., Wang, H., & Xie, X. (2020). Altruistic behaviors relieve physical pain. *Proceedings of the National Academy of Sciences*, *117*(2), 950-958.

Wheeler, J. A., Gorey, K. M., & Greenblatt, B. (1998). The beneficial effects of volunteering for older volunteers and the people they serve: A meta-analysis. *The International Journal of Aging and Human Development*, *47*(1), 69-79.

Wiwad, D., & Aknin, L. B. (2017). Motives matter: The emotional consequences of recalled self-and other-focused prosocial acts. *Motivation and Emotion*, *41*, 730-740.

Xia, W., van Wijngaarden, J. D., Huijsman, R., & Buljac-Samardžić, M. (2024). The effect of long-term (im)balance of giving versus receiving support with nonrelatives on subjective well-being among home-dwelling older people. *The Journals of Gerontology, Series B: Psychological Sciences and Social Sciences*, *79*(4), gbad198.

Yin, Y., & Wang, Y. (2023). Is empathy associated with more prosocial behaviour? A meta-analysis. *Asian Journal of Social Psychology*, *26*(1), 3-22.

Yan, Z. Q., Su, J. L., & Su, Y. J. (2017). Increasing empathy in Chinese college students: a cross-temporal meta-analysis. *Psychology of Technology Applied*, *5*(10), 578-585.

Yang, F. (2024). Being good and feeling good: What happiness means to children. *Child Development Perspectives*. https://doi.org/10.1111/cdep.12522

Yuan, M., Spadaro, G., Jin, S., Wu, J., Kou, Y., Van Lange, P. A., & Balliet, D. (2022). Did cooperation among strangers decline in the United States? A cross-temporal meta-analysis of social dilemmas (1956–2017). *Psychological Bulletin*, *148*(3-4), 129.

Wu, F., & Nugent, J. B. (2024). Mitigating Life Challenges to Subjective Well-being through Civic Engagement: Insights from a Global Perspective (No. 1489). GLO Discussion Paper.

Zhang, W., Pan, J., Liu, J., Zhang, Y., & Chen, M. (2021). Recipients' happiness in prosocial spending: The role of social ties. *Journal of Consumer Affairs*, *55*(4), 1333-1351.

Chapter 3

Sharing meals with others

How sharing meals supports happiness and social connections

Jan-Emmanuel De Neve
University of Oxford

Andrew Dugan
Gallup

Micah Kaats
Harvard Kennedy School

Alberto Prati
University College London

The authors are grateful for the use of the Gallup World Poll and, in particular, the special world poll survey module on food and meal sharing funded by the Ajinomoto Group on which this chapter relies. We are also grateful for edits and feedback from John Helliwell, Lara Aknin, Richard Layard, Barry Grimes, and the participants of the WHR workshop in Rome.

doi.org/10.18724/whr-g119-bv60

This version was last updated on 13 March 2025. Please see worldhappiness.report for the latest data.

The connection between food and social relationships is far from new. In French, *copain* (friend) and in Italian *compagno* (mate) come from the Latin *cum+pañis*, literally “with-bread”.

Key Insights

For over a decade, the *World Happiness Report* has shown that social connections are important drivers of happiness, both at the individual and national level, and across cultures.

In this chapter, we present new evidence on an understudied measure of social connection – sharing meals. Given the relatively objective way in which it is measured, sharing meals is uniquely comparable across countries and cultures, between individuals, and over time.

Using novel data for 142 countries and territories collected by Gallup in 2022 and 2023, we find stark differences in rates of meal sharing around the world. While residents of some countries share almost all of their meals with other people, residents of other countries eat almost all of their meals alone. These differences are not fully explained by differences in income, education, or employment.

Sharing meals proves to be an exceptionally strong indicator of subjective wellbeing – on par with income and unemployment. Those who share more meals with others report significantly higher levels of life satisfaction and positive affect, and lower levels of negative affect. This is true across ages, genders, countries, cultures, and regions.

In the United States, using data from the American Time Use Survey, we find clear evidence that Americans are spending more and more time dining alone. In 2023, roughly 1 in 4 Americans reported eating all of their meals alone the previous day – an increase of 53% since 2003. Dining alone has become more prevalent for every age group, but especially for young people.

Meal sharing also appears to be closely related to some, but not all, measures of social connectedness. Most notably, countries where people share more meals have higher levels of social support and positive reciprocity, and lower levels of loneliness.

Nevertheless, there remain vast gaps in our understanding of the causal dynamics of meal sharing, subjective wellbeing, and social connections. We point to a number of promising avenues for future research and discuss implications for policy.

Introduction

Social connections are critically important for human health, happiness, and prosperity. People who are more socially connected tend to be happier, less stressed, more satisfied with their lives, less prone to depression, more engaged in their communities, and less likely to suffer from disease or disability.[1] In their professional lives, people with more social connections are more creative, cooperative, trusting, and likely to be promoted.[2] They are less likely to commit crimes, earn higher levels of income, and live longer lives.[3]

At the same time, social isolation and loneliness are strongly associated with negative life outcomes. The absence of social ties has been linked to higher rates of disease, shorter life expectancies, lower levels of subjective wellbeing, higher rates of criminality, and greater support for authoritarianism.[4] One widely cited meta-analysis estimated that the negative health consequences of loneliness and isolation were roughly equivalent to smoking 15 cigarettes a day.[5] In short, to paraphrase Dr. Chris Peterson, one of the founding fathers of positive psychology - other people matter.

Social connections are not only important for individual health and happiness, but also for societal health and happiness writ large. People who are more connected to each other are more trusting of others and have more faith in institutions.[6] They are more likely to donate to charity, be more politically engaged, and report higher levels of pride in their communities.[7] They tend to be more considerate and compassionate, not only towards friends and family, but also towards strangers. They are more likely to volunteer time to help those in need and share resources with others.[8]

In this chapter, we explore links between sharing meals, social connections, and wellbeing. Although the topic of sharing meals has remained relatively understudied in the academic literature, the connection between food and social relationships is far from new. In French, *copain* (friend) and in Italian *compagno* (mate) come from the Latin *cum+pa-nis*, literally "with-bread". The Chinese term for companion/partner, 伙伴, stems from a similar term (火伴) which literally translates to "fire mate", a reference to sharing meals over a campfire.

Recently, an emerging body of empirical evidence has begun to point to potential links between sharing meals and a range of social benefits. One review of the literature found that adolescents who ate more meals with family members had better diet and nutritional habits, lower levels of obesity, fewer eating disorders, and greater academic achievement.[9] Another experiment found strong links between meal sharing and positive affect, although these effects were diminished with increased smartphone use during meal times.[10] Yet another study of roughly 9,000 older adults in China found that sharing meals with others was associated with lower rates of depression.[11]

In this chapter, we extend this body of work by looking at the relationship between sharing meals and wellbeing using novel data collected on a global scale. We present evidence from the first-ever global dataset on social eating, collected in 2022 as part of the Ajinomoto module on the Gallup World Poll. In 2023, Gallup asked these questions again in 17 countries. More than 150,000 people from around the world answered the following two questions: "Thinking about the last 7 days ... (i) On how many days did you eat lunch with someone you know? (ii) On how many days did you eat dinner with someone you know?" In addition to this new dataset, we present new evidence from the American Time Use Survey (ATUS) in the United States on the association between sharing meals, social connections, and wellbeing over time.

Our aim is twofold. First, we explore the extent to which sharing meals can serve as an indicator of social connectedness. In this respect, the number of meals shared with others has a number of advantages compared to existing proxies. The act of sharing a meal is relatively objective and straightforward to report. Even if it is self-reported (as are all survey questions), the number of shared meals is an observable and objective aspect of people's lives. Conversely, many other measures of social connectedness rely on more subjective assessments. For example, survey respondents may be asked to report how close they feel to friends and family, how many close relationships they have in general, or how often they feel isolated or left out.

While undeniably valuable, these sorts of questions present a series of challenges for researchers studying the relationship between subjective measures of wellbeing and subjective indicators of social connection. For one, statistical correlations between all subjective measures tend to be artificially high to begin with. This can make it seem like subjectively reported variables are more closely related to each other than they really are. Moreover, it is often difficult to assess whether one person's self-report of a close social connection is directly comparable to someone else's. These problems are exacerbated when trying to make comparisons across countries and cultures, or over long periods of time.

A metric based on the number of shared meals is poised to address these issues. Sharing meals is a cross-cultural social ritual, practiced every day by millions of people. It is a universal practice. This is particularly useful when studying social connections and wellbeing on a global scale as it allows for relatively reliable international and intercultural comparisons. The number of meals shared with others is also much more objectively comparable over time than related measures of social engagement. While interpretations of closeness or belonging may evolve and change over time, the number of meals shared with others is not expected to. In this way, our approach is operationally similar to other well-established questions to measure related aspects of human capital. For example, the question "How many books were there in your home when you were 16?" is routinely used by international surveys to measure parental cultural capital.[12]

Our second aim in this chapter is to consider the relationship between sharing meals and subjective wellbeing. Given the strong link between wellbeing and social connections, sharing meals with others may be an important indicator of positive wellbeing. This indeed turns out to be the case. We present the largest and most robust evidence to date showing that sharing meals with others is strongly predictive of greater life evaluations, increased positive affect, and decreased negative affect. We also find that dining alone is at least as (if not even more) strongly associated with low levels of wellbeing.

We present the largest and most robust evidence to date showing that sharing meals with others is strongly predictive of greater life evaluations, increased positive affect, and decreased negative affect.

However, studying the correlation between sharing meals and wellbeing raises an important issue of causality. Does sharing meals make people happier? Or do people who are happy to begin with share more meals? Or, perhaps even more likely, is the relationship bi-directional? These are important questions with significant implications for research and policy. We do not conclusively resolve them here. While we discuss preliminary evidence and efforts to get at the underlying causal dynamics of meal sharing and subjective wellbeing, arriving at a full answer to this question is a task that remains open to future research.

To begin, we present new global evidence on the variation in meal sharing and dining alone around the world. We then turn to the relationship between sharing meals and subjective wellbeing. Specifically, we consider the extent to which sharing meals with others is associated with life evaluations, positive affect, and negative affect. Next, we generate novel indicators of meal sharing and dining alone using data from the American Time Use Survey (ATUS) to take a closer look at links between sharing meals and subjective wellbeing over time in the United States. Finally, we consider associations between meal sharing and a range of related social indicators. We conclude with a discussion of policy implications and point to a number of promising avenues for future research.

Sharing meals around the world

In 2022 and 2023, the Gallup World Poll asked representative samples in 142 countries and territories how often they ate lunch or dinner with family, friends, or anyone else they knew.[13] In Figure 3.1, we present regional differences in meal sharing, broken down by lunches and dinners. Additional regional descriptive statistics are provided in Table A1 of the online appendix.

Overall, we find stark differences in the frequency of dining with others and dining alone around the world. Latin America and the Caribbean emerges as the global leader in meal sharing. On average, residents of these countries share approximately 9 meals with other people per week. At the bottom of the list is South Asia, where people report eating fewer than 4 meals with others per week.

The relatively low levels of meal sharing in both South and East Asian countries is particularly notable. Past research has found that dining alone is on the rise in East Asian countries, most notably in Japan and the Republic of Korea.[14] Two of the most commonly cited explanations are the rise of single-person households and demographic ageing. However, differences in the interpretation of the survey items used to measure meal sharing may also play a role. There are some indications that East and South Asian respondents may be less likely to consider family members or other members of their household as "someone you know."[15] Whatever the underlying explanation, the considerably low rates of meal sharing in these regions clearly warrant further investigation.

However, these regional differences also mask significant variation across countries. In Figure 3.2, we present rates of meal sharing for all countries. Full country rankings are provided in Table A2 of the online appendix. Senegal tops the list, where residents report sharing 11.7 meals with others per week on average. Gambia, Malaysia, and Paraguay come next, where residents report sharing approximately 11 meals with others per

Figure 3.1: Meal sharing by region
Gallup World Poll (2022–2023)

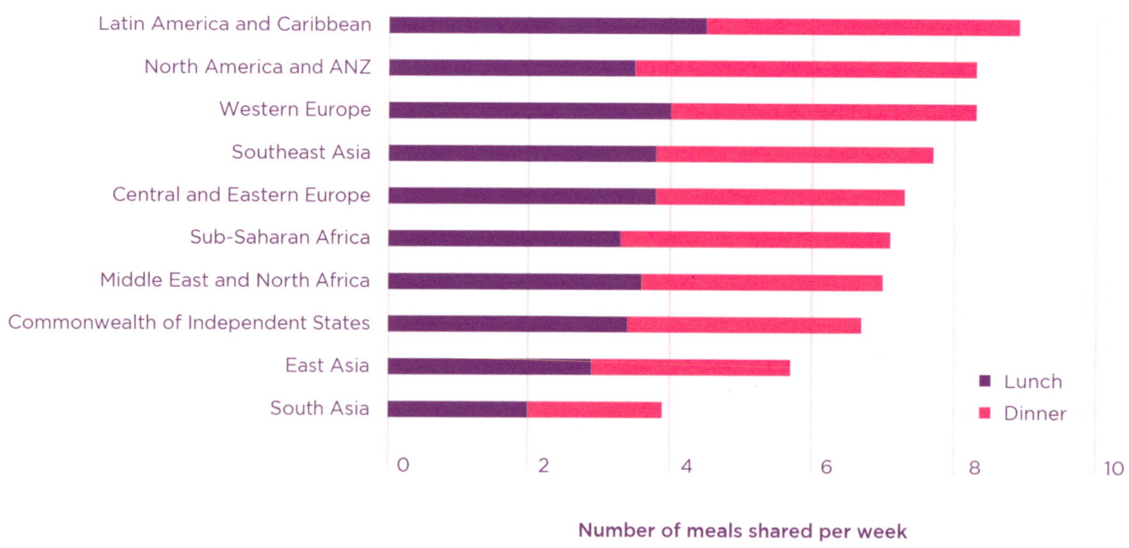

Figure 3.2: Meal sharing by country
Gallup World Poll (2022–2023)

Average number of lunches and dinners eaten "with someone you know" in the past 7 days

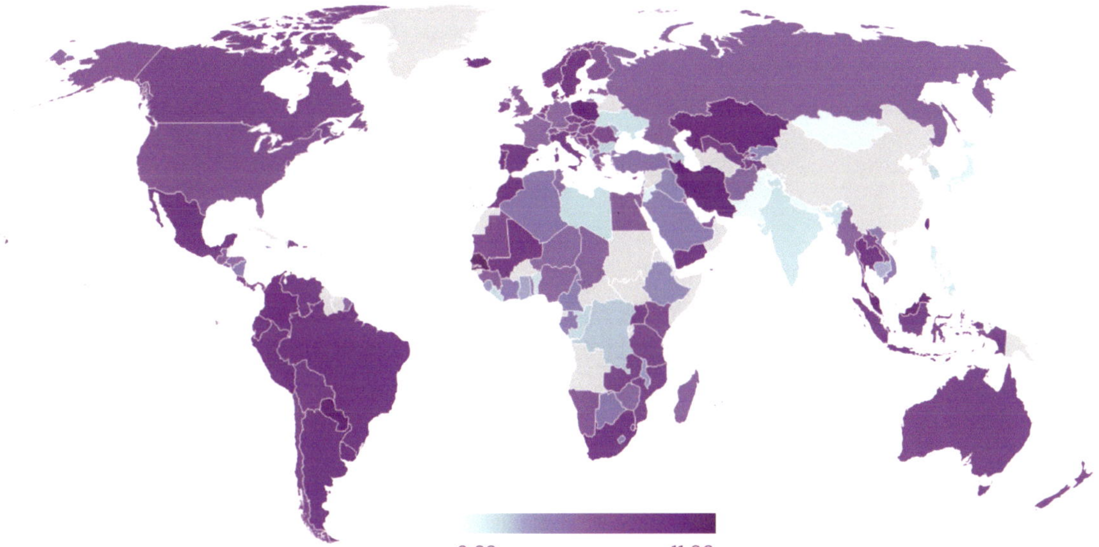

Note: Grey regions denote places for which there is no data

week. Iceland is the only country from Europe or North America represented in the top 10 with an average of 10 meals shared per week. Canada ranks 53rd with 8.4 meals shared per week, the United States ranks 69th, and the United Kingdom ranks 81st. Germany appears in 91st place, while India ranks 132nd with 4 meals shared per week. At the very bottom of the list are Bangladesh and Estonia, where residents report sharing only 2.7 meals per week.

We present additional maps with lunches and dinners considered separately in Figures A1 and A2 of the online appendix. Overall, the dynamics are broadly consistent with Figure 3.2. Senegal and Gambia continue to rank highly in both categories, Iceland jumps to second place for shared dinners, and Middle Eastern countries including Iran and Morocco move closer to the top for shared lunches. Residents in the United States, Canada, New Zealand, and Australia are much more likely to eat dinner with others than they are to share lunches. On average, residents of these countries share roughly 5 dinners per week – more than twice as much as residents in the Republic of Korea, Japan, and Mongolia where respondents report sharing just 1 to 2 dinners per week.

The underlying explanations for these differences are sure to be complex and multifaceted. Nevertheless, explanations that appeal to differences in income alone seem unlikely. For example, one potential interpretation of our results could be that people who eat more meals overall also share more meals with other people. If so, one might expect that residents of high-income countries would eat more meals overall, and therefore eat more meals with other people than residents of low-income countries.

However, the fact that low-income countries in Sub-Saharan Africa and Latin America report such high levels of meal sharing casts doubt on the assumption that more meals eaten with others is simply a function of more meals eaten overall. Although we do observe a moderate and statistically significant correlation of 0.2 between income and meal sharing at the country level, this association explains only 4.6% of the global variation in meal sharing.[16] Explaining the other 95.4% represents a rich opportunity for future research.

In Figure 3.3, we extend our analysis by plotting the number of shared meals for all regions broken down by age. In almost every region, younger people share more meals with others. This is an important difference worth further study. If meal sharing is a strong proxy for (and potentially a causal contributor to) subjective wellbeing, then age-related differences in meal sharing may shed new light on differences and changes in wellbeing across the lifespan, and over time.

Considering gender, men and women report similar numbers of meals shared per week around the world. Across all regions, we find that gender

Figure 3.3: Meal sharing by region and age
Gallup World Poll (2022–2023)

differences in meal sharing are statistically insignificant. These results are presented in Figure A3 in the online appendix.

Sharing meals and wellbeing

In the previous section, we documented considerable differences in meal sharing around the world. In this section, we consider what, if anything, these variations can tell us about corresponding differences in subjective wellbeing. Our focus throughout this section will be specifically on life evaluations, positive affect, and negative affect. Once again, we rely on survey responses from the Gallup World Poll in 2022 and 2023.

In Figure 3.4, we present the overall relationship between life evaluations measured using the Cantril Ladder[17] and the total number of meals shared with others in the previous week. We calculate country averages for both variables so that each point on the graph represents a different country.

Overall, we find a positive relationship between sharing meals and life evaluations. Across countries, sharing one more meal per week is associated with an average increase of roughly 0.2 points on a scale from 0 to 10. This difference is both statistically significant and practically meaningful. A difference of 0.2 points is roughly equivalent to a difference of five places in the global happiness rankings presented in Chapter 2 of this report.

However, to better understand the relationship between sharing meals and subjective wellbeing, it is worth diving deeper into the data to consider differences in meal sharing and subjective wellbeing across individuals.

Fig. 3.4: Meal sharing and life evaluations around the world
Gallup World Poll (2022–2023)

Figure 3.5 shows average life evaluations by the number of meals shared per week based on individual comparisons rather than country or regional averages. In general, we see an upward trend - albeit a subtle, uneven one - in average life evaluations as the number of shared meals increases. The largest difference in life evaluations is between those who eat all meals alone and those who eat one meal with someone else. People who shared just one meal in the past week have notably higher life evaluations (5.2) than those who ate all meals alone (4.9). This 0.3-point difference is again statistically and practically significant. For context, it is about half as large as the decline in life evaluation associated with unemployment, which is consistently found to be one of the largest effects documented in the wellbeing literature.[18]

From there, life evaluations tend to increase as the number of shared meals increases.[19] Life evaluations are broadly constant (5.2 to 5.3) for people who shared 1–5 meals in the past week and uptick slightly (5.5 to 5.6) for those who ate 5–8 meals with others. People who shared 9–10 meals reported average life evaluations of 5.7, which increases to 5.8 for those who shared 11 meals per week. The peak occurs for those who shared 13 meals with others in the previous week, reporting average life evaluations of 6.1.

In Figures 3.6 and 3.7, we plot relationships between meal sharing and wellbeing by gender and age. Even at this fine-grained level of analysis, we find strong and significant associations between sharing meals and subjective wellbeing. Sharing meals not only predicts more positive life evaluations, but also higher levels of positive affect and lower levels of negative affect. The relationship between sharing meals and positive affect is particularly strong - even stronger than the relationship between sharing meals and life evaluations. Overall, we estimate the correlation between positive affect and meal sharing to be 0.44. Correlations for life evaluations and negative affect are 0.34 and -0.21, respectively.

Figure 3.5: Life evaluations by number of meals shared per week
Gallup World Poll (2022–2023)

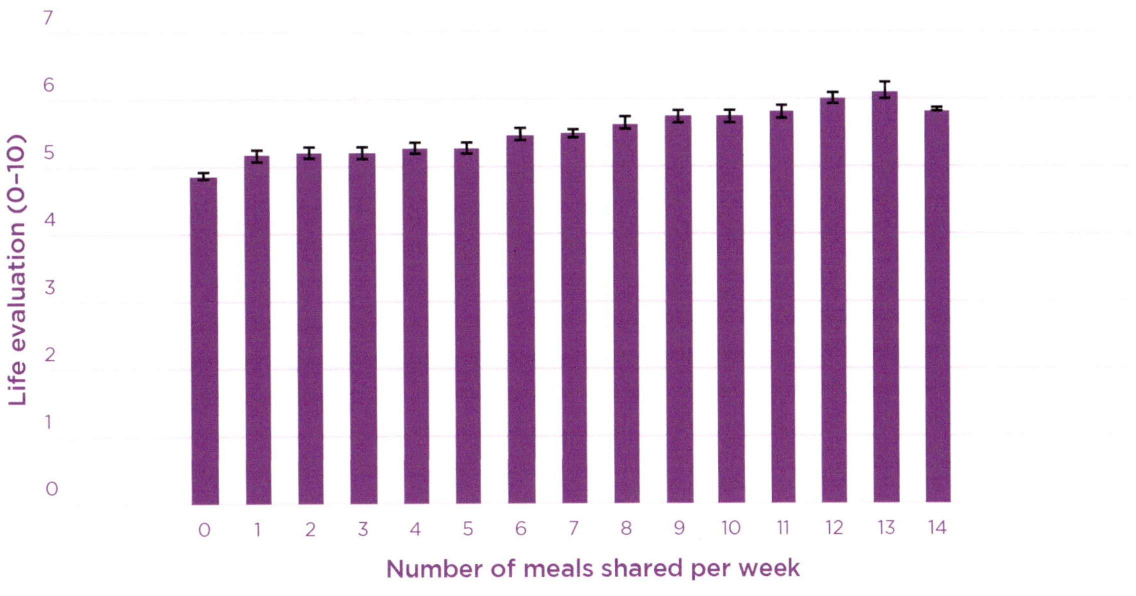

Figure 3.6: Shared meals and subjective wellbeing by gender
Gallup World Poll (2022–2023)

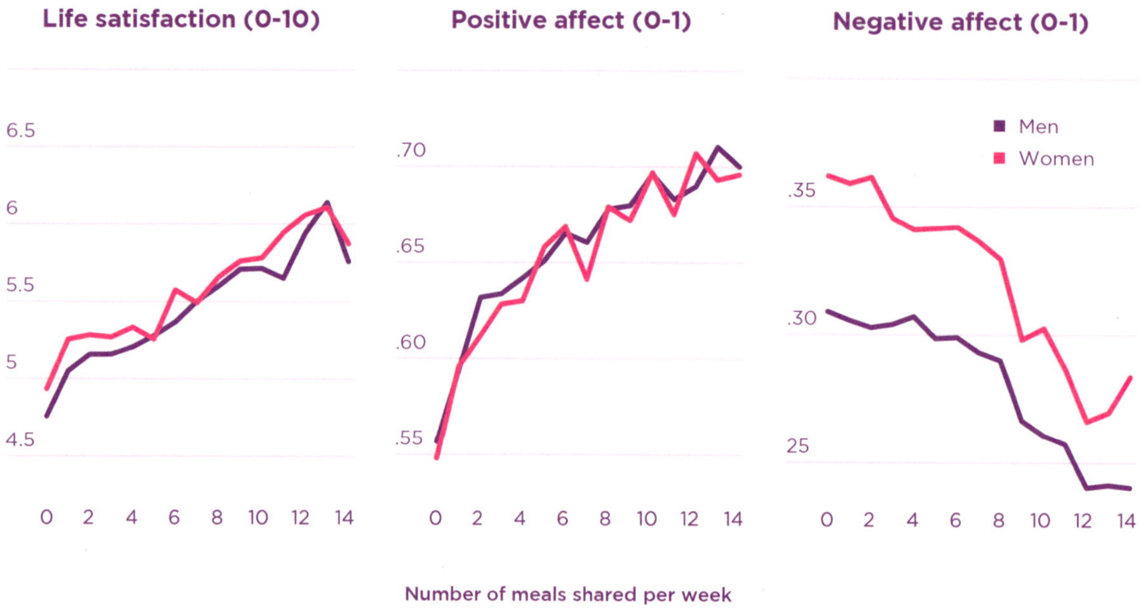

When we consider men and women separately in Figure 3.6, we find no statistically significant difference in the link between sharing meals and life evaluation by gender, nor do we find strong evidence of gender differences in the association between positive affect and meal sharing. In other words, sharing meals appears to be just as important for men and women in terms of how they evaluate their lives and how often they experience positive emotions.

However, when we consider negative emotions, the story begins to change. First, it is worth noting that men report considerably lower levels of negative affect than women overall. This is consistent with evidence presented in this and previous editions of the *World Happiness Report*. Second, we also find that sharing meals appears to be more closely related to negative emotions for women than for men. Women who spend more time dining alone report much higher levels of negative affect than women who spend more time dining with others. This is also true for men, but the difference is smaller than it is for women. This is indicated by the steeper slope of the line for women in the third panel of Figure 3.6.

In Figure 3.7, we present the relationship between sharing meals and wellbeing for younger (age 16–24) and older (age 65+) adults. Results for all age groups are presented in Figure A4 of the online appendix. There are two clear takeaways.

The first is the overall difference in subjective wellbeing across the two age groups, represented by the gap between the purple (16–24) and pink (65+) lines. Older adults report higher life evaluations than young people overall, but lower levels of positive affect. In other words, they are more likely to report being satisfied with their lives as a whole but less likely to report feeling happy the previous day. We observe no significant differences in levels of negative affect between younger and older adults.

The second feature – perhaps more relevant for this chapter – is the difference in the slope of the lines, where steeper lines represent stronger relationships between meal sharing and subjective wellbeing. Overall, we find the relationship between sharing meals and life evaluations, as well as the relationship between sharing meals and positive affect, to be stronger for younger people than for older people. When we compare young people who dine alone to young people who share meals, we find much greater differences in life evaluations and positive affect than we do for older adults. We do not observe similar patterns for negative affect – for both young and old, eating more meals alone is equally predictive of higher levels of negative affect.

As a final note, not only do people who share more meals report more positive emotions overall, they also seem to enjoy their food more. In Figure 3.8, we plot average levels of reported enjoyment while cooking and eating for those dining alone compared to those dining with others. We find a clear positive trend in both cases. The more meals we share with other people, the more we seem to enjoy them.

Put simply, across regions, countries, and cultures, for men and women, young and old, sharing more meals is associated with greater subjective wellbeing.

Recent research has also suggested that individuals who feel more positively about different aspects of their eating experience tend to have higher life evaluations and experience more positive emotions.[20] Taken together, this may suggest that dining experiences are an

Figure 3.7: Shared meals and subjective wellbeing by age
Gallup World Poll (2022–2023)

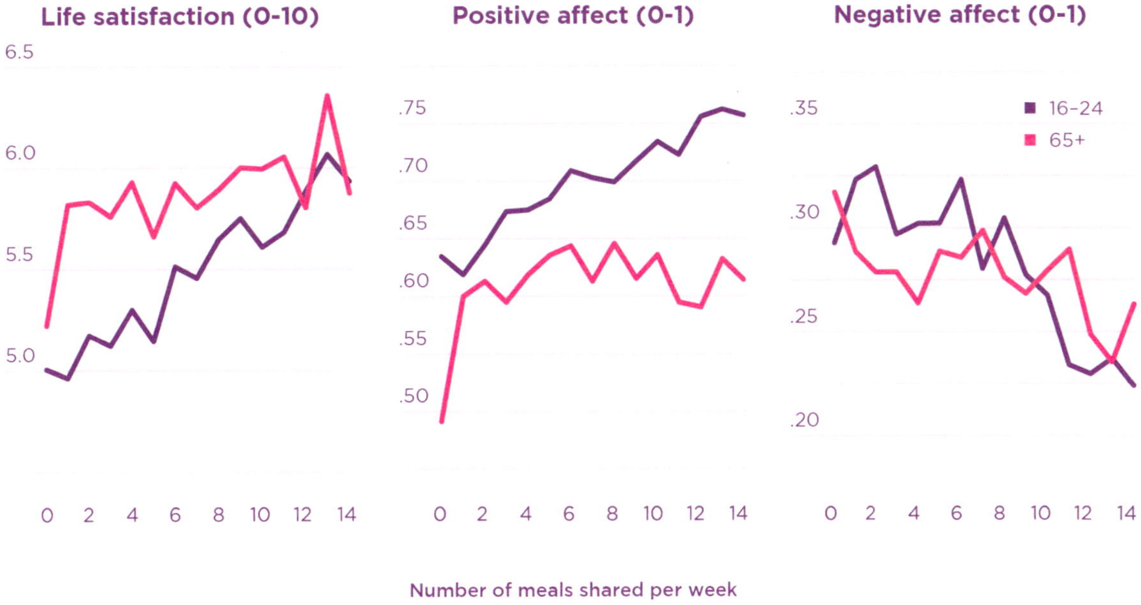

Figure 3.8: Sharing meals and enjoyment of food
Gallup World Poll (2022)

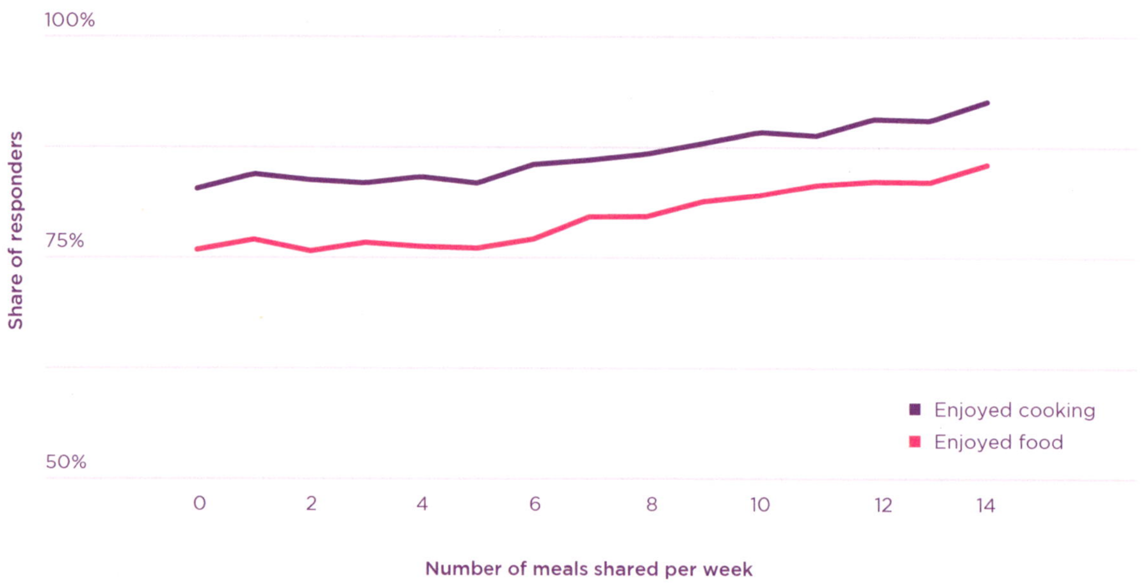

important, if sometimes overlooked, ingredient in shaping overall wellbeing.

To briefly summarise, in the previous section, we observed sizable differences in rates of meal sharing around the world. This was true across regions, countries, and individuals. While we did not find significant differences in rates of meal sharing for men and women, younger people appear to eat more of their meals with others in almost every region of the world.

In this section, we found that differences in sharing meals are also closely related to differences in subjective wellbeing. This is true across multiple levels of analysis. At the country level, countries where residents share more meals report greater average life evaluations. At the individual level, men and women who eat more meals with others report greater life evaluations, increased positive affect, and decreased negative affect – although the relationship between meal sharing and negative affect appears to be stronger for women than it is for men. We observed similar dynamics across age cohorts. Both younger and older adults who share more meals report higher levels of wellbeing, but these links are stronger for the young than the old. Finally, we noted that sharing meals is particularly important for positive affect, more so than for life evaluations or negative affect. Put simply, across regions, countries, and cultures, for men and women, young and old, sharing more meals is associated with greater subjective wellbeing.

Testing possible explanations

A key question to emerge from these results is whether sharing meals with others is merely an indicator of wellbeing, or a direct causal contributor to it. While we cannot conclusively answer this question here, we can begin to consider some potential explanations.

For example, eating alone may simply be more affordable than eating with others. This seems particularly likely in high-income countries, where sharing meals with others may be more common in restaurants. If so, perhaps the reason we find such a strong relationship between happiness and sharing meals is simply because people who share more meals have more money.

Or consider the related case of (un)employment. It is plausible to imagine that employed adults who eat at work are more likely to share meals than those who are unemployed. If so, then the link between sharing meals and wellbeing may be partially, or even substantially, reducible to differences in employment.

To address these concerns, Figure 3.9 presents the results of multivariate linear regressions which estimate the relationships between meal sharing, life evaluation, positive affect, and negative affect for all regions while controlling for a variety of other potentially relevant factors. Specifically, we control for gender, age, education, employment, income, household size, and country fixed effects. We also control for people's ability to meet basic needs for food, as measured by the question "Have there been times in the past 12 months where you did not have enough money to buy the food that you or your family needed?" With these controls included, if we continue to observe a significant relationship between sharing meals and subjective wellbeing, we can be more confident that this relationship is important in its own right, and not merely attributable to other factors.

In fact, this is precisely what we observe. Even after accounting for income, education, employment, and other key indicators, we continue to find strong and robust relationships between sharing meals and subjective wellbeing around the world. In almost all regions, sharing more meals with others proves to be highly predictive of higher life evaluations, more positive affect, and less negative affect.

However, there are notable differences in the magnitude of these relationships across regions. The relationship between sharing meals and wellbeing appears to be particularly strong in North America, Australia, and New Zealand. In these countries, the differences in wellbeing between those who eat more or fewer meals alone is greater than for any other region. One potential interpretation of this result is that the importance of sharing meals with others may be driving the relatively high levels of meal sharing we observe in this region (Figure 3.1). However, we do not observe similar dynamics for Latin America and the Caribbean. While meal sharing is most common in this region, it does not appear particularly important for wellbeing relative to other parts of the world.

We find similarly complex relationships between levels of meal sharing and its importance for wellbeing at the opposite end of the spectrum. In East Asia – where meal sharing is relatively rare (Figure 3.1) – we find strong links between sharing meals and negative affect, but weak links when it comes to life evaluation or positive affect. At the same time, sharing meals with others appears to be particularly important for life evaluations and negative affect in South Asia – another region with relatively low levels of meal sharing overall – but less so for positive affect. Here again, all of these effects are estimated after controlling for age, gender, income, education, and employment. Taken together, the relationship between how often meals are shared and how important meal sharing is for wellbeing is clearly neither simple nor straightforward. Examining and identifying

Figure 3.9: The link between sharing meals and subjective wellbeing by region
Gallup World Poll (2022–2023)

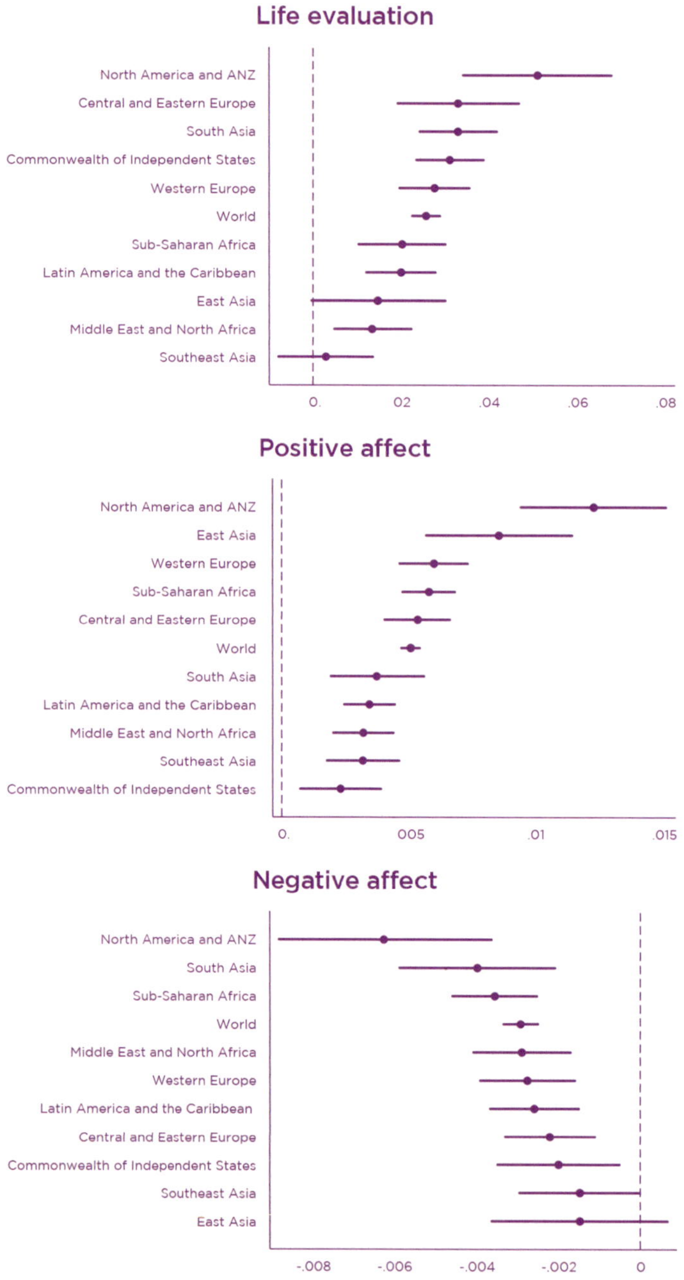

Note: Coefficients plotted from 11 separate regressions per image. N ≈ 150,000. Controls: country fixed effects, income quintile, household size, gender, age, age-squared, education group, employment group, people's ability to meet basic needs for food. Life evaluation is expressed on a 0–10 scale, positive and negative affect on a 0–1 scale. Data includes the 2023 sample. Predictive effect of the number of shared meals is assumed to be linear. Regressions use country-level survey weights.

potential explanations for these differences can provide a rich opportunity for future research.

Nevertheless, although the magnitude of these relationships may differ across regions, their direction does not. Even after controlling for a wide variety of other factors, sharing meals continues to be strongly and consistently associated with better life evaluations, increased positive affect, and decreased negative affect around the world.

Assessing practical significance

The fact that we observe such strong and consistent links between sharing meals and wellbeing is striking. But how significant are these relationships practically? Given the impressive size and scope of our dataset, it is certainly possible to find statistically significant relationships that are nonetheless relatively small, and therefore perhaps not useful for real-world applications or matters of policy.

One way to assess the practical significance of meal sharing is to consider how much this variable reveals about wellbeing compared to other important social indicators. To that end, we briefly return to income and employment. Decades of research has found strong and sizable links between income, unemployment, and subjective wellbeing.[18] In particular, the dramatic decline in life evaluation associated with unemployment is one of the largest and most consistent effects to emerge from empirical wellbeing research. If we compare the significance of these associations with sharing meals, how do they stack up?

In Figure 3.10, we present the results of a series of regression analyses testing the extent to which income, unemployment, dining alone, and sharing meals explain variation in subjective wellbeing around the world. In each panel, we estimate four separate regressions in which we relate differences in subjective wellbeing (considered as the dependent variable) to differences in sharing meals, dining alone, income quintile, and unemployment (considered as independent variables). In Panel A, we consider relationships with life evaluation, Panel B with positive affect, and Panel C with negative affect. In each case, we measure the extent to which differences in the independent variable of interest can explain differences in the dependent variable of interest. This is estimated empirically by the R-squared value produced by each regression. By implication, the size of the bars in each figure represent the extent to which differences in e.g., shared meals can explain differences in e.g., life evaluation.

In our view, the results of these analyses are the most striking so far. Not only do we find sharing meals and dining alone to be important predictors of wellbeing compared to income and employment, but in many cases, they seem to be even more so. That is, asking people if they shared at least one meal last week can tell us more about their overall life evaluation than knowing if they are unemployed. Or relatedly, knowing how many meals someone shared last week can tell us more about their positive emotions than their income.[21]

Again, the links with positive affect are particularly notable. When explaining variation in positive emotions, the extent to which people share meals with others is a more important predictor than both income and unemployment combined. At the same time, sharing meals also does a better job explaining variation in life evaluations around the world than income or unemployment. Dining alone is a more important predictor of differences in life evaluations than unemployment, but not income. For negative affect, income continues to be a crucially important indicator, yet sharing meals and dining alone are not far behind. Indeed, both prove to be more powerful predictors of negative emotions than unemployment.

Not only do we find sharing meals and dining alone to be important predictors of wellbeing compared to income and employment, but in many cases, they seem to be even more so.

Figure 3.10: Explaining differences in subjective wellbeing around the world
Gallup World Poll (2022–2023)

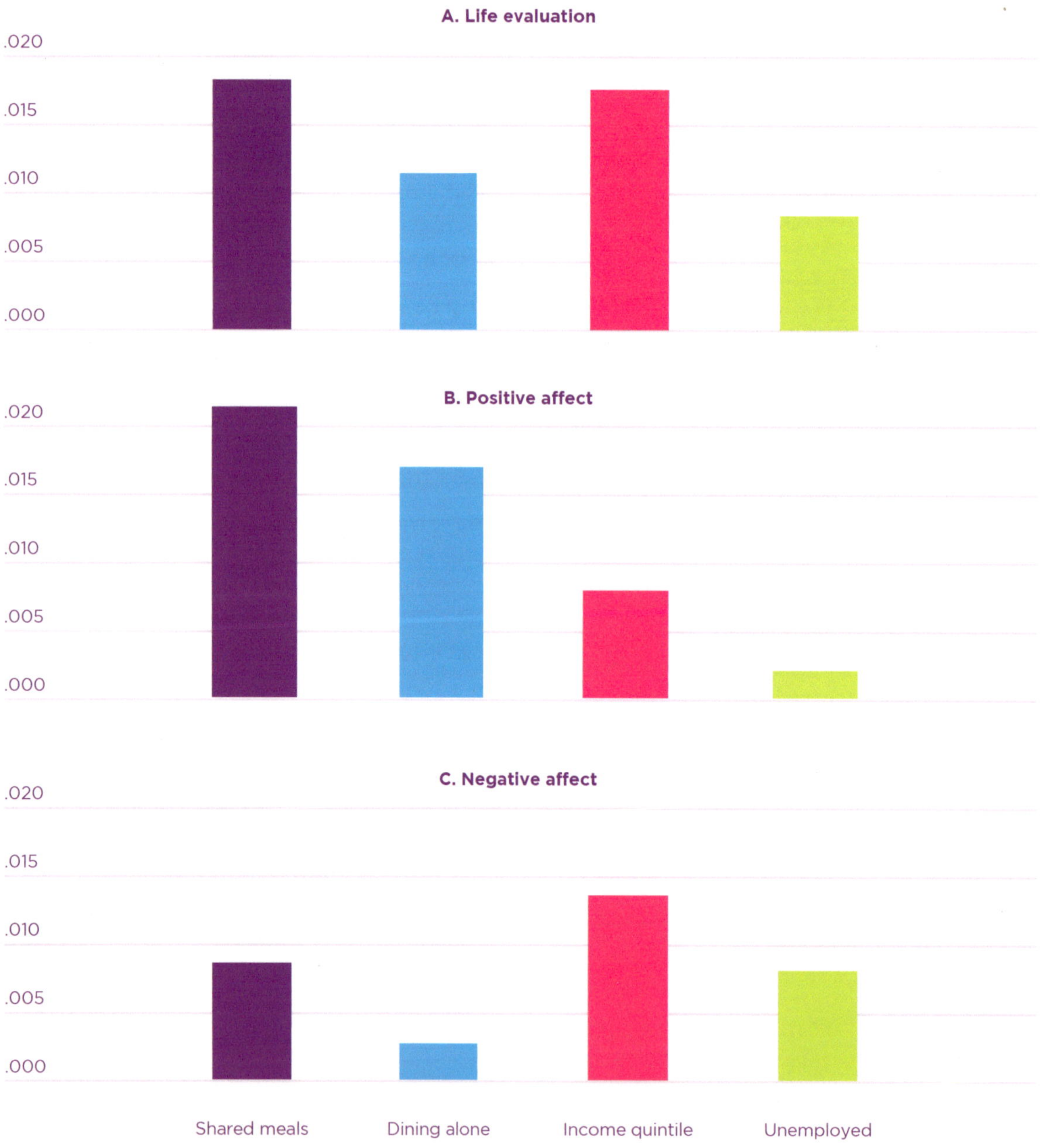

Note: Bars represent R^2 estimates from four separate regressions per image ($N \approx 150{,}000$). No controls. Dependent variables are all expressed on a 0–100 scale and sharing meals is measured using five categories for comparability (0–2; 3–5; 6–8; 9–11; 12–14). Data includes the 2023 sample. The predictive effect of the number of shared meals is not assumed to be linear. Regressions use country-level survey weights.

Taken together, these results underscore the importance and usefulness of sharing meals as an indicator of subjective wellbeing. The fact that we observe such strong links between sharing meals, life evaluations, positive affect, and negative affect suggests that meal sharing should be given much more weight and attention by researchers and policymakers around the world.

However, our discussion in this section is not intended to resolve the thorny matter of causation. While we find strong and significant associations between sharing meals and wellbeing, this could indicate that sharing meals itself causes people to be happy, or that happy people are more likely to share meals with others. Even after controlling for a range of related variables, we cannot conclusively rule out either explanation. In all likelihood, both dynamics are probably true, at least to some extent. The question of which pathway is stronger is nevertheless important and we will return to it in the final section. Before that, we turn to the United States to take a deeper look at changes in meal sharing over time.

Meal sharing over time: a case study of the United States

So far, our analysis has been based on data from the Gallup World Poll, collected from over 150,000 survey respondents in more than 140 countries. The size and scope of this data allowed us, for the first time, to compare differences in meal sharing and its relationship to subjective wellbeing for more people and more countries than ever before.

However, this survey module was only introduced in 2022, so we are unable to examine how meal sharing has changed over time. Given the close association between sharing meals and subjective wellbeing, this is an important perspective to consider as it may provide an objective yardstick for thinking about longitudinal changes in subjective wellbeing.

To this end, this section turns to data collected by the American Time Use Survey (ATUS) in the United States from 2003 to 2023. Each year, the ATUS asks a representative sample of roughly

12,000 Americans how they spend their time on a day-to-day basis. Survey respondents fill out long and detailed questionnaires about what they did the day before, who they did it with, and how they felt while doing it.[22] This data has been used and referenced extensively in research, media, and policy circles. Nevertheless, to our knowledge, the extent to which Americans spend time eating and cooking alone or with others has remained relatively unexplored.

Longitudinal trends in meal sharing are particularly important when considered against the backdrop of declining social capital and connection in the United States. These trends were starkly documented by Robert Putnam in his landmark 2000 book, *Bowling Alone*. Drawing on a truly expansive array of datasets, Putnam found that Americans were spending more and more time alone, while civic institutions and social organisations including religious groups, labour unions, veterans' associations, and even dinner parties were declining. These declines were occurring alongside similar declines in political participation, voting rates, trust in other people, faith in institutions, indicators of physical and mental health, rates of educational achievement, social mobility, and economic opportunity.

More recent studies have reinforced this general story of social decline in the United States and found evidence of similar trends in other countries.[23] One large-scale study looking at data from 1990 to 2012 in 30 European countries found that participation in social groups was falling while distrust in political institutions was rising.[24]

Meal sharing over time in the United States

With this context in mind, we begin our analysis by plotting the number of people dining alone in the United States over time in Figure 3.11. Each dot represents the percentage of survey respondents who reported eating all of their meals alone the previous day. Our sample includes approximately 235,000 American adults from 2003 to 2023 and is weighted to be representative of the general population. Importantly, this measure of meals shared the previous day is also distinct from prior sections in which we considered meal sharing over the course of an entire week.

The trendline is unmistakable. There has been a sharp rise in the number of Americans dining alone since 2003. Perhaps unsurprisingly, some of the highest levels of dining alone were recorded in 2019 and 2020 during the height of the COVID-19 pandemic. However, rates of dining alone were increasing long before the pandemic, and they have not come down since. In 2023, the most recent year for which data is available, rates of dining alone in the United States were even higher than they were during the pandemic. Roughly 1 in 4 American adults (26%) now report eating all of their daily meals alone - an overall increase of more than 50% since 2003.

In Figure A5 of the appendix, we present a series of robustness checks which expand our definition of dining to include cooking and food preparation, and consider the percentage of total meals eaten alone, rather than a binary indicator of every meal eaten alone. In each case, the results all point in the same direction. In one of the most reliable, reputable, and widely used time-use datasets in

Figure 3.11: Dining alone in the United States
ATUS (2003-2023)

Note: Data from the American Time Use Survey weighted to be representative of the general population (n = 234,185 individuals). 95% confidence intervals displayed. Dining alone measured as the share of respondents in each survey year reporting eating all meals alone in the previous day.

the world, we find clear and consistent evidence that, with every passing year, Americans are spending more and more time dining alone.

At this point, it is worth considering one obvious potential explanation for these trends – the rise in living alone. It is well-documented that Americans have become increasingly likely to live by themselves.[25] In Chapter 4 of this report, we see similar results for European countries. There are many reasons for this. Some of the most widely cited explanations are declines in family size, delays in marriage and parenthood, and increased economic opportunity for women. These, and related dynamics, have led to a considerable rise in the share of one-person households in the United States – a trend that we confirm using ATUS data in Figure A6 of the online appendix.

In one of the most reliable, reputable, and widely used time-use datasets in the world, we find clear and consistent evidence that, with every passing year, Americans are spending more and more time dining alone.

We conducted a series of analyses to see if the rise in living alone can explain the rise in dining alone in the US. In Figure A7, we find that people who live alone are considerably more likely to eat alone. This was particularly true during the COVID-19 pandemic, yet has barely come down in the years since. In 2023, roughly 70% of those living alone reported eating all of their meals alone the previous day, compared to 20% of those who live with others.

However, when it comes to *changes* in dining alone, there has been a greater relative increase in dining alone among those who live with others (Figure A8). To be specific, in 2023, roughly 18% of Americans who live with others ate all of their meals alone the previous day, compared to 12% in 2003 – an increase of 50%. Among people who live alone, the corresponding figures are 69% in 2023 and 55% in 2003 – an increase of 25%.

So, does living alone explain dining alone? To some extent it does. In regression analyses, we estimate that recent increases in living alone explain 15–20% of associated increases in dining alone (Figure A9).[26] Nevertheless, even after controlling for household size, we continue to find sizable and significant increases in dining alone since 2003. This remains true even after controlling for age, sex, gender, and income. By implication, while we do find evidence that the rise in living alone is, at least partly, to blame for the rise in dining alone, there is clearly much more to the story.

In Figures 3.12 and 3.13, we present similar trends in dining alone broken down by gender and age. We find that men have generally been more likely to eat all of their meals alone on the previous day than women since 2003. However, we observe sharp and similar increases in dining alone for both genders. Today, both men and women are eating more meals alone than ever before.

Results for different age groups are another story entirely (Figure 3.13). As we observed in Gallup's global data, older people are much more likely to spend time dining alone than younger people. In this case, we present results for 10-year age cohorts from 18-year-olds to over-65s. In every year since 2003, over-65s report eating more meals alone than their younger counterparts. Rates of dining alone for those under the age of

Figure 3.12: Dining alone in the United States by gender
ATUS (2003–2023)

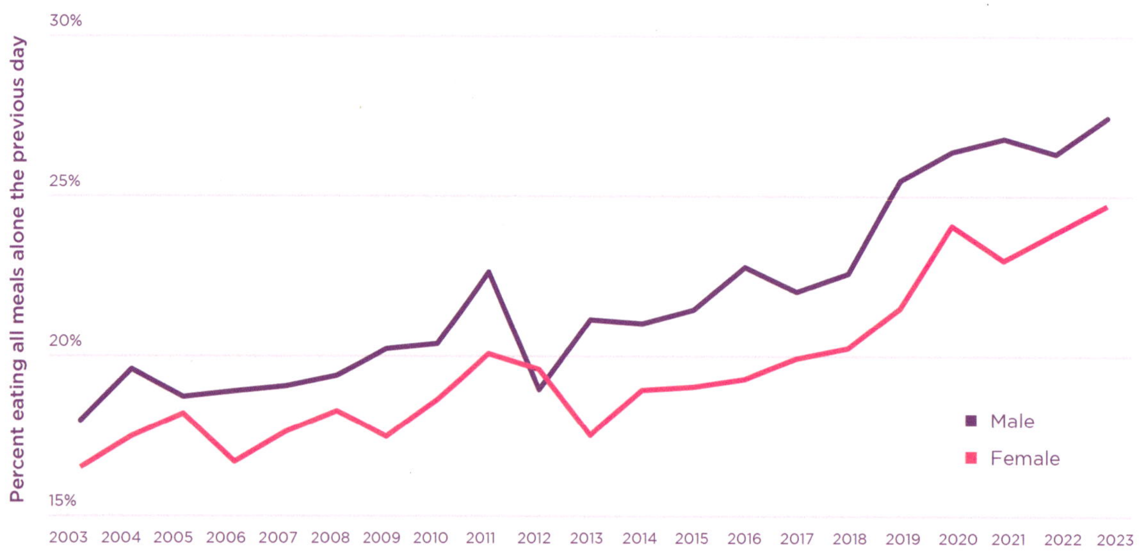

Note: Data from the American Time Use Survey weighted to be representative of the general population (n = 234,185). Dining alone measured as the share of respondents in each survey year reporting eating all meals alone in the previous day.

Figure 3.13: Dining alone in the United States by age
ATUS (2003–2023)

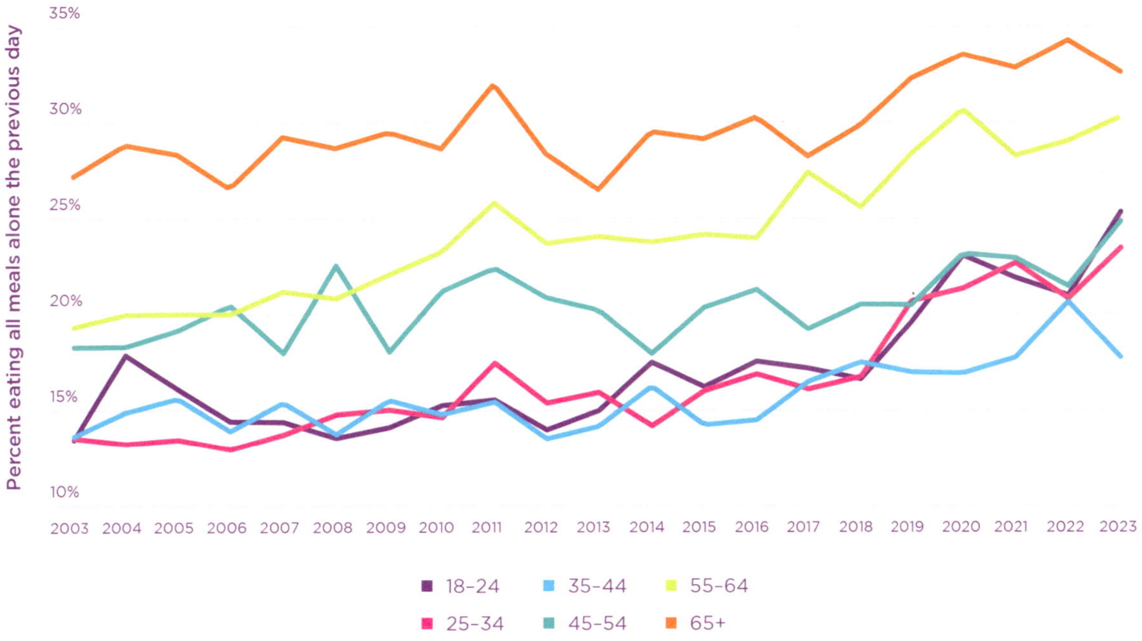

Note: Data from the American Time Use Survey weighted to be representative of the general population (n = 234,185). Dining alone measured as the share of respondents in each survey year reporting eating all meals alone on the previous day.

45 are among the lowest recorded and remained largely consistent and comparable – that is, until 2018. Beginning in 2018, we observe sharp increases in dining alone for almost every age group. The trendlines for those under 35 are particularly stark.

In Figure 3.14, we normalise rates of dining alone for all age groups to their 2003 levels and plot the overall changes for each cohort in the years since. We see that levels of dining alone have increased for every age group since 2003, even among older cohorts who were already much more likely to dine alone 20 years ago. Nevertheless, the largest and most dramatic changes are for those under 35. The extent to which 25- to 34-year-olds report eating all of their meals alone on the previous day has increased by more than 180% in two decades. We observe a similarly dramatic increase for 18- to 24-year-olds.

Such sizable increases in rates of dining alone among young adults in the US clearly and urgently warrant further research and policy attention. We are not the first to document concerning levels of

Such sizable increases in rates of dining alone among young adults in the US clearly and urgently warrant further research and policy attention.

Figure 3.14: Changes in dining alone in the United States by age
ATUS (2003-2023)

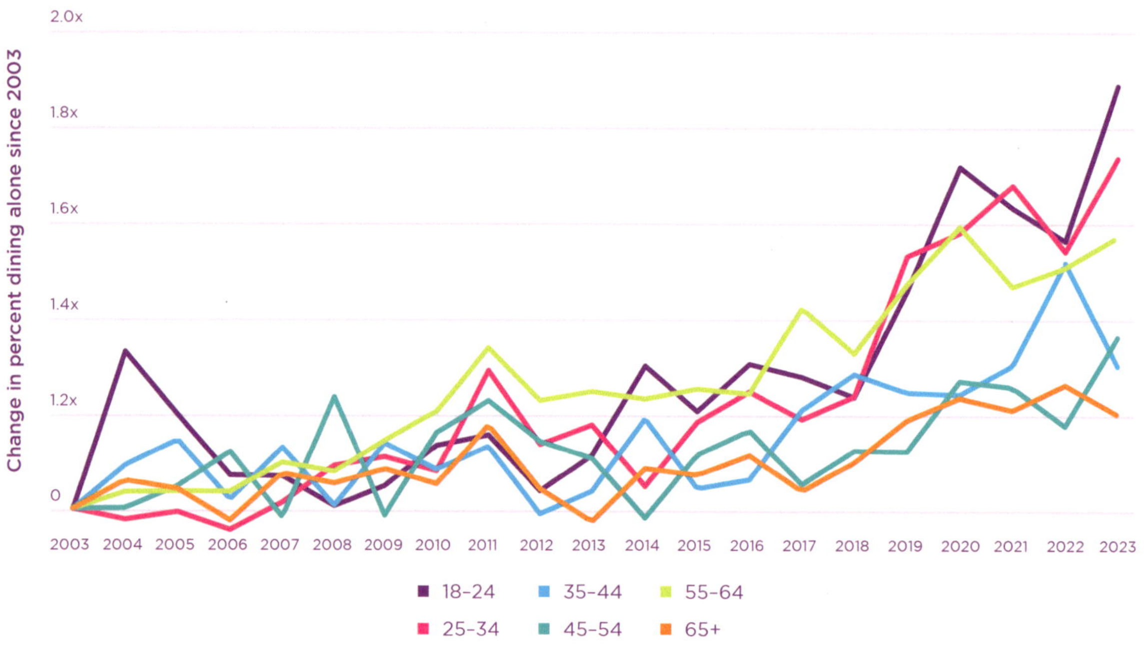

Note: Data from the American Time Use Survey weighted to be representative of the general population (n = 234,185). Dining alone measured as the share of respondents in each survey year reporting eating all meals alone on the previous day.

isolation among young people,[27] but many of the explanations emanating from public discourse and academic literature are somewhat unsatisfying in the present context.

Most notably, the rise of smartphones and social media is often credited with observable declines in young people's wellbeing. However, when it comes to dining alone, the timelines do not line up as neatly as one might expect. With the launch of Facebook in 2004 and the introduction of the iPhone in 2007, you might expect that the sharpest increases in dining alone would emerge around these times. Instead, we observe a relatively steady and consistent rise in young people dining alone from 2003 to 2015, followed by increasingly steeper inclines in the years since.

Another common explanation for (or at least contributor to) the decline in young people's mental health has been the COVID-19 pandemic. Indeed, we do see some of the highest recorded rates of dining alone during the pandemic in 2020 and 2021. However, rates of dining alone were increasing long before the pandemic began to spread in the United States. During the pandemic, rates of dining alone among young people even appear to have declined slightly – potentially reflecting more meals eaten at home with family members. If the pandemic was the whole story, we may also expect rates of dining alone among young adults to have declined in more recent years. In fact, we observe the highest levels of dining alone among those under 35 in 2023.

While we certainly do not reject the notion that COVID-19 may have contributed to higher levels of isolation among young adults, the timing suggests that it has not been the primary driver of increases in dining alone.

Meal sharing and subjective wellbeing in the United States

For the remainder of this section, we turn from the overall levels and trends in meal sharing to the relationship between meal sharing and subjective wellbeing. In 2010, 2012, 2013, and 2021, the American Time Use Survey included a special wellbeing module in which all respondents were asked about their daily emotions and their overall satisfaction with life (using the Cantril Ladder). In Figure 3.15, we use this data to compare average differences in life evaluation between Americans who reported eating all of their meals alone in the previous day relative to those who shared meals.

We find that Americans who dine alone reported life evaluations that are, on average, 0.5 points lower than those who dine with others. We observe similarly large differences in life evaluations when we split our sample by age cohorts and gender. Dining alone is strongly associated with substantial differences in subjective wellbeing for both men and women, young and old alike.

In Figure 3.16, we present analogous results for happiness, tiredness, stress, pain, and sadness. Again, we split our sample by Americans who reported eating all meals alone on the previous day compared to those who ate at least one meal with someone else and plot the average levels of daily emotions for each group.

Figure 3.15: Meal sharing and life evaluation in the United States
ATUS (2010, 2012, 2013, 2021)

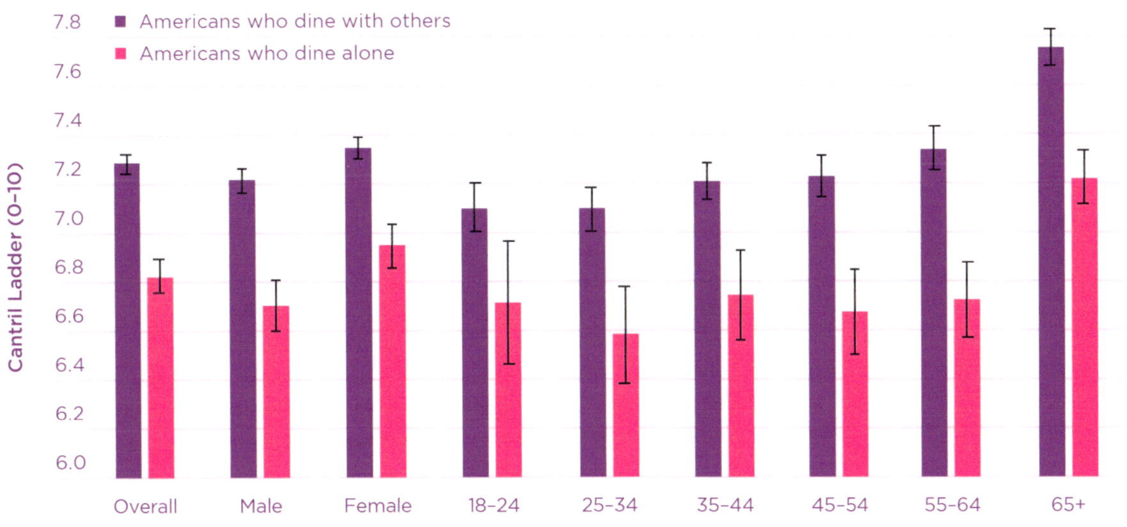

Note: Bars represent average life evaluations of those who reported eating all meals alone on the previous day relative to those who shared at least one meal. Data weighted to be representative of the general population. 95% confidence intervals (n = 28,811).

In line with prior results, we observe notable differences in emotions between those who share meals and those who dine alone. Differences in happiness, pain, and sadness are particularly large, although we also find slight differences in self-reported levels of stress. In each case, Americans who eat more meals with others report higher levels of positive affect and lower levels of negative affect than those who dine alone. We find no significant differences for tiredness.[28]

In Tables A3, A4, and A5 of the online appendix, we conduct a series of robustness checks to estimate the size and strength of the relationship between dining alone and subjective wellbeing using linear regressions. In every instance, we find that differences in life evaluations, positive affect, and negative affect between Americans who dine alone and Americans who share meals are statistically significant at a 99% confidence level. This is true even when controlling for age, sex, geographic location, marital status, and race. In an echo of earlier results, relationships with positive affect are strongest. In Table A5, using standardised measures of all wellbeing outcomes, we find that gaps in happiness, in particular, are larger than those estimated for life evaluations or any other affect measure under consideration.

Taken together, our results in this section point to concerning declines in how often Americans share meals with each other. Unlike other social indicators, such as loneliness or depression, the relatively objective nature of sharing meals makes it a uniquely reliable metric by which to compare differences over time. The fact that fewer Americans report sharing meals with others is particularly concerning given the close relationship

Figure 3.16: Meal sharing and emotions in the United States

ATUS (2010, 2012, 2013, 2021)

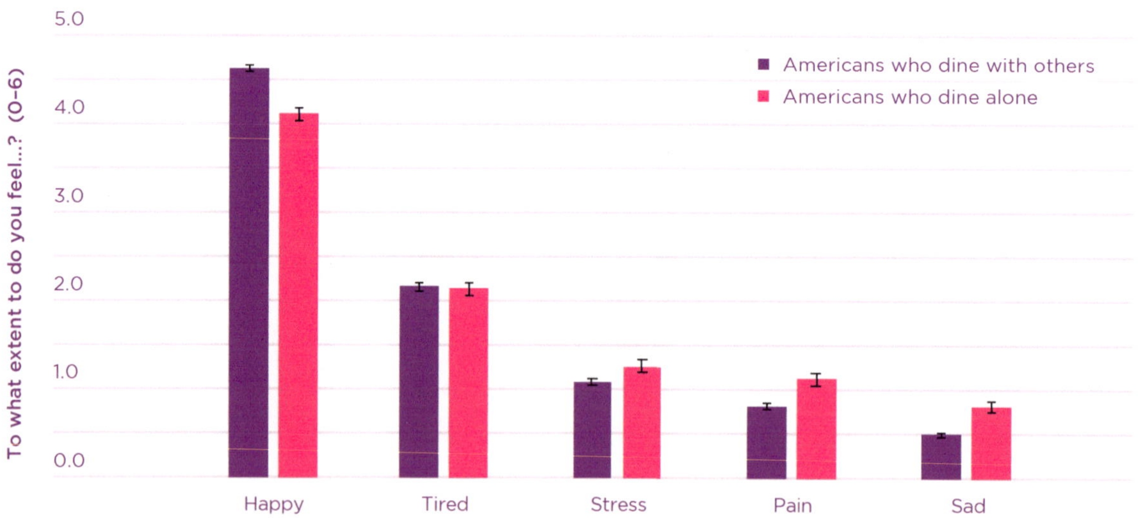

Note: Bars represent average levels of daily affect reported by those who ate all meals alone the previous day vs. those who ate at least one meal with someone else. Data weighted to be representative of the general population. 95% confidence intervals (n = 28,811).

between meal sharing and subjective wellbeing – a relationship that we have now confirmed in two separate large-scale representative datasets. In the final section of this chapter, we take one last look at the link between sharing meals and social connections, and what this link may tell us about societal health and stability writ large.

Sharing meals and social connections

In this final section, we turn our gaze to the relationship between sharing meals and social connections. One plausible interpretation of the importance of sharing meals is that it promotes and sustains social ties. Given the widely documented, well-established links between positive social relationships and subjective wellbeing,[29] this could help to explain our results in previous sections.

Moreover, if sharing meals really does help to build and sustain social ties, it is not only academically interesting but politically important. As decades of social science research has demonstrated – and this year's *World Happiness Report* highlights – social connections are not only important for individual health and happiness, but for societal health and happiness writ large.[30]

With this backdrop in mind, we return to the country-level data provided by the Gallup World Poll (GWP) and incorporate additional data from the Global Preferences Survey (GPS) administered by Gallup in 2012. The GPS captured detailed information regarding risk and time preferences, positive and negative reciprocity, altruism, and trust from a large sample of roughly 80,000 individuals in 76 countries, representing more than 90% of the world's population. It remains one of the most reliable, robust, and expansive datasets on political and economic opinions to date.[31] Additional details and variable descriptions for GPS and GWP data are provided in Table A7 of the online appendix.

In Figure 3.17, we present country-level correlations between GPS indicators of trust and reciprocity in 2012 with rates of meal sharing collected by

Gallup in 2022 and 2023. These are presented numerically in Table A6 of the online appendix. Overall, we find positive relationships between levels of meal sharing, trust, and reciprocity. We observe the strongest relationships for indicators of positive reciprocity, social support, and loneliness. Associations with measures of negative reciprocity, trust, and altruism tend to be more modest.

In Figure 3.18, we plot analogous associations for dining alone. In this case, relationships between dining alone and indicators of social connectedness appear stronger. Dining alone is negatively correlated with all measures of social capital under consideration, except for loneliness, where we find a positive correlation. Links with trust, social support, and reciprocity again seem to be the most robust. Many of these relationships – calculated at the country level, rather than respondent level – are also statistically significant. We find significantly negative correlations between dining alone and various measures of reciprocity, trust, and altruism.

Why is sharing meals so strongly predictive of subjective wellbeing but only moderately related to indicators of social trust, reciprocity, and altruism?

Nevertheless, it is worth noting that relationships between sharing meals and dining alone with many of the social indicators under consideration are relatively weak and, in some cases, statistically insignificant. This is surprising when evaluated in the context of the results presented in previous

Figure 3.17: Sharing meals, trust, altruism, and reciprocity
Gallup World Poll (2022–2023), Global Preferences Survey (2012)

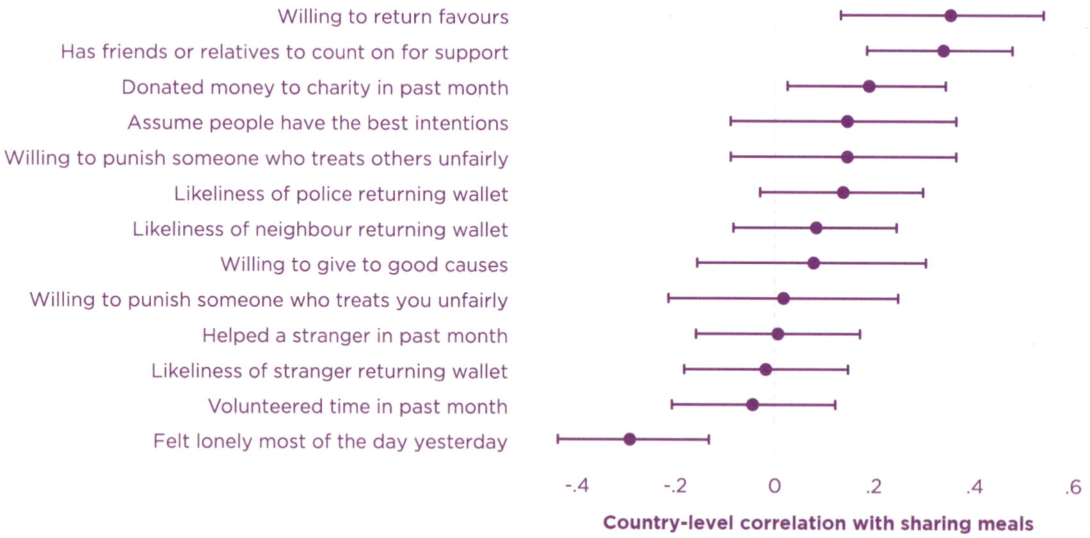

Note: Pearson correlation coefficient is calculated using the weighted country-level results for each indicator. Number of observations (countries) ranges between 71–141.

Figure 3.18: Dining alone, trust, altruism, and reciprocity
Gallup World Poll (2022–2023), Global Preferences Survey (2012)

Note: Pearson correlation coefficient is calculated using the weighted country-level results for each indicator. Number of observations (countries) ranges between 71–141.

sections of this chapter and raises an important question. Why is sharing meals so strongly predictive of subjective wellbeing but only moderately related to indicators of social trust, reciprocity, and altruism?

One answer to this question is somewhat technical. Although our analysis is based on data collected from individual survey respondents, Figures 3.17 and 3.18 compare averages across countries. This is unavoidable as several of the indicators under consideration were measured using different surveys at different times. As the survey respondents did not respond to every question, we're unable to make individual-level comparisons. However, this higher level of analysis comes at a cost. Most notably, the number of observations is now much smaller as we are considering countries rather than individuals. This smaller sample size could help to explain the large confidence intervals we observe in the figures above.

Many of these indicators are also measured using different time scales. For example, sharing meals and loneliness are asked in terms of the past week or previous day, but questions on trust and reciprocity are asked in much broader terms e.g., time spent helping strangers or volunteering in the past month. This too could help to explain why we find relatively weak relationships between sharing meals and some indicators of social connectedness.

Another potential explanation is that the positive benefits of sharing meals may operate through channels that have little or less to do with social connections than one might expect. We have already commented on this dynamic with regard to income, education, and living alone. While we found some supportive evidence of all three channels – that is, people who share more meals are more likely to be employed, more likely to earn higher levels of income, and more likely to live with others – none could fully account for the differences in rates of meal sharing we observe around the world, nor could they fully explain the relationship between sharing meals and subjective wellbeing.

Nevertheless, these may not be the only mechanisms at play. People who share more meals with others may also be more likely to eat healthily, be more physically active, spend more time outdoors, spend less time behind screens, live in more densely populated areas, and so forth. Any or all of these factors could help to explain why sharing meals is so strongly related to subjective wellbeing without needing to appeal to any role it may or may not play in promoting social connections.

Yet another potential explanation is even more subtle. While social connections, broadly construed, are generally taken to be a good thing for individuals and societies writ large, recent research has begun to add a few important asterisks to the story. Not all social attitudes and behaviours are equally important for wellbeing, nor are they necessarily related to each other.

It is entirely plausible to imagine that people who have others to rely on in times of need may still lack trust in societal institutions. People who feel strongly connected to their communities may, nevertheless, spend little time volunteering or helping strangers in need. Those who donate money to charity may still strongly believe that people who treat others unfairly ought to be punished. Even more importantly, not all of these social attitudes and behaviours are likely to be equally important for subjective wellbeing.

To draw this point out, consider the case of social support and loneliness. Fortunately, in the 2022 and 2023 waves of the Gallup World Poll, many of the same survey respondents were asked how often they shared meals with others, how often they felt lonely, and if they had others to count on in times of need. In Figure 3.19, we group these individuals into five categories depending on how often they share meals and present average levels of loneliness and social support for each group.

We find clear evidence that individuals who share more meals with others are significantly less likely to feel lonely (Panel A) and significantly more likely to experience social support (Panel B). Now that we can analyse individual-level responses, as opposed to country averages, our sample size is considerably larger and our resulting estimates are much more precise. This analysis suggests that sharing meals may indeed strengthen and support social ties. Even so, it may still suggest little to nothing at all about levels of trust.

To return to our earlier question, the fact that we find such strong relationships between sharing meals and wellbeing, while simultaneously finding modest or even mixed links between sharing meals and trust or reciprocity, may simply suggest that meal sharing is more conducive or supportive of certain types of social attitudes and behaviours than others. This interpretation could explain why sharing meals is so closely related to social support and loneliness – both of which have consistently been shown to be strong predictors of subjective wellbeing in their own right. Nevertheless, there is clearly much more to be explored in future research regarding the correlational and causal relationships between sharing meals, dining alone, and social connections.

Conclusion

In this chapter, we have presented new evidence on the global variation in meal sharing and what it implies for subjective wellbeing and social connections. Unlike most indicators of social relationships, and all indicators of subjective wellbeing, the number of meals shared with others is relatively objective and directly comparable across individuals, between countries, and over time. This feature makes sharing meals (and its counterpart, dining alone) uniquely valuable and well-positioned to reveal new insights into the nature and dynamics of human wellbeing.

Overall, we find stark differences in meal sharing around the world. These differences prove to be closely tied to age – on average, younger people share more meals with others than older adults – but mostly unrelated to gender. Global

Figure 3.19: Differences in loneliness and social support by number of meals shared
Gallup World Poll (2022–2023)

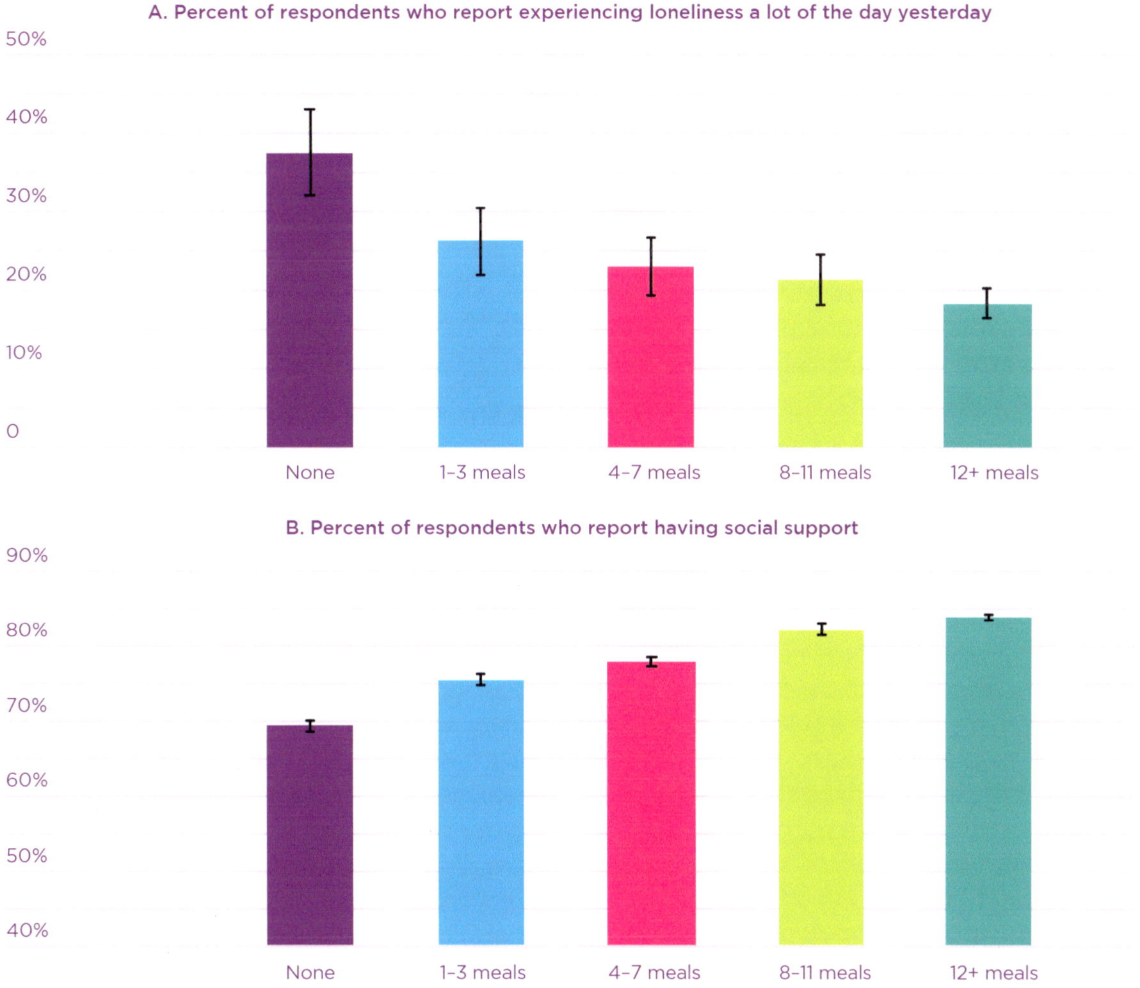

Note: Loneliness and social support measured on scale from 0 to 1. Social support is measured as having someone to count on in times of need. Individual-level data for Panel A from 13 countries in 2023 (n=6,375). Individual-level data for Panel B from 142 countries and territories in 2022–2023 (n = 147,678). See the online appendix for additional details regarding variable descriptions and coverage.

differences in meal sharing are also not fully explained by regional or individual characteristics such as income, employment, or household size.

When we examine the link between sharing meals and subjective wellbeing, we find that individuals who share more meals with others report higher levels of life satisfaction, lower levels of negative affect, and especially higher levels of positive affect.

In the United States, rates of meal sharing appear to be in stark decline, while rates of dining alone are on the rise. This is true for both genders and all ages, but particularly for young adults. Finally, we find rates of meal sharing to be closely tied with a handful of related social indicators - most notably, social support and loneliness - while being only modestly related to others including reciprocity and trust.

However, our analysis also has some limitations. First, and most importantly, we are unable to answer the key question of causation. There is already considerable evidence that social connections contribute to greater wellbeing,[32] and early suggestive evidence that sharing meals with others may promote social connection.[33] However, more research is clearly warranted. Second, in our analysis of Gallup World Poll data, we are unable to properly control for the fact that some respondents might skip either lunch or dinner. This element might be particularly relevant in countries where many citizens are exposed to food insecurity. Third, although our analyses converge in highlighting the association between meal sharing and wellbeing, they do not delve deeply into the mechanisms driving this relationship. Factors such as the quality of social interactions, the type of meal, or other confounding variables (e.g., work schedules and working from home) may play a significant role but are not explored.

Despite these limitations, the strong and robust relationships between sharing meals, wellbeing, and social connections that we have documented in this chapter cry out for future research and exploration. One of the most fruitful avenues for future research relates to the key question of causation. At present, it remains unclear whether sharing meals leads to greater wellbeing or whether greater wellbeing leads to more shared meals. In all likelihood, both are probably true, at least to some extent. Large-scale experiments such as the Health and Happiness Study, which will administer daily surveys and collect real-time smartphone and smartwatch data from a global sample population, may help shed new light on the underlying causal dynamics of sharing meals, social connection, and wellbeing.[34] Small-scale experiments conducted in more controlled environments may also help to pin down the causal nature of these relationships and would represent a meaningful and important contribution to research.

As a final point, it is worth commenting on the possible policy implications of these findings. While researchers and policymakers have long lamented declines in social connectedness across modern societies, isolating the key contributing factors and identifying solutions has proven frustratingly difficult. If sharing meals is just as important for promoting social ties and subjective wellbeing as the evidence in this chapter suggests, it may serve as a uniquely valuable, actionable, and cost-effective policy tool by which to facilitate and promote societal welfare. Several initiatives including Project Gather in the United States have already begun to chart exciting new pathways along these lines by providing financial support for shared meals.[35] As Dr. Vivek Murthy, the sponsor of the program and former US Surgeon General eloquently put it, "When we gather with others around food, we not only feed our bodies but also nourish our spirits."

Endnotes

1 Barger (2013); Chetty et al. (2016); Diener et al. (2018b); Kawachi and Berkman (2014); Rodríguez-Pose and von Berlepsch (2014).

2 Diener et al. (2018a); Kansky (2017); Lauricella et al. (2022); Sözbilir (2018); Tenney et al. (2016).

3 Chetty et al. (2016); Diener et al. (2018a); Diener et al. (2018b); Diener and Chan (2011); Johnson et al. (2018); Robison and Siles (1999); Shen and Bian (2018).

4 Diener et al. (2018a); Diener and Chan (2011); Holt-Lunstad et al. (2015); Johnson et al. (2018); Knapp (1976); Leigh-Hunt et al. (2017); Osborne et al. (2023); Robison and Siles (1999); Roßteutscher (2010); Tung et al. (2019).

5 Holt-Lunstad et al. (2015); U.S. Department of Health and Human Services. (2023).

6 Glanville et al. (2013); Sønderskov and Dinesen (2016).

7 Austin and Baba (1990); Brehm and Rahn (1997); Brown and Ferris (2007); Forrest and Kearns (2001); Glanville et al. (2016); Lyubomirsky et al. (2005a); McClurg (2003); Wang and Graddy (2008).

8 Brown and Ferris (2007); Glanville et al. (2016); Liang and Meng (2023); Lyubomirsky et al. (2005a); Wang and Graddy (2008); Wilson and Musick (1998).

9 Dwyer et al. (2018).

10 Glanz et al. (2021).

11 Wang et al. (2016).

12 For example, the World Inequality Study, the OECD's Programme for the International Assessment of Adult Competencies (PIAAC), and the Life in Transition Survey (LITS) run by the European Bank for Reconstruction and Development all employ versions of this question.

13 In 2023, respondents were asked: "Thinking about the past 7 days, on how many days did you eat LUNCH/DINNER with someone you know; including family, friends, or anyone else you know." In 2022, the question wording did not include the language "including family, friends, or anyone else you know." Countries surveyed in 2023 include: Albania, Cambodia, Czechia, Denmark, Egypt, El Salvador, Germany, Japan, Kazakhstan, Malaysia, Paraguay, Republic of Moldova, Sénégal, Sierra Leone, Togo, Tunisia, and Zambia. The approximate sample size in each country is 500.

14 Kimura et al. (2020); Lee et al. (2020).

15 Following the revised wording in 2023 (see Footnote 13), in two countries - Japan and Cambodia - the average number of meals shared rose by a substantial and statistically significant margin. However, even in this case, Japan's rate of meal sharing remained well below average. At the same time, in other countries including Malaysia, the average number of meals shared was essentially static in 2023 compared to 2022.

16 This is the estimated R^2 value in a binary OLS linear regression with the number of shared meals eaten per week as the dependent variable and log GDP per capita as the independent variable.

17 The specific question wording is as follows: "Please imagine a ladder with steps numbered from zero at the bottom to 10 at the top. The top of the ladder represents the best possible life for you and the bottom of the ladder represents the worst possible life for you. On which step of the ladder would you say you personally feel you stand at this time?"

18 For thorough discussions, see Layard & Ward (2020), Clark et al. (2018), and Chapter 6 of the 2017 edition of the *World Happiness Report* (De Neve and Ward, 2017).

19 It is perhaps also worth noting that we observe slight declines in life evaluations and increases in negative affect for those who eat 14 meals with others per week. One potential explanation for this result is that individuals who report sharing all of their weekly meals with others may be uniquely different from other groups. They may, for example, be more likely to live with children or in group settings. They could also be younger or more prone to financial difficulties than those with more flexibility in the number of meals they share with others. Alternatively, this may also be due to measurement error or rounding issues as it represents the upper bound of the scale. We cannot conclusively distinguish between these explanations. A full investigation of this dynamic remains open to future research.

20 This research comes from the Ando Foundation/Nissin Food Products Satisfaction With Food Enjoyment and Variety Survey. See Gallup (2024).

21 This accords with prior work suggesting that more volitional activities (e.g., eating with others or dining alone) matter more than life circumstances (e.g., income or age) in predicting wellbeing (Lyubomirsky et al. 2005b).

22 A module to capture subjective wellbeing was introduced in the *American Time Use Survey* in 2010, 2012, 2013, and 2021. When we consider relationships between meal sharing and subjective wellbeing, we focus on these years in particular.

23 Iglič et al. (2021); Sarracino (2010).

24 Sarracino and Mikucka (2017).

25 U.S. Census Bureau (2023).

26 Estimated as the difference in coefficients predicting the increase in dining alone across two separate regression models in which we include and exclude a control variable for household size. These coefficients are plotted in Figure A9.

27 For relevant discussions, see Kirwan et al. (2024) and U.S. Department of Health and Human Services (2023).

28 In an echo of earlier results using Gallup data, here again, we find that differences in negative affect - specifically sadness and pain - between those who report eating all meals alone the previous day and those who shared meals are slightly larger for women than for men. However, we do not observe similar gender differences for happiness, stress, or tiredness when comparing men and women who eat alone to those who share meals.

29 See Chapter 2.

30 For discussions, see Joshanloo et al. (2018), Siedlecki et al. (2014), and Waldinger and Shultz (2023).

31 For more information, see: Falk et al. (2018).

32 Diener et al. (2018a, 2018b); Holt-Lunstad et al. (2010); Howick et al. (2019); Kawachi and Berkman (2014).

33 Putnam (2000); Dwyer et al. (2018); Glanz et al. (2021); Wang et al. (2016).

34 For more information, see: www.healthandhappinessstudy.com.

35 For more information, see: www.projectgather.org.

References

Austin, D. M., & Baba, Y. (1990). Social determinants of neighborhood attachment. *Sociological Spectrum, 10*(1), 59–78.

Barger, S. D. (2013). Social integration, social support and mortality in the US National Health Interview Survey. *Psychosomatic Medicine, 75*(5), 510–517.

Brehm, J., & Rahn, W. (1997). Individual-level evidence for the causes and consequences of social capital. *American Journal of Political Science, 41*(3), 999–1023.

Brown, E., & Ferris, J. M. (2007). Social capital and philanthropy: An analysis of the impact of social capital on individual giving and volunteering. *Nonprofit and Voluntary Sector Quarterly, 36*(1), 85–99.

Chetty, R., Stepner, M., Abraham, S., Lin, S., Scuderi, B., Turner, N., & Cutler, D. (2016). The association between income and life expectancy in the United States, 2001–2014. *JAMA, 315*(16), 1750–1766.

Clark, A. E., Fleche, S., Layard, R., Powdthavee, N., & Ward, G. (2018). *The origins of happiness*. Princeton University Press.

De Neve, J. E., & Ward, G. (2017). Happiness at work. *World Happiness Report 2017.* New York: Sustainable Development Solutions Network. https://worldhappiness.report/ed/2017/

Diener, E., & Chan, M. Y. (2011). Happy people live longer: Subjective well-being contributes to health and longevity. *Applied Psychology: Health and Well-Being, 3*(1), 1–43.

Diener, E., Oishi, S., & Tay, L. (2018a). Advances in subjective well-being research. *Nature Human Behaviour, 2*(4), 253–260.

Diener, E., Seligman, M. E. P., Choi, H., & Oishi, S. (2018b). Happiest people revisited. *Perspectives on Psychological Science, 13*(2), 176–184. https://doi.org/10.1177/1745691617697077

Dwyer, R. J., Kushlev, K., & Dunn, E. W. (2018). Smartphone use undermines enjoyment of face-to-face social interactions. *Journal of Experimental Social Psychology, 78*, 233–239. https://doi.org/10.1016/j.jesp.2017.10.007

Falk, A., Becker, A., Dohmen, T., Enke, B., Huffman, D., & Sunde, U. (2018). Global evidence on economic preferences. *The Quarterly Journal of Economics, 133*(4), 1645–1692.

Gallup (2024). *Nourishing wellbeing: A global perspective on food enjoyment, healthy eating, and choices in food: 2023 edition*. Gallup, Inc. https://www.gallup.com/analytics/512390/eating-behavior-wellbeing.aspx

Glanville, J. L., Andersson, M. A., & Paxton, P. (2013). Do social connections create trust? An examination using new longitudinal data. *Social Forces, 92*(2), 545–562.

Glanville, J. L., Paxton, P., & Wang, Y. (2016). Social capital and generosity: A multilevel analysis. *Nonprofit and Voluntary Sector Quarterly, 45*(3), 526–547.

Glanz, K., Metcalfe, J. J., Folta, S. C., Brown, A., & Fiese, B. (2021). Diet and health benefits associated with in-home eating and sharing meals at home: A systematic review. *International Journal of Environmental Research and Public Health, 18*(4), Article 4. https://doi.org/10.3390/ijerph18041577

Holt-Lunstad, J., Smith, T. B., & Layton, J. B. (2010). Social relationships and mortality risk: A meta-analytic review. *PLOS Medicine, 7*(7), e1000316. https://doi.org/10.1371/journal.pmed.1000316

Holt-Lunstad, J., Smith, T. B., Baker, M., Harris, T., & Stephenson, D. (2015). Loneliness and social isolation as risk factors for mortality: A meta-analytic review. *Perspectives on Psychological Science, 10*(2), 227–237.

Howick, J., Kelly, P., & Kelly, M. (2019). Establishing a causal link between social relationships and health using the Bradford Hill Guidelines. *SSM-population health*, 8, 100402.

Iglič, H., Rözer, J., & Volker, B. G. M. (2021). Economic crisis and social capital in European societies: The role of politics in understanding short-term changes in social capital. *European Societies, 23*(2), 195–231. https://doi.org/10.1080/14616696.2020.1765406

Johnson, B. R., Pagano, M. E., Lee, M. T., & Post, S. G. (2018). Alone on the inside: The impact of social isolation and helping others on AOD use and criminal activity. *Youth & Society, 50*(4), 529–550. https://doi.org/10.1177/0044118X15617400

Joshanloo, M., Sirgy, M. J., & Park, J. (2018). Directionality of the relationship between social well-being and subjective well-being: Evidence from a 20-year longitudinal study. *Quality of Life Research, 27*(8), 2137–2145.

Kansky, J. (2017). Benefits of well-being: Health, social relationships, work, and resilience. *Journal of Positive Psychology and Wellbeing, 1*(2), 129–169.

Kawachi, I., & Berkman, L. F. (2014). Social capital, social cohesion, and health. In L. F. Berkman, I. Kawachi, & M. M. Glymour (Eds.), *Social Epidemiology* (2nd ed., pp. 290–319). Oxford University Press.

Kimura, Y., Wada, T., Okumiya, K., Ishimoto, Y., Fukutomi, E., Kasahara, Y., & Matsubayashi, K. (2020). Eating alone among community-dwelling Japanese elderly: Association with depression and food diversity. *The Journal of Nutrition, Health and Aging, 24*(6), 687–693.

Kirwan, E. M., Burns, A., O'Súilleabháin, P. S., Summerville, S., McGeehan, M., McMahon, J., Gowda, A., & Creaven, A.-M. (2024). Loneliness in emerging adulthood: A scoping review. *Emerging Adulthood, 12*(1), 47–63.

Knapp, R. J. (1976). Authoritarianism, alienation, and related variables: A correlational and factor-analytic study. *Psychological Bulletin, 83*(2), 194–212. https://doi.org/10.1037/0033-2909.83.2.194

Lauricella, T., Parsons, J., Schaninger, B., & Weddle, B. (2022). Network effects: How to rebuild social capital and improve corporate performance. *McKinsey Quarterly, 2.*

Layard, R., & Ward, G. (2020). *Can we be happier?: Evidence and ethics*. Penguin UK.

Lee, E. J., Lee, K. R., & Kim, J. Y. (2020). Analysis of differences in eating alone attitude of Koreans by dietary habits and age. *Appetite, 152*, Article 104695.

Leigh-Hunt, N., Bagguley, D., Bash, K., Turner, V., Turnbull, S., Valtorta, N., & Caan, W. (2017). An overview of systematic reviews on the public health consequences of social isolation and loneliness. *Public Health*, *152*, 157–171. https://doi.org/10.1016/j.puhe.2017.07.035

Liang, P., & Meng, J. (2023). Paying it forward: An experimental study on social connections and indirect reciprocity. *Review of Economic Design*, *27*(2), 387–417.

Lyubomirsky, S., King, L., & Diener, E. (2005a). The benefits of frequent positive affect: Does happiness lead to success? *Psychological Bulletin*, *131*(6), 803.

Lyubomirsky, S., Sheldon, K. M., & Schkade, D. (2005b). Pursuing happiness: The architecture of sustainable change. *Review of General Psychology*, *9*(2), 111–131.

McClurg, S. D. (2003). Social networks and political participation: The role of social interaction in explaining political participation. *Political Research Quarterly*, *56*(4), 449–464.

Osborne, D., Costello, T. H., Duckitt, J., & Sibley, C. G. (2023). The psychological causes and societal consequences of authoritarianism. *Nature Reviews Psychology*, *2*(4), 220–232. https://doi.org/10.1038/s44159-023-00161-4

Putnam, R. D. (2000). *Bowling alone: The collapse and revival of American community*. Simon & Schuster.

Robison, L. J., & Siles, M. E. (1999). Social capital and household income distributions in the United States: 1980, 1990. *The Journal of Socio-Economics*, *28*(1), 43–93. https://doi.org/10.1016/S1053-5357(99)00012-8

Rodríguez-Pose, A., & von Berlepsch, V. (2014). Social capital and individual happiness in Europe. *Journal of Happiness Studies*, *15*(2), 357–386. https://doi.org/10.1007/s10902-013-9426-y

Roßteutscher, S. (2010). Social capital worldwide: Potential for democratization or stabilizer of authoritarian rule? *American Behavioral Scientist*, *53*(5), 737–757. https://doi.org/10.1177/0002764209350835

Sarracino, F. (2010). Social capital and subjective well-being trends: Comparing 11 Western European countries. *The Journal of Socio-Economics*, *39*(4), 482–517. https://doi.org/10.1016/j.socec.2009.10.010

Sarracino, F., & Mikucka, M. (2017). Social capital in Europe from 1990 to 2012: Trends and convergence. *Social Indicators Research*, *131*(1), 407–432. https://doi.org/10.1007/s11205-016-1255-z

Shen, J., & Bian, Y. (2018). The causal effect of social capital on income: A new analytic strategy. *Social Networks*, *54*, 82–90. https://doi.org/10.1016/j.socnet.2018.01.004

Siedlecki, K. L., Salthouse, T. A., Oishi, S., & Jeswani, S. (2014). The relationship between social support and subjective well-being across age. *Social Indicators Research*, *117*, 561–576.

Sözbilir, F. (2018). The interaction between social capital, creativity, and efficiency in organizations. *Thinking Skills and Creativity*, *27*, 92–100. https://doi.org/10.1016/j.tsc.2017.12.006

Sønderskov, K. M., & Dinesen, P. T. (2016). Trusting the state, trusting each other? The effect of institutional trust on social trust. *Political Behavior*, *38*, 179–202.

Tenney, E. R., Poole, J. M., & Diener, E. (2016). Does positivity enhance work performance?: Why, when, and what we don't know. *Research in Organizational Behavior*, *36*, 27–46.

Tung, E. L., Hawkley, L. C., Cagney, K. A., & Peek, M. E. (2019). Social isolation, loneliness, and violence exposure in urban adults. *Health Affairs*, *38*(10), 1670–1678. https://doi.org/10.1377/hlthaff.2019.00563

U.S. Census Bureau. (2023). More than a quarter of all households have one person. *U.S. Department of Commerce*. https://www.census.gov/library/stories/2023/06/more-than-a-quarter-all-households-have-one-person.html

U.S. Department of Health and Human Services. (2023). *Our epidemic of loneliness and isolation: The U.S. Surgeon General's advisory on the healing effects of social connection and community*. U.S. Public Health Service. https://www.hhs.gov/sites/default/files/surgeon-general-social-connection-advisory.pdf

Waldinger, R., & Schulz, M. (2023). *The good life: Lessons from the world's longest scientific study of happiness*. Simon & Schuster.

Wang, L., & Graddy, E. (2008). Social capital, volunteering, and charitable giving. *Voluntas: International Journal of Voluntary and Nonprofit Organizations*, *19*, 23–42.

Wang, X., Shen, W., Wang, C., Zhang, X., Xiao, Y., He, F., Zhai, Y., Li, F., Shang, X., & Lin, J. (2016). Association between eating alone and depressive symptoms in elders: A cross-sectional study. *BMC Geriatrics*, *16*(1), 19. https://doi.org/10.1186/s12877-016-0197-2

Wilson, J., & Musick, M. (1998). The contribution of social resources to volunteering. *Social Science Quarterly*, *79*(4), 799–814.

Chapter 4

Living with others

How household size and family bonds relate to happiness

Mariano Rojas
Tecnológico Nacional de México

Lina Martínez
Universidad ICESI, Observatory of Public Policies (POLIS), Cali - Colombia

Gerardo Leyva Parra
Research Institute for Development with Equity (EQUIDE), Universidad Iberoamericana

Roberto Castellanos
Faculty of Political and Social Sciences, National Autonomous University of Mexico

Margarita Tarragona
ITAM Center for Wellbeing Studies, Instituto Tecnológico Autónomo de México

We are grateful to the editors of the *World Happiness Report* for the opportunity to share the research and perspectives on wellbeing and happiness that is being carried out in different regions and countries of the world. We would also like to thank John Helliwell, Jan-Emmanuel De Neve, and particularly Lara Aknin, Barry Grimes, and Leoni Boyle for their valuable comments and suggestions on earlier versions of this chapter. Ultimate responsibility for its content rests, of course, with the authors.

doi.org/10.18724/whr-5xep-q992

This version was last updated on 13 March 2025. Please see worldhappiness.report for the latest data.

Happiness is nurtured in relational spaces and the family is at the heart of these connections.

4

Key Insights

For most people in the world, family relationships are an important source of happiness. This chapter explores how the size and configuration of households – where most family interactions take place – are associated with people's happiness.

A household size of about four members is predictive of higher happiness levels. People in these households enjoy abundant and very satisfactory relationships.

People who live on their own often experience lower levels of happiness, primarily due to lower levels of relational satisfaction. People in very large households can also experience less happiness, probably linked to diminished economic satisfaction.

Governments should consider how economic policies may have secondary effects on relationships, hence affecting the wellbeing in families. National statistical offices should prioritise the development of metrics that assess the quantity and quality of interpersonal relationships and the bonds that underpin them.

Latin American societies, characterised by larger household sizes and strong family bonds, offer valuable lessons for other societies that seek higher and sustainable wellbeing.

Introduction

Caring and sharing - sustained by warm, close, and enduring relational bonds - are crucial to human happiness.[1] In particular, family bonds promote lasting relationships, and households provide a context where these bonds develop and, in many cases, thrive.[2] Thus, the field of wellbeing science should pay more attention to household configurations and intra-household relationships.

This chapter examines the relationship between happiness, household size, and family configuration. We make extensive use of the rich data provided by INEGI, the National Statistical Office of Mexico, through its ENBIARE 2021 survey, as well as additional information from Colombia. Our analyses contrast the situation in Mexico with that of European countries, drawing on data from the European Social Survey 2020.

We hypothesise that a small number of household members may limit affective connections, which negatively impacts happiness, while a large number of members may impose economic burdens that also threaten wellbeing. Consequently, the chapter explores the potential existence of an inverted U-shaped relationship between life satisfaction and the number of household members. An in-depth analysis of satisfaction across different life domains suggests that the number of household members is associated with certain economic costs but also offers broad relational benefits, such as increased satisfaction with affective life, family, and personal relationships.[3]

We also investigate the association between life satisfaction and various family configurations. The findings indicate that these configurations significantly influence happiness. For example, two-parent households are associated with higher levels of life satisfaction among adult members, while adults living in single-person and single-parent households tend to experience lower levels of happiness. The presence of additional family members in single-parent households appears to mitigate some of the negative effects of single-parenthood on happiness.

Literature review

The family as a central relational space

Happiness is nurtured in relational spaces and the family is at the heart of these connections.[4] Caring and sharing are practices that inherently rely on the presence of and interaction with others, beginning with family members. The family is where people first learn to care for and share with others, creating the foundation for broader social interactions and for wellbeing.[5] The family works as a reference for how people interact with others in their life.[6] Families are associated with close, warm, and genuine relationships that last for long stretches in people's journey in life.

Wellbeing researchers have recently been encouraged to adopt a systemic perspective,[7] which has roots in psychology since the late 1950s, particularly in family therapy and family studies.[8] Systemic approaches assert that the whole is greater than the sum of its parts and that emergent phenomena arise between people who are in relationships, rather than from out-of-context individuals. This perspective examines how a family's structure and dynamics influence its members, recognising that people exist within an emotional ecosystem where family bonds shape identity and wellbeing. It emphasises that happiness is not solely individual enjoyment but the shared joy and caring experienced within relationships.[9]

Research on subjective wellbeing often emphasises social cohesion, community involvement, and prosocial behaviours within broader civic spaces, but the family's foundational role in shaping these behaviours is frequently overlooked. By

Happiness is not solely individual enjoyment but the shared joy and caring experienced within relationships.

acknowledging the influence of family dynamics on prosocial development, we gain a more comprehensive understanding of how to enhance wellbeing in societies. Before civic engagement and charitable activities appear, the family is the space where people start to build interpersonal relationships.

From a sociological perspective, a family may be conceived as a social unit or group of people who are related by blood, marriage, adoption, or other long-term commitments who typically live together and share economic, emotional, and social activities.[10] From a psychological perspective, families are understood in terms of caring, sharing gratifying activities, and nurturing and supporting their members which foster a sense of belonging and identity that significantly contributes to people's wellbeing.[11] This perspective emphasises emotional bonds, interpersonal dynamics within the family, and the shared joys and challenges in life.[12]

Families and households

Families and households are different both conceptually and empirically. On the one hand, the concept of family tends to emphasise kinship ties, socialisation, social roles, nurturing, and the transmission of culture and values across generations. On the other hand, a household commonly refers to a group of people, regardless of their type of relationship, who live together in the same dwelling and share living arrangements. A household may consist of one person who lives alone, persons who are not related, or several people with family bonds. In most cases, households are constituted based on family ties.

Household size and configuration are commonly determined by the number of children and the type of coresident family group.[13] Household size and its possible configurations are important for the family dynamics that may emerge from them, including those relating to caring and sharing practices and their relationship with happiness.

Household size, family configurations, and happiness

Mexican data shows the existence of an inverted U-shaped relationship between life satisfaction and the number of people in a household.[14] In Colombia, the heads of households, spouses in two-parent households, and especially those who are married, work, and have medium-sized families (four persons), report the highest levels of happiness.[15] Convergent results have been found in Mexico where university students from two-parent households show greater levels of self-esteem and life satisfaction than those living in single-parent households.[16] The impact of family configuration on wellbeing points to the importance of the quality of interpersonal relationships that are established and developed in families. In particular, relationships based on affection, close communication, repeated contact, and mutual support are a source of family satisfaction and, in turn, life satisfaction as a whole.[17]

Research on the relationship between family configuration, household size, and wellbeing offers insights that are relevant to the analyses in this chapter. First, the size and configuration of families and the dynamics within them are not innocuous, they affect the happiness of their members. This is particularly true for marriage bonds,[18] parenthood bonds,[19] and the number of family members.[20] Second, there are family characteristics, such as the presence of one or two parents, that have a particularly significant effect on the wellbeing of family members. Third, life satisfaction is related to the contextual circumstances of the family – such as the material living conditions and the family's life cycle – and to specific characteristics of the parents or heads of household (age, education, etc.). All these factors are interrelated and it is not always easy to disentangle their links and their effects on wellbeing.

While family relationships have historically been viewed as traditional sources of support,[21] it is also important to highlight the emotional depth and sense of value in family relationships. They are rooted in mutual affection and companionship that transcend mere supportive roles and imply

Relationships based on affection, close communication, repeated contact, and mutual support are a source of family satisfaction and, in turn, life satisfaction as a whole.

person-based relationships, where people know each other well and where the purpose of the relationship is the relationship itself. This kind of relationship is central to the joint enjoyment of life.[22] The intrinsic value of such relationships lies in the warmth, closeness, and genuine affection that family members share with each other over long periods of time. Hence, family relationships are valued not only for what they provide but, fundamentally, for the quality of the emotional and meaningful bonds involved. The abundance and quality of family relationships contribute to people's happiness and we expect household size and configuration to contribute to both the quantity and quality of family relationships.[23]

Household characteristics in Latin America and Europe

We now turn to the specific and contrasting characteristics of households in Latin America and Europe. Figure 4.1 presents the average household size for many Latin American and European countries taken from the CORESIDENCE Database.[24] Substantial differences are observed between the two regions. Except for Cuba, Uruguay, and Puerto Rico, the average household size in Latin America exceeds three people per household, a size only reached in two European countries: Montenegro and Slovakia. The average household size exceeds five in Nicaragua and it is fewer than two in Finland and Denmark.

Linear thinking may suggest that living alone may be better than living with others, even if it's more expensive. However, this view leaves aside the value of interpersonal relationships, which are an important source of wellbeing and are fostered by sharing the same house with other people that we care about.[25]

Figure 4.1: Average household size in Latin America and Europe
CORESIDENCE Database (2023)

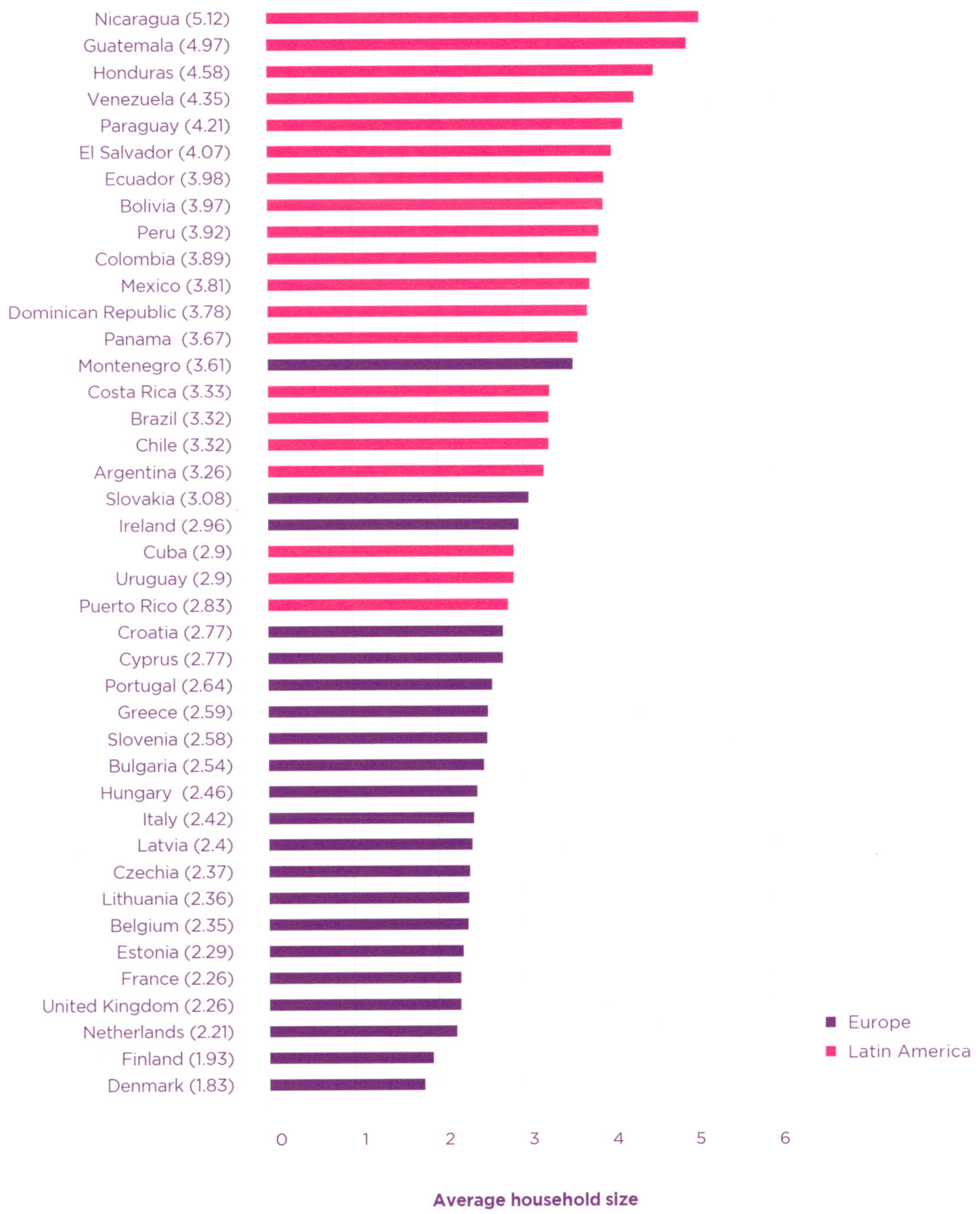

Figure 4.2: Single-person households in Latin America and Europe
CORESIDENCE Database (2023)

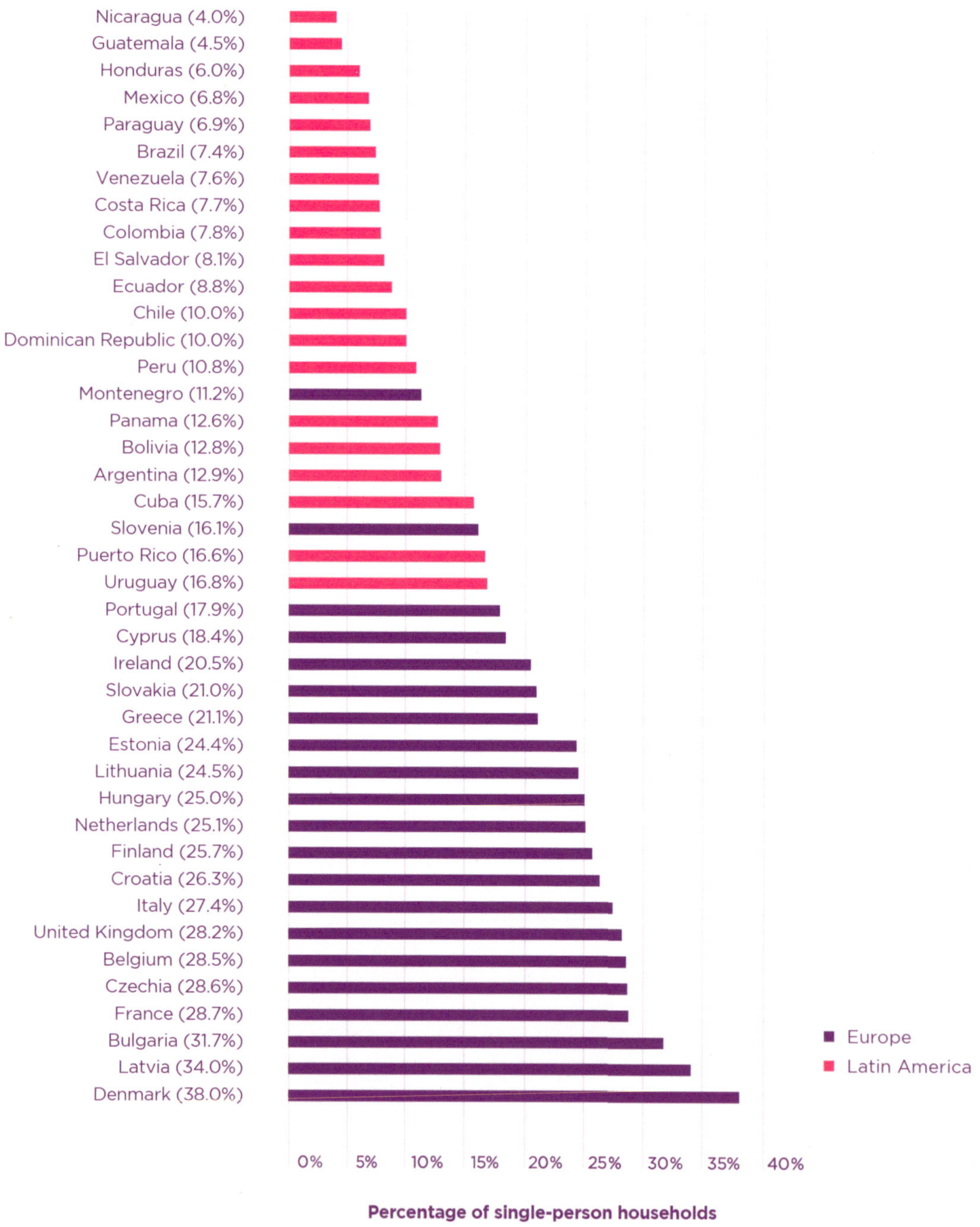

Figure 4.2 shows the proportion of single-person households in Latin American and European countries. The proportion ranges from 4% in Nicaragua to 38% in Denmark.

In Figure 4.3, we compare the distribution of household size between Mexico and European countries. We observe that single-person households make up 23% of European households, but that figure is only 11% in Mexico. In addition, households with two members comprise 34% of European households and less than 20% in Mexico. Thus, 55% of households in Europe have two members or fewer, but this figure is about 30% in Mexico. Furthermore, almost half of Mexican households include four people or more, while this figure is about 24% for Europe. Mexico is economically poorer than the average European country. However, larger households imply a potential advantage to build positive social interactions within the household, which could partially counterbalance the differences in income with Europe. This is one of several plausible explanations for why most Latin American countries report higher wellbeing than predicted by their GDP per capita.

Figure 4.3: Distribution of household size in Mexico and Europe
ENBIARE (2021), European Social Survey (2020)

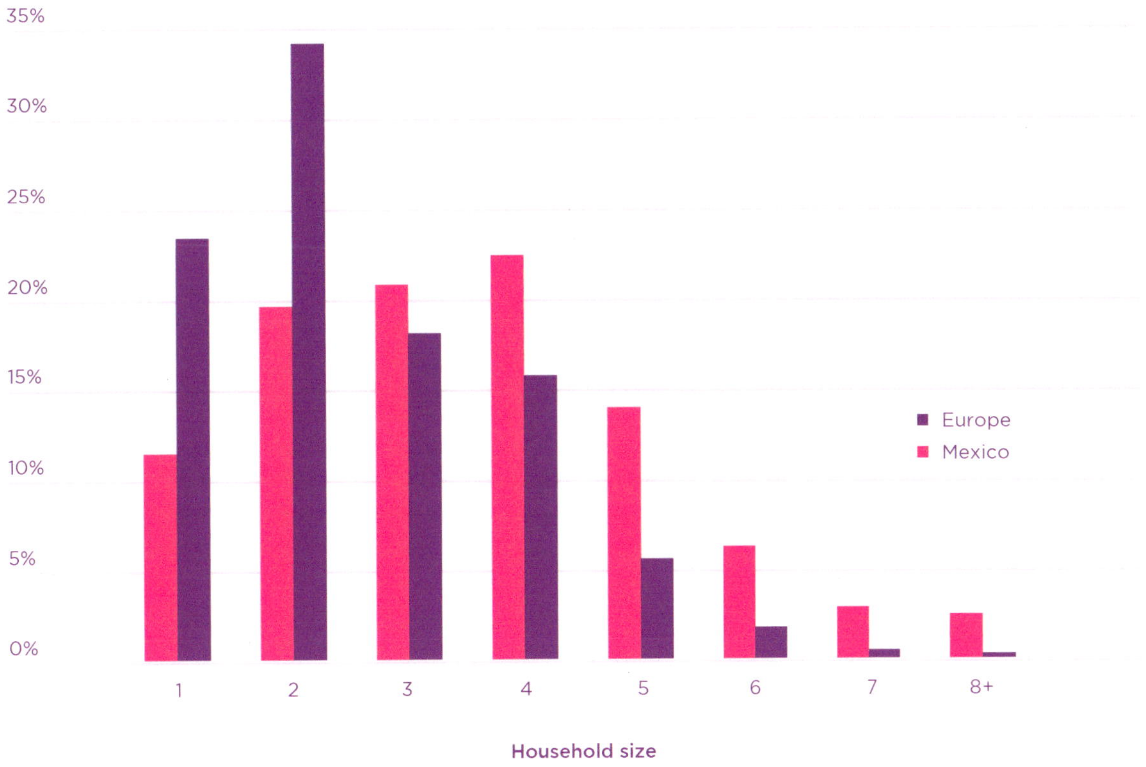

Box 4.1: Trends in family configurations

Over the past 50 years, the composition of families and households worldwide has undergone significant change.[26] The emerging trends raise significant concerns from a wellbeing perspective as they suggest increasing threats to both the quantity and quality of person-based relationships and the role that households play in contributing to happiness. It is unclear if economic growth can adequately compensate for these detrimental effects.

Some of the most relevant trends are:

- **Decrease in family size:** Between 1970 and 2020, the majority of households have decreased by approximately 0.5 persons per decade on average.[27] This global trend is largely explained by the reduction in the number of children. Fertility rates across the globe have halved, from 4.84 in 1950 to 2.23 in 2021.[28]
- **Rise in single-person households:** They are becoming widespread in Europe and are growing rapidly in Latin America, but they are still rare in Africa and most Asian countries.[29] Single-person households range from 2.6% in Cambodia to 38% in Switzerland.
- **Rise in single-parent families:** Since the 1990s, single-mother households have been on the rise in all developing regions, while single-father households have remained stable.[30]
- In many parts of the world, **population aging has led to more multigenerational households**, in which elderly parents live with their adult children and grandchildren. This change is particularly prevalent in regions and social sectors where economic restrictions and cultural norms favour family care for the young and old.[31]
- More people are choosing to **live without children**. Couples without children are prevalent in OECD countries, ranging from 15% in Poland and Slovenia to 26% in Canada. In the United States, couples without children (25%) are slightly more prevalent than couples with children (24%).[32]

The relationship between household size and happiness

Descriptive statistics of household size

What is the relationship between household size and happiness? This is the first empirical question we tackle in this chapter. We use surveys from Mexico, Colombia, and several European countries to address this question, taking advantage of the data available in Mexico to deepen our understanding of the relationship between life satisfaction and household size. Figure 4.4 presents average life satisfaction by the number of members in the household for both Mexico and Europe (note that these figures differ from the rankings in Chapter 2 as they come from different data sources).[33] Our objective is to analyse how life satisfaction varies with the number of household members in each region, rather than comparing the regions directly.

In both Mexico and Europe, the highest average life satisfaction is reported by people who live in households with four to five members. We also observe an inverted U-shaped relationship. Average life satisfaction is lower for people in single-person households as well as households with six or seven members. In Europe, there is a high wellbeing cost for people in single-person households. While the average life satisfaction reaches 7.5 for people in households with five members, it is only 6.6 for people in single-person households. We should not forget that almost

Figure 4.4: Life satisfaction by household size in Mexico and Europe
ENBIARE (2021), European Social Survey (2020)

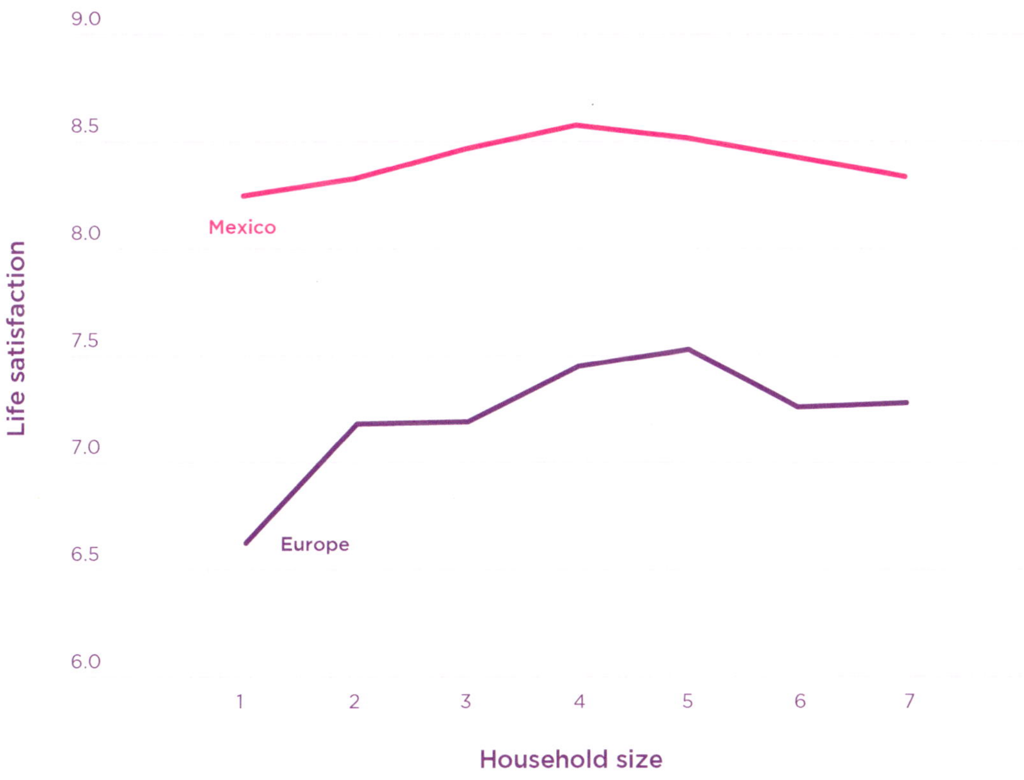

24% of households in Europe are single-person. The wellbeing cost for people in single-person households is also observed in Mexico, but to a lesser degree than in Europe. It seems that living in single-person households has a wellbeing cost, but it depends on the context so this cost varies across regions. For example, it may not be the same to live in a single-person household when you have close relatives in the neighbourhood or good friendship ties.[34]

Statistical analysis of household size

We ran regression analyses to delve deeper into the relationship between household size and life satisfaction in Mexico, taking socio-economic and demographic characteristics into account. The regression specification is flexible enough to test the existence of an inverted U-shaped relationship. The statistical findings are presented in Table 4.1 in the online appendix.[35] It is worth noting that the coefficients for the number of household members are statistically significant, although the goodness of fit of the entire model is low, suggesting that it would be very risky to predict the life satisfaction of a particular person from their information on assets, education, gender, age, and number of members in the household.[36]

Figure 4.5 presents the predicted life satisfaction for different household sizes based on the estimated coefficients presented in Table 4.1. The predicted value is computed for a woman with average age, education, and assets, and an inverted U-shaped

Figure 4.5: Predicted life satisfaction by household size in Mexico
ENBIARE (2021)

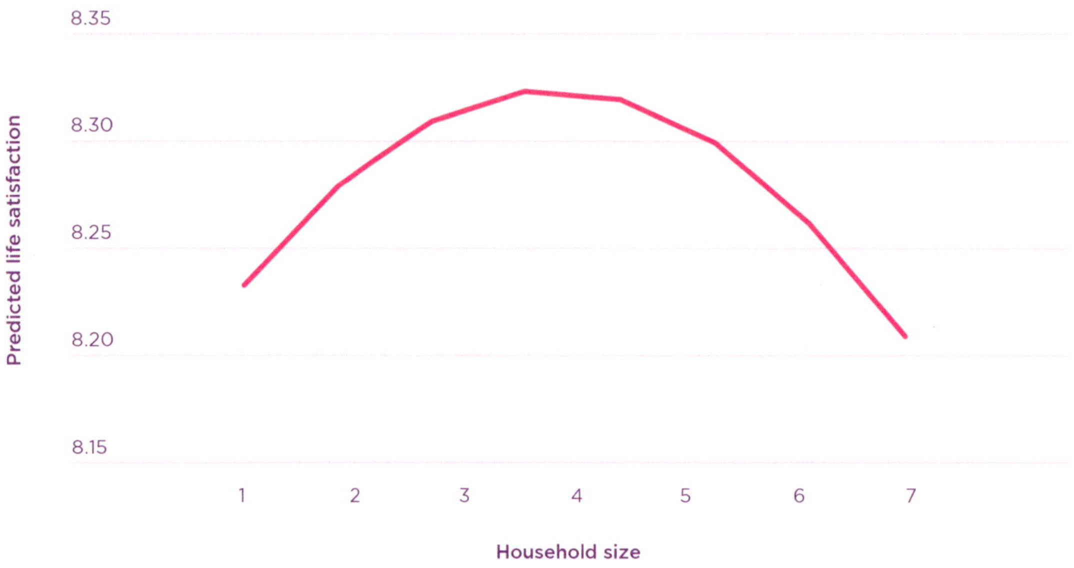

Note: See Table 4.1 of the online appendix for details of the statistical analysis

relationship is confirmed. We estimate that the highest life satisfaction is reached in households of 4 to 5 members, keeping socio-economic and demographic characteristics constant.[37] This figure is higher than the average household size in Mexico, which is 3.5. Therefore, from a wellbeing perspective, the current average household size in Mexico is suboptimal.

We conducted similar regression analyses for the European region using data from the European Social Survey. Table 4.2 in the online appendix presents the estimated coefficients for two models.[38] Figure 4.6 presents the predicted life satisfaction for different household sizes based on a woman with average age, education, income, and living in Belgium.

Figure 4.6 uses the estimated coefficients for the model that incorporates more countries and observations. Again, we observe an inverted U-shaped relationship, regardless of the estimated model. The highest life satisfaction is achieved with a household size of four members, well above the current average for European households which is 2.5. Therefore, from a wellbeing perspective, the current average household size in European countries seems to be suboptimal too.

This analysis shows two important results for both Mexico and Europe. First, the current average size of households is below the size associated with the highest predicted level of life satisfaction for adults within the region. Second, people who live in single-person or very large households tend to report lower wellbeing. The reasons for suboptimal household size are not addressed in this chapter. A deeper understanding of the inverted U-shaped relationship is obtained by studying how satisfaction in different life domains relates to household size.

Figure 4.6: Predicted life satisfaction by household size in Europe
European Social Survey (2020)

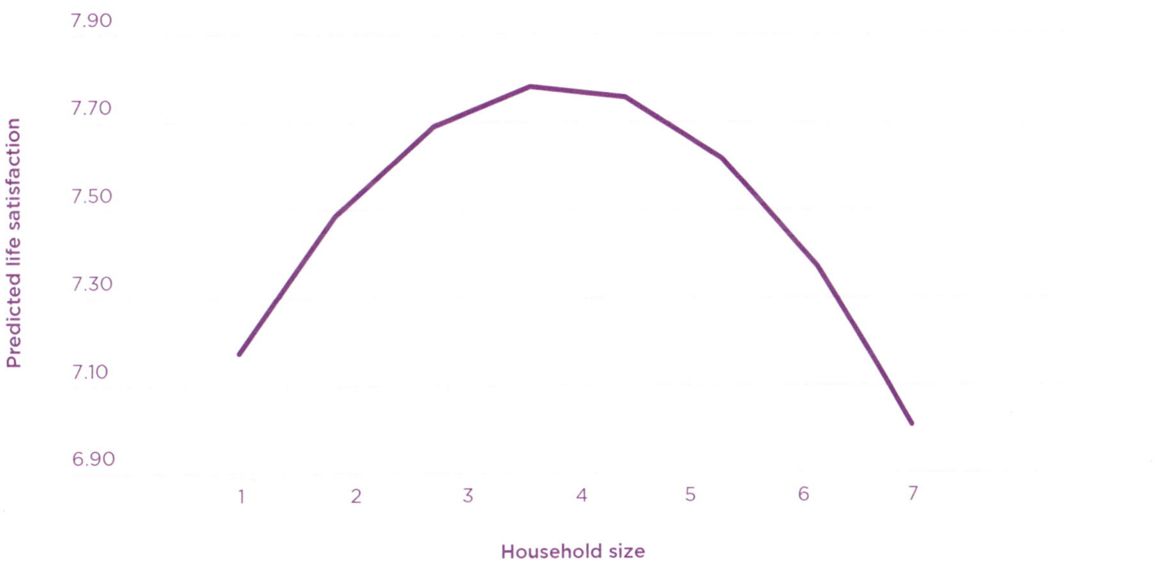

Note: See Table 4.2 of the online appendix for details of the statistical analysis

Domains-of-life explanation for the inverted U-shaped relationship

The domains-of-life approach understands life satisfaction as emerging from satisfaction in specific realms of life.[39] The Mexican ENBIARE survey asks people about their satisfaction in several domains of life. Here, we consider seven domains: personal relationships, social life, affective life, family, economic situation, health, and occupational situation. The first four domains are clearly relational. Figure 4.7 shows average satisfaction by domain of life. We observe that satisfaction with family life is quite high in Mexico, with an average satisfaction above 9, on a scale ranging from 0 (totally unsatisfied) to 10 (totally satisfied). Satisfaction with the economic situation is relatively low.

How is household size related to satisfaction in these different domains of life?[40] Figure 4.8 presents the predicted satisfaction in each domain for different household sizes, based on the estimated coefficients presented in Table 4.3 in the online appendix. The predicted satisfaction is computed for a woman with sample average age, education, and assets.

The key insight portrayed by Figure 4.8 is in the relational domains. An inverted U-shaped relationship is observed in family satisfaction, where maximum family satisfaction would be reached with a household size of about six members. This suggests that there are substantial benefits from living in large households in terms of family satisfaction. The same situation is observed for satisfaction with affective life, where maximum satisfaction is reached for about five to

Figure 4.7: Satisfaction across domains of life in Mexico
ENBIARE (2021)

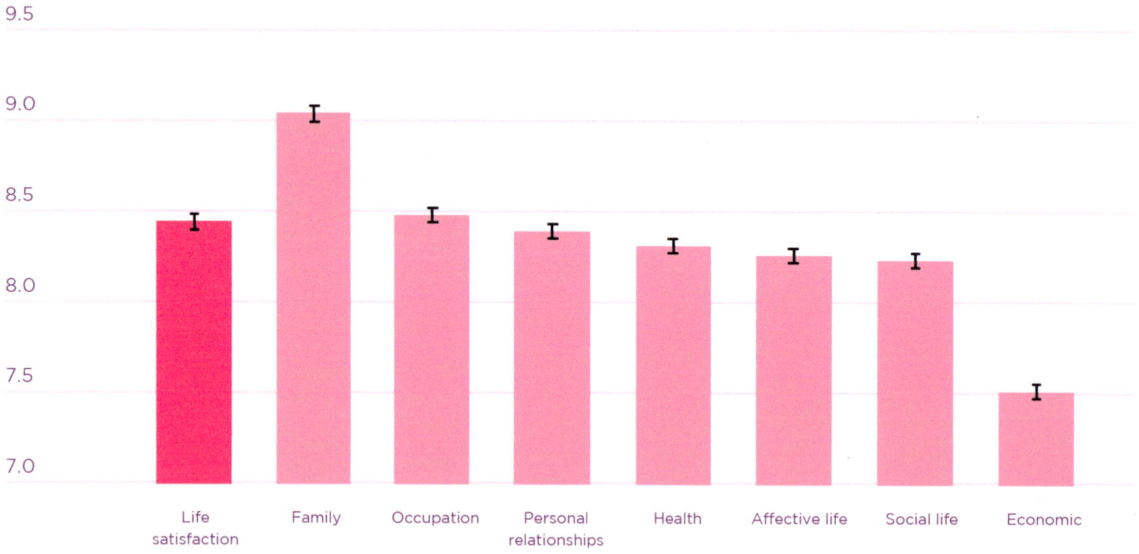

six household members. Similarly, for the domain of personal relationships, maximum satisfaction is reached with household members close to five.

These results suggest there are relational benefits associated with living in households between four and six members, which is well above the average household size for Mexico of 3.5 members. Recent research[41] shows that, to a certain degree, the size of the household contributes to generating quality relationships which may be associated with greater satisfaction with personal relationships, affective life, and family. There is also evidence that the main variable explaining family satisfaction is its relational foundation and, in particular, its affective component, more so than economic factors.[42] Thus, the life satisfaction gains from living in large households seem to be associated with the important relational benefits associated with large households.

Our empirical analysis also shows that economic satisfaction is inversely associated with the number of household members. This finding suggests that the number of members in the household implies an economic burden that reduces economic satisfaction, possibly signalling that the benefits from material resources have to be distributed among a greater number of household members as household size increases.

The findings presented in Figure 4.8 (and in Table 4.3) pose a dilemma: small households – and even single-person households – tend to report higher levels of economic satisfaction. However, small households report lower satisfaction with their interpersonal relationships. If economic satisfaction is prioritised, then a small household size offers advantages. However, people are much more than mere consumers and their human relationships play a central role in their lives and their wellbeing.

Figure 4.8: Predicted satisfaction in domains of life by household size in Mexico
ENBIARE (2021)

Note: See Table 4.3 of the online appendix for details of the statistical analysis.

Household configuration and life satisfaction

Households are relational spaces and the different types of family bonds within the household may influence the nature of relationships and, in consequence, the life satisfaction of household members. This section focuses on the nature of family bonds within households and how they are associated with life satisfaction.[43]

We consider six basic household configurations in the following analyses (see right).

Single-person households

Couples without children

Couples with at least one child

Couples with at least one child and other relatives

Single-parent households

Single-parent households and other relatives

Figure 4.9: Life satisfaction and distribution of household configurations in Mexico
ENBIARE (2021)

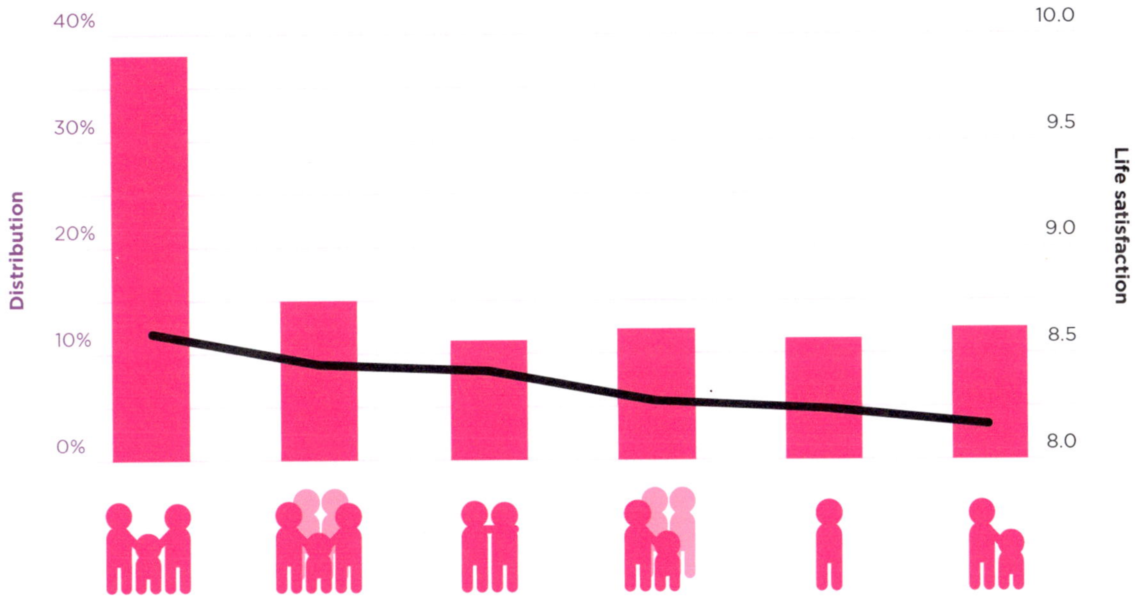

Note: Percentages are estimated on frequencies over expanded data. Households composed of non-family members and people who live with family members but without a partner or children are left out as both cases are rare.

The number of households inhabited by two or more people with no family ties at all is negligible so they are not considered in this analysis. All six configurations involve some type of family bond and we examine the life satisfaction reported by adults who live in one of these household configurations and were selected to respond to the ENBIARE or European Social Survey.

Descriptive statistics of household configuration (Mexico)

In Figure 4.9, we present the distribution of the Mexican population by household configuration and the average life satisfaction associated with each configuration. Couples with at least one child are the most frequent household configuration in Mexico, representing almost 38% of households in the country. People in this type of household report the highest life satisfaction with an average of 8.6. Life satisfaction is also relatively high for those who live in households of couples without children and for those who live in households with a couple, children, and other relatives. In these cases, the average life satisfaction is around 8.4. Life satisfaction is relatively low for people who live in single-person households, in single-parent households with children, and in single-parent households with children and other relatives. The presence of extended family has a favourable effect for single parents with children but seems to be detrimental for couples with children.

These results suggest that households based around couples report the highest levels of life satisfaction. This high level of life satisfaction, combined with a high percentage of these household configurations, clearly contributes to increased life satisfaction in Mexico. This is a wellbeing driver that is not frequently considered in the happiness literature.

Statistical analysis of household configuration (Mexico)

We ran regression analyses to study the association between life satisfaction and family configurations further. The analysis controls by age, age squared, gender, and educational level of the interviewee, as well as by assets in the household (as a proxy for the household's economic situation). Table 4.4 in the online appendix presents the estimated coefficients. The category of reference corresponds to a couple with at least one child.

Figure 4.10 shows the relevant estimated coefficients. The lowest levels of life satisfaction are associated with single-person and single-parent households. We also observe significantly lower life satisfaction among single-parent households, even after adjusting for the economic situation of the household. However, this observation is for single-parent households where there are no other family members, which suggests that the presence of other relatives can mitigate the wellbeing cost for people who live in single-parent households. Couples without children report levels of life satisfaction that are statistically similar to people who live in couples with children.[44]

Figure 4.10: Life satisfaction and household configurations in Mexico
ENBIARE (2021)

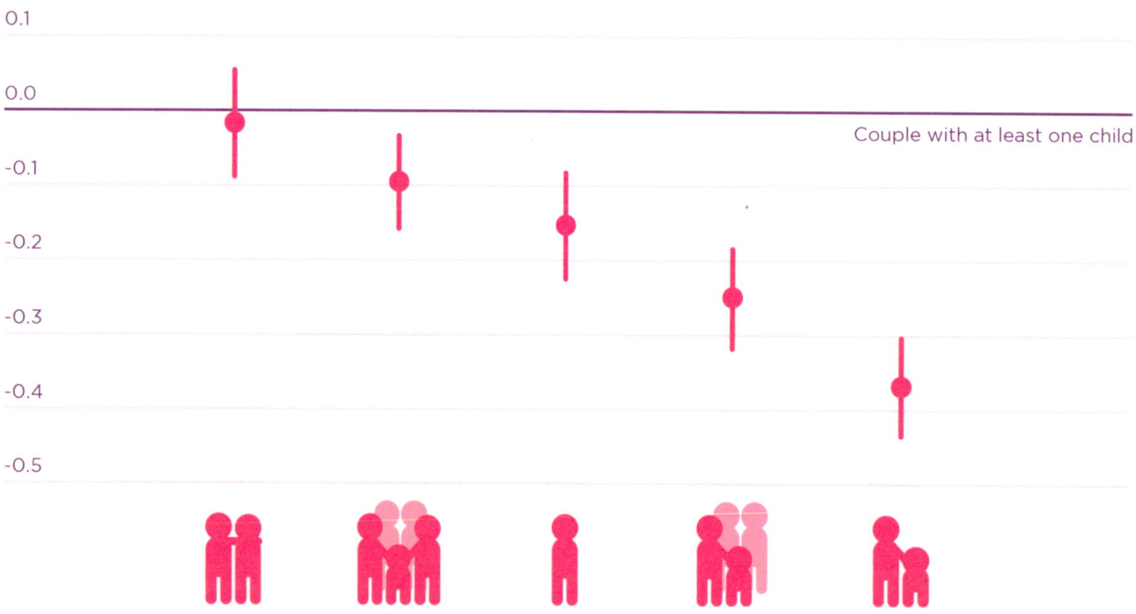

Note: See Table 4.4 of the online appendix for details of the statistical analysis.

Domains-of-life explanation (Mexico)

The domains-of-life approach allows us to delve into the origin of the wellbeing costs associated with single-person and single-parent households. We ran regression analyses to study the association of satisfaction in each domain of life with family configurations, controlling by age, age squared, gender, education level, and assets in the household. Table 4.5 in the online appendix shows the estimated coefficients from the quantitative analyses.

The relevant estimates are presented in Figure 4.11. In comparison to people living in a couple with at least one child, people living in single-person households report lower satisfaction in their personal relationships, affective life, and family life. They also report higher economic satisfaction but this is not enough to compensate for lower satisfaction in relational domains. This explains the lower life satisfaction reported by people in single-person households seen in Figure 4.10.

Compared to couples with children, people who live in couples without children have greater economic and affective satisfaction, but lower satisfaction with health and family. Overall, their life satisfaction is no different than people who live in couples with children.

People living in couples with children and other relatives report lower life satisfaction than people living in couples with children. This lower life satisfaction is explained by lower satisfaction in almost all the domains of life that were studied.

People who live in single-parent households (with or without other relatives) report lower life satisfaction than couples with children and they have lower satisfaction in all the domains of life under consideration. It is important to note that people in single-parent households with other relatives report greater life satisfaction than people in single-parent households with no relatives. This is mostly explained by their greater satisfaction with affective life and personal relationships.

Figure 4.11: Satisfaction in domains of life by household configuration in Mexico
ENBIARE (2021)

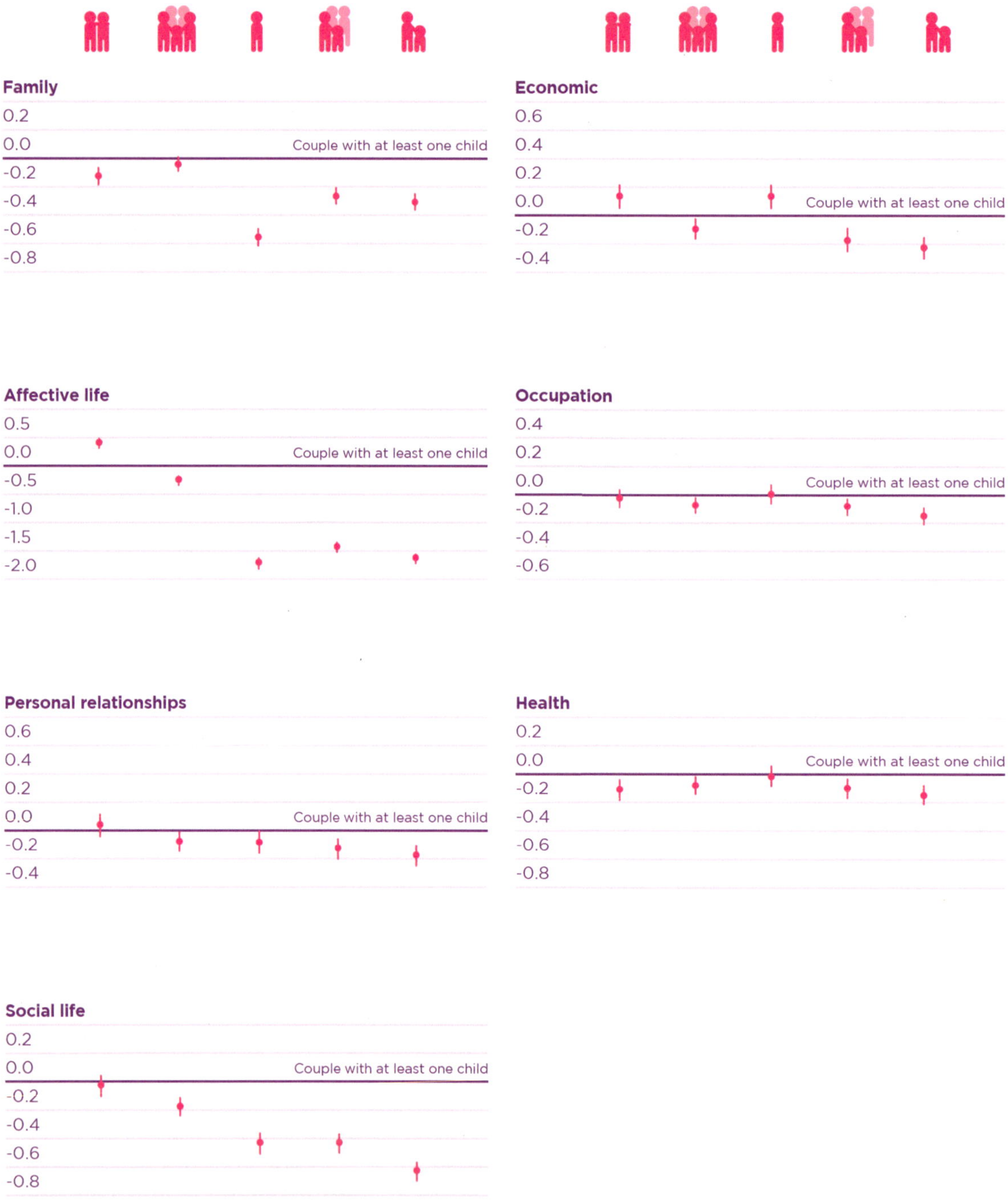

Note: See Table 4.5 of the online appendix for details of the statistical analysis.

Descriptive statistics of household configuration (Europe)

The distribution of household configurations in European countries (Figure 4.12) differs from Mexico (Figure 4.9). The percentage of single-person households in Europe (24%) is more than double the number in Mexico (11%). The situation is similar for couples without children, 28% for Europe and 11% for Mexico. Couples with children represent 25% of households in Europe, while in Mexico it reaches almost 38%. Couples with children and other relatives are less than 4% of households in Europe, while this figure is almost 15% in Mexico. The percentage of single-parent households is relatively small in Europe (5%) compared to Mexico (12%), while the percentage of single-parent households with other relatives is slightly higher in Europe (15%) than in Mexico (12%).

As we saw in Mexico, there is a life satisfaction cost for single-person households. A similar situation is observed for single-parent households with no other relatives. Couples, with or without children, enjoy the highest level of life satisfaction.

The difference in life satisfaction between people who live in couples with children and people living alone is much smaller in Mexico than in Europe (a decline of 0.37 in Mexico vs. a decline of 0.84 in Europe). In addition, there is a higher percentage of people living alone in Europe. This combination of factors – the high cost of living in single-person households and the large percentage of households in this category – is clearly detrimental to the life satisfaction of Europeans.[45]

Figure 4.12: Life satisfaction and distribution of household configurations in Europe
European Social Survey (2020)

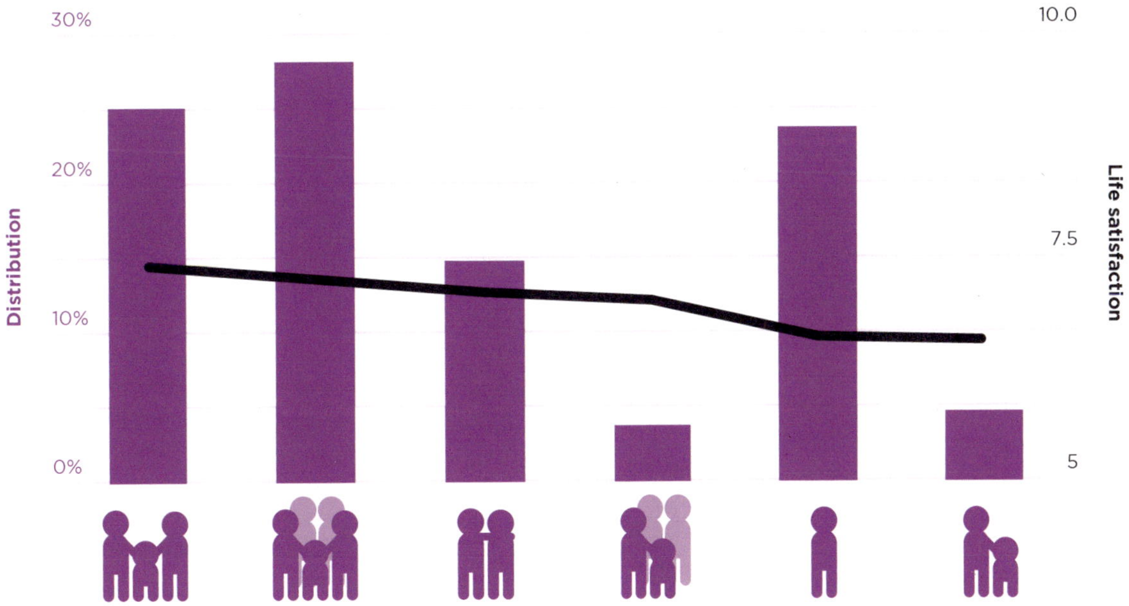

Note: Percentages are estimated on frequencies over expanded data. Households composed of non-family members and people who live with family members but without a partner or children are left out as both cases are rare.

Statistical analysis of household configuration (Europe)

We ran regression analyses for European countries using data from the European Social Survey. The analysis studies the relationship between life satisfaction and household configurations and controls by age, age squared, gender, education level, and country fixed effects. Table 4.6 in the online appendix presents the estimated coefficients.

Figure 4.13 shows the relevant coefficients for household configurations in Europe. Similarly to Mexico, people living in single-person and single-parent households have lower life satisfaction than people who live in couples with children. However, this cost is larger in Europe than in Mexico. Contextual factors such as the role of extended family and friendships may explain this difference. People who live in couples with children and other relatives also have lower life satisfaction than couples with children and no other relatives. People who live in single-parent households with other relatives have greater life satisfaction than single parents who live with no other relatives. In general, the life satisfaction pattern across household configurations is very similar in Mexico and Europe.

Figure 4.13: Life satisfaction and household configurations in Europe
European Social Survey (2020)

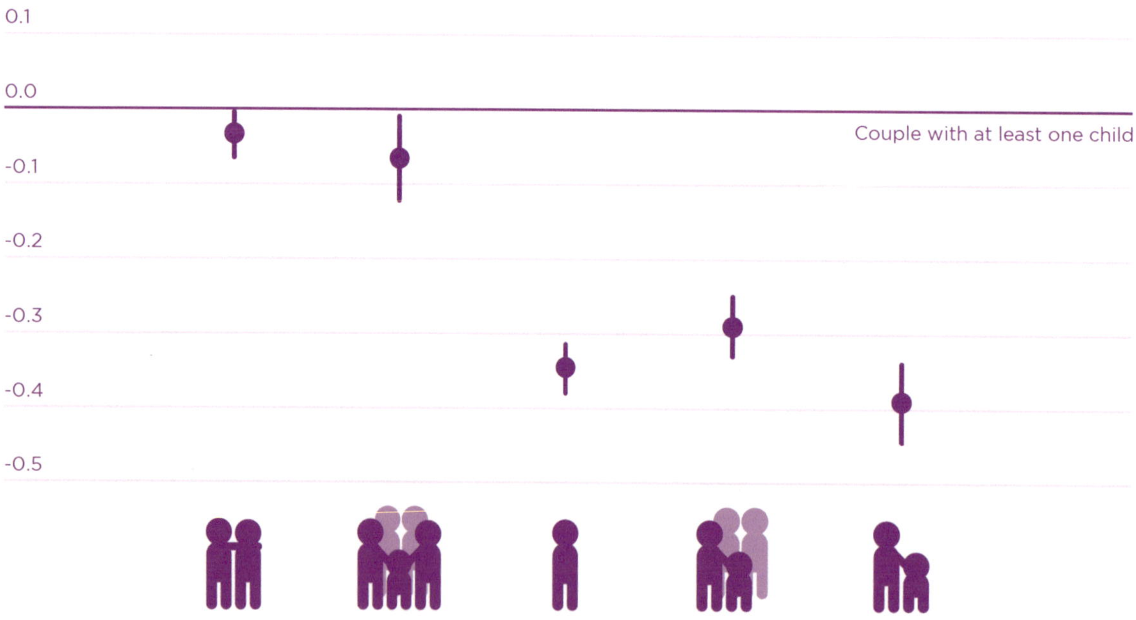

Note: See Table 4.6 of the online appendix for details of the statistical analysis.

Final considerations

The size and configuration of households are highly relevant for people's wellbeing. The household is not only a dwelling. It is a space for coexistence that favours the emergence of high-quality relationships which may significantly contribute to life satisfaction. The study of households often emphasises economies of scale in the use of resources as well as specialisation in the division of labour. However, the wellbeing benefits of living together in households are not limited to economic aspects.

Households are relational spaces – a community of caring and sharing – where members create strong interpersonal relationships that contribute to their life satisfaction. Households of two or more people frequently foster close, genuine, and long-lasting relationships, with subsequent benefits for life satisfaction.

In this chapter, we show that household size and configuration are statistically associated with life satisfaction in Mexico and Europe. We find that people who live on their own report lower levels of life satisfaction, and this is not associated with economic reasons. On the contrary, single-person households report greater economic satisfaction but lower life satisfaction due to relational deprivation. Controlling for economic resources, this effect is smaller in Mexico than in Europe, which suggests that the wellbeing cost experienced

In this chapter, we show that household size and configuration are statistically associated with life satisfaction in Mexico and Europe. We find that people who live on their own report lower levels of life satisfaction, and this is not associated with economic reasons.

by those who live on their own may be context dependent. Presumably, higher average levels of income may help to hide the relatively larger cost of loneliness to people's wellbeing.

We present evidence of an inverted U-shaped relationship between household size and life satisfaction. A household of around four people has the highest life satisfaction in both Mexico and Europe. Information from Latin American countries indicate that the quantity (i.e., time spent with family members) and quality (i.e., sharing emotions, manifesting affection, communicating, and giving support when facing challenges) of family relationships is positively associated with household size. We also find

Globalisation and the geographical relocation of production impose considerable strain on the social fabric, destabilising families and weakening familial and social bonds.

that people who live in single-parent households report lower life satisfaction. This is mostly explained by their lower satisfaction in relational domains of life, as well as their lower economic satisfaction.

Further analyses indicate that the relationship between household size and life satisfaction is influenced by the extent to which family members engage in caring and sharing activities.[46] Thus, the time spent together as a family, along with positive emotional exchanges, affective bonds, genuine interest, communication, and mutual support largely accounts for the positive link between household size and life satisfaction. Additional analyses also indicate that household size is positively associated with access to support.[47] This allows us to conclude that, at least in part, the association between household size and life satisfaction is mediated by the relevance of household size on caring and sharing activities.

Household configurations differ across regions, countries, and decades. Due to these differences, and because household configurations matter for people's wellbeing, researchers should take these differences into account when contrasting life satisfaction between countries and across time. It is also important to consider these differences even when contrasting the wellbeing of people living in the same country, as their household configuration may vary.

Some prevailing social trends are detrimental to the kinds of household configurations that promote life satisfaction. Indeed, these trends are often linked to the erosion of relational spaces. Such trends are intricately tied to economic policies and development strategies that have, in recent decades, prioritised economic growth while neglecting the vital role of family relationships and broader social connections.

These policies aim at economic targets such as investment, exports, and infrastructure development, yet their consequences extend well beyond the economic sphere, often being inadequately acknowledged. For instance, globalisation and the geographical relocation of production impose considerable strain on the social fabric, destabilising families and weakening familial and social bonds. The reallocation of resources is frequently linked to heightened job insecurity which, in turn, can undermine the quality of family relationships. Moreover, the deregulation of labour markets and capital movements, while attracting foreign investment, can also exacerbate job vulnerability and disrupt work-life balance. Similarly, an educational focus on enhancing human capital, while crucial for productivity, may overlook the development of socio-emotional skills that are essential for fostering positive social interaction.

In this context, social policy has often been conceived as a palliative measure, designed to address the social problems generated or, at best, not sufficiently addressed by pro-growth policies. It is within the realm of social policy to introduce pro-family initiatives, albeit as a reactive measure rather than as part of a proactive strategy. Specific pro-family policies, such as initiatives to improve work-life balance, promote gender equity, and provide maternity or childcare benefits, attempt to mitigate some of the familial challenges that negatively impact wellbeing. However, these policies are far less effective when implemented within a broader and hostile socio-economic context. Consequently, there is a pressing need for a more comprehensive reassessment of economic policies – one that recognises the critical importance of family relationships for overall wellbeing.

Ultimately, societal progress should not be measured by income levels, but rather by the wellbeing experienced by its members. A more holistic approach to policy-making, which acknowledges the relevance of relations and family bonds as a core element of prosperity, is essential for fostering sustainable, long-term wellbeing.

Box 4.2: Family, the social fabric, and the crisis of violence in Ciudad Juárez

At the beginning of the 21st century, Ciudad Juárez, a Mexican city on the border with the United States, suffered from an acute crisis of violence to the point that, with 229 homicides per 100,000 inhabitants in 2008, it was considered "the most violent city in the world".[48] With only 1% of the country's population, Ciudad Juárez concentrated 28% of the total homicides committed in Mexico. At the same time, it's youth were victims and participants in an enormous escalation of violence that included organised crime, robberies, kidnappings, 'rent' collection, homicides, femicides, and serious problems of domestic violence and child abuse.

This crisis resulted from the confluence of multiple factors, one of which was the deterioration of family relationships and community life. This was a city that based its growth on the export maquiladora industry, whose viability demanded that large volumes of the working population migrate from other parts of the country. Those who came to work in Ciudad Juárez were mostly poorly educated and did not have the traditional family support networks that they left in their places of origin. They lacked various social services (e.g., institutional support for care activities) which encouraged their children to grow up without adequate guardianship and with emotional deficiencies that made them more vulnerable to being co-opted by criminal gangs.

This situation coincided with several demographic characteristics of Ciudad Juárez, in relative terms, such as high frequency of working women, single mothers, absent fathers, both parents working, smaller households, and more homes made up of people without family ties. The growth model followed in Ciudad Juárez focused on the conditioning of the town for the proper functioning of businesses without considering the many aspects involved in a more humane model of development. Hence, there was not an effective policy for promoting healthy family relationships nor proper institutional and community development programs that could counterbalance the tendencies towards the deterioration of the social fabric.[49]

Perhaps because of the deeply socio-relational origin of the problem, police actions to address it were ineffective. It was not until the authorities adopted an holistic approach to rebuild the social fabric through community work and social participation that the crisis was contained and reversed.[50] This approach focused on education, culture and sports, the construction or rehabilitation of public spaces, physical and mental health services, social security protection, support for local businesses, and attention to addictions among many other initiatives.

All this highlights the importance of maintaining healthy social ties so that economic and community progress can be sustained in the long run. Likewise, it shows there are links between family, community, public safety and economic spheres, that leaders must be very attentive to. In turn, this implies that statistical monitoring should not be limited to material progress. It should also monitor indicators of subjective wellbeing, with a focus on the quality of family and social ties. Specifically, it is a reminder to national statistical offices around the world of the importance of systematically tracking these types of variables and communicating them in a timely and relevant way to citizens and policy makers.

Endnotes

1 Aknin et al. (2015); Rojas (2024).

2 Bengtson (2001); Botha and Booysen (2014); Millán and Esteinou (2021); Wall and Gouveia (2014).

3 An alternative explanation of reverse causality - happier people forming larger households - is explored; retrospective life satisfaction information from Mexico is used. The analyses suggest that the main causal directionality is from household size to happiness, rather than from happiness to household size.

4 Beytía (2016): Russo et al. (2023); Velásquez (2016).

5 McHale et al. (2003); Newman et al. (2016).

6 Bartholomew and Horowitz (1991).

7 Lomas et al. (2021).

8 Beavin Bavelas (2021); Watzlawick et al. (1967).

9 Referring to caring communities, Chapter 2 addresses a similar idea to this one.

10 Oliva and Villa (2014); Rojas (2006).

11 Rojas et al. (2023).

12 Brown and Brown (2014); Thomas et al. (2017).

13 Global studies on family configurations consider the household as the unit of enumeration for data collection. The United Nations defines a household as "a small group of persons who share the same living accommodation, who pool some or all of their income and wealth, and who consume certain types of goods and services collectively, mainly housing and food" (United Nations, 1993).

14 Leyva (2014).

15 Velázquez and Ramírez (2012).

16 Montoya and Landero (2008).

17 Millán and Esteinou (2021); Rojas (2024).

18 Beytía (2017); Diener and Diener (2008); Florenzano and Dussaillant (2011); Moyano and Ramos (2007); Velázquez and Ramírez (2012); Wadsworth (2016).

19 Nelson et al. (2014); Nomaguchi and Milkie (2020); Rojas (2018); Twenge et al. (2003).

20 You et al. (2018).

21 Barrera (1986); Barrera et al. (1981); Inglehart (2010, 2018); Inkeles and Smith (1976).

22 Rojas (2024).

23 Beytía (2017); Millán and Esteinou (2021); Rojas (2018, 2024).

24 Esteve et al. (2023).

25 Botha and Booysen (2014); Thomas et al. (2017).

26 These trends emerge out of a variety of social, economic, and cultural factors, including, among others: declining fertility rates, longer life spans, later marriage and nonmarriage, higher divorce rates, delayed childbearing and childlessness, widowhood, and in general, a growing desire for autonomy and independence.

27 Esteve et al. (2024).

28 Fertility rates are projected to further decline worldwide, reaching a total fertility rate of 1.83 in 2100 (Bhattacharjee et al, 2024).

29 Cohen (2021); Esteve et al. (2024); Ortiz-Ospina (2024).

30 According to the United Nations (2021), since the mid-1990s, single-mother parenting has increased from 5% to 8% in Northern Africa and Western Asia; from 7% to 10% in sub-Saharan Africa, and from 8% to 10% in Latin America and the Caribbean, while the proportion of single-father households has remained between 1% to 2%. In selected OECD countries, 5% to 10% of households have children living with a single-mother and 1% to 3% of households have children living with a single-father.

31 According to an analysis by Pew Research Centre (Cohn et al., 2022), the share of the US population in multi-generational households has increased from 7% in 1971 to 18% in 2021 (US Census Bureau, 2022). Data gathered by the United Nations (2019) indicates that multi-generation households are extremely common in many countries or areas in Asia (more than half in India and Pakistan), as well as several in sub-Saharan Africa. This type of household accounts for at least one in four households in all countries of Latin America and the Caribbean. In Europe and Northern America, multi-generation households represent more than 30% of all households in countries like Greece, Poland, Spain and Ukraine, but are rare in Canada and the UK.

32 OECD (2016).

33 The ENBIARE 2021 survey, generated by the national statistics office of Mexico (INEGI), asks the following question: *"Could you tell me, on a scale of 0 to 10, how satisfied you are currently with your life? 0 means "totally dissatisfied" and 10 "totally satisfied"*. Round 10 of the European Social Survey, applied in 2020 in 25 countries, asks the following question regarding life satisfaction: *"All things considered, how satisfied are you with your life as a whole nowadays? 0 means "extremely dissatisfied" and 10 means "extremely satisfied"*. The countries included in the ESS 2020 are: Belgium, Bulgaria, Switzerland, Cyprus, Czechia, Denmark, Estonia, Finland, France, United Kingdom, Greece, Croatia, Hungary, Ireland, Iceland, Italy, Lithuania, Latvia, Montenegro, North Macedonia, Netherlands, Norway, Portugal, Slovenia, and Slovakia. Both questions and the response scales are quite similar. In any case, the empirical exercise does not compare across regions but explores the variability within regions.

34 In a study in the Caldas department of Colombia, Velázquez and Ramírez (2012) arrived at similar results. The authors found higher levels of happiness among household heads and partners who live in households of four members. In this group of households, 69.5% of those surveyed rated their happiness as high or very high, which exceeds the levels reached in households with two, three, five and more members (between 60.0% and 65.8%) and particularly higher than single-person households (51.5%).

35 OLS regressions are run. Life satisfaction is treated as cardinal. An alternative ordered-probit model was tested. The main conclusions of the research sustain when life satisfaction is treated as ordinal rather than cardinal.

36 All the econometric models are run based on raw (unweighted) sample data. For the case of Mexico, the models were also run based on a weighted (expanded) version of the data. Both versions show very similar results.

37 The control variables are age, age squared, gender, education level, and a constructed assets variable, which refers to the existence of the following goods and services in the household: refrigerator, washing machine, automobile, television, computer, video game, access to the internet, and access to any music and/or video streaming service.

38 In this case, household income information is available for a group of European countries. Therefore, two models are run, one that includes income but relies on a smaller number of observations, and another that does not include income but uses a greater number of observations.

39 Recommendations about the domains of life and their demarcation can be found in the OECD Guidelines on Measuring Subjective Well-being (OECD, 2013).

40 Regression analyses similar to the one presented in Figure 4.5 (Table 4.1 of the online appendix) are followed. Satisfaction in the different life domains is the dependent variable. A quadratic specification for the number of household members is included and control variables in the regression are: age, age squared, gender, education level, and assets.

41 Rojas (2024).

42 Millán and Esteinou (2021).

43 The ENBIARE survey defines a household as one in which one or more people live and share the costs and preparation of their meals. There may be cases where people physically share the same household, but do not participate in either spending on or preparing food (or do not do so all the time). There may also be households in which several families live, each sharing the cost and preparation of their food independently, but not necessarily together (or not all the time). Given the number of cases in these situations, the percentages are considered to be negligible for the purposes of the analysis conducted here.

44 It is important to note that the same econometric exercise was applied to information that comes from two other reliable sources: the quarterly subjective wellbeing survey applied by the national statistics office of Mexico (INEGI) since 2014, known as BIARE Básico or Basic BIARE, and the quality-of-life survey applied by the national statistics office of Colombia (DANE) in 2023. In both cases, the results obtained are similar to those presented in Figure 4.10.

45 The proportion of single-person households in Europe (24%) may be one of the factors behind the so-called loneliness epidemic in some European countries (Baarck et al., 2022). Despite its importance, to explore this possibility lies outside the scope of our analysis here.

46 These analyses rely on information from the Understanding Happiness in Latin America survey, which involves three Latin American countries: Colombia, Costa Rica, and Mexico.

47 These analyses are based on information from ENBIARE 2021 and consider the following variables: 1) considering that one can always count on the help of family members, 2) caring for or attending to family members who could not take care of themselves, 3) helping household members with their schoolwork or taking them to school, 4) frequency with which one usually has meetings with family members and 5) frequency with which one usually has meetings with friends of one's partner.

48 El Economista (2009).

49 Barraza et al. (2009).

50 In early 2010, after the killing of 14 children and young adults (and 15 wounded) in Ciudad Juárez, the authorities issued, and started the implementation of a new strategy to tackle the violence in Juárez, with an approach that went beyond policing and military action and included many social measures. The strategy was called *Todos somos Juárez* or 'We are all Juárez'. See: Gobierno Federal et al. (2010).

References

Aknin, L. B., Broesch, T., Hamlin, J. K., & Van de Vondervoort, J. W. (2015). Prosocial behavior leads to happiness in a small-scale rural society. *Journal of Experimental Psychology: General*, *144*(4), 788–795. https://doi.org/10.1037/xge0000082

Baarck, J., d'Hombres, B., & Tintori, G. (2022). Loneliness in Europe before and during the COVID-19 pandemic. *Health Policy*, *126*(11), 1124–1129. https://doi.org/10.1016/j.healthpol.2022.09.002

Barraza, L., et. al. (2009). *Diagnóstico sobre la realidad social, económica y cultural de los entornos locales para el diseño de intervenciones en materia de prevención y erradicación de la violencia en la región norte: el caso de Ciudad Juárez, Chihuahua.* Comisión Nacional para Prevenir y Erradicar la Violencia contra las Mujeres, Secretaría de Gobernación, Gobierno de México. http://www.conavim.gob.mx/work/models/CONAVIM/Resource/pdf/JUAREZ.pdf

Barrera, M. (1986). Distinctions between social support concepts, measures, and models. *American Journal of Community Psychology*, *14*, 413-445.

Barrera, M. Jr., Sandler, I. N., & Ramsey, T. B. (1981). Preliminary development of a scale of social support: Studies on college students. *American Journal of Community Psychology*, *9*, 435-447.

Bartholomew, K., & Horowitz, L.M. (1991) Attachments Styles Among Young Adults: A Test of a Four-Category Model. *Journal of Personality and Social Psychology*, *61*(2), 226-244. https://doi.org/10.1037/0022-3514.61.2.226

Beavin Bavelas, (2021). Pragmatic of Human Communication 50 years later. *Journal of Systemic Therapies*, *40*(2), 3–25.

Bengtson, V.L. (2001). Beyond the Nuclear Family: The Increasing Importance of Multigenerational Bonds. *Journal of Marriage and Family*, *63*, 1-16. https://doi.org/10.1111/j.1741-3737.2001.00001.x

Beytía, P. (2016) The singularity of Latin American patterns of happiness. In: M. Rojas (ed) *Handbook of Happiness Research in Latin America*. (17-29) Springer.

Beytía, P. (2017) Vínculos Familiares: Una Clave Explicativa De La Felicidad. In: Reyes, C. & Muñoz, M., *La familia en tiempos de cambio*. Santiago: Ediciones UC. https://ssrn.com/abstract=3089321

Botha, F., & Booysen, F. (2014). Family functioning and life satisfaction and happiness in South African households. *Social Indicators Research*. *119*(1), 163-182. https://doi.org/10.1007/s11205-013-0485-6

Brown, Roy I., & Brown, I. (2014) Family Quality of Life. In A. Michalos (ed) *Encyclopedia of Quality of Life and Wellbeing Research*. (2194-2201). Springer Reference.

Bhattacharjee, N. V., Schumacher, A. E., Aali, A., Abate, Y. H., Abbasgholizadeh, R., Abbasian, M., Abbasi-Kangevari, M., Abbastabar, H., Abd ElHafeez, S., Abd-Elsalam, S., Abdollahi, M., Abdollahifar, M.-A., Abdoun, M., Abdullahi, A., Abebe, M., Abebe, S. S., Abiodun, O., Abolhassani, H., Abolmaali, M., … Vollset, S. E. (2024). Global fertility in 204 countries and territories, 1950–2021, with forecasts to 2100: A comprehensive demographic analysis for the Global Burden of Disease Study 2021. *The Lancet*, *403*(10440), 2057–2099. https://doi.org/10.1016/S0140-6736(24)00550-6

Cohen, P. N. (2021). The Rise of One-Person Households. Socius, 7. https://doi.org/10.1177/23780231211062315

Cohn, D., Horowitz, J., & Arditi, T. (2022) *Financial Issues Top the List of Reasons U.S. Adults Live in Multigenerational Homes*. https://www.pewresearch.org/wp-content/uploads/sites/20/2022/03/PSDT_03.24.22_multigenerationalhouseholds.report.pdf (Retrieved: August 2024).

Diener M. L., & Diener, M. B. (2008) What makes people happy? A Developmental Approach to the Literature on Family Relationships and Wellbeing. In M. Eid & R. J. Larsen (eds) *The Sciences of Subjective Wellbeing*. (347-375). The Guilford Press. NY.

El Economista (2009) Juárez, la ciudad más violenta del mundo. August, 26. https://www.eleconomista.com.mx/politica/Juarez-la-ciudad-mas-violenta-del-mundo-20090826-0102.html

Esteve, A., Pohl, M., Becca, F., Fang, H., Galeano, J., García-Román, J., Reher, D., Trias-Prats, R., & Turu, A. (2024) A global perspective on household size and composition, 1970–2020. *Genus* 80, 2. https://doi.org/10.1186/s41118-024-00211-6

Esteve, A., Galeano, J., Turu, A., García-Román, J., Becca, F., Fang, H., Pohl, M. L. C., & Trias Prat, R. (2023). The CORESIDENCE Database: National and Subnational Data on Household and Living Arrangements Around the World, 1964-2021 [Data set]. Zenodo. https://doi.org/10.5281/zenodo.8142652

European Social Survey (2020) *European Social Survey European Research Infrastructure Consortium*. Round 10. https://www.europeansocialsurvey.org/

Florenzano, R., & Dussaillant, F. (2011) Felicidad, salud mental y vida familiar. In M. Rojas (coord) *La medición del progreso y del bienestar. Propuestas desde América Latina* (247-257). Foro Consultivo Científico y Tecnológico.

Gobierno Federal, Gobierno de Chihuahua, & Gobierno Municipal de Juárez (2010) *Estrategia Todos somos Juárez. Reconstruyamos la Ciudad*. https://www.sep.gob.mx/work/models/sep1/Resource/889/2/images/todossomos-juarezb%281%29.pdf

INEGI (2021) *Encuesta Nacional de Bienestar Autorreportado (ENBIARE)*. México. https://www.inegi.org.mx/programas/enbiare/2021/

Inkeles, A., & Smith, D. H. (1976). *Becoming modern: Individual change in six developing countries* (2. print). Harvard Univ. Press.

Inglehart, R. (2010). Faith and freedom: Traditional and modern ways to happiness. In E. Diener, J.F. Helliwell, & D. Kahneman (Eds.), *International Differences in Well-Being*. (pp. 351-397) Oxford University Press.

Inglehart, R. (2018). *Cultural Evolution*, Cambridge University Press.

Leyva, G. (2014) Los hogares pequeños viven mejor… ¿o no? *Coyuntura Demográfica*, Num. 6, July, 21-27.

Lomas, T., Waters, L., Williams, P., Oades, L. G., & Kern, M. L. (2021). Third wave positive psychology: Broadening towards complexity. *The Journal of Positive Psychology*, *16*(5), 660-674. https://doi.org/10.1080/17439760.2020.1805501

McHale, S., Crouter, A., & Whiteman, S. (2003). The family contexts of gender development in childhood and adolescence. *Social Development*, *12*(1), 125-148. https://doi.org/10.1111/1467-9507.00225

Millán, R, & Esteinou, R. (2021). Satisfacción familiar en América Latina: ¿importan las relaciones? *Perfiles Latinoamericanos*, 29(58), 1-12. DOI: dx.doi.org/10.18504/pl2958-012-2021

Montoya, B. I., & Landero, R. (2008) Satisfacción con la vida y autoestima en jóvenes de familias monoparentales y biparentales. *Psicología y Salud*, *18*(1), 117-122.

Moyano, E. & Ramos, N. (2007) Bienestar subjetivo: midiendo satisfacción vital, felicidad y salud en población chilena de la región del Maule. *Universum. Revista de Humanidades y Ciencias Sociales*, 2(22), 184-200, Universidad de Talca, Chile, http://www.redalyc.org/articulo.oa?id=65027764012.

Nelson, S. K., Kushlev, K., & Lyubomirsky, S. (2014). The pains and pleasures of parenting: When, why, and how is parenthood associated with more or less well-being? *Psychological Bulletin*, *140*(3), 846-895. doi: 10.1037/a0035444.

Newman, L., Arthur, L., Staples, K., & Woodrow, C. (2016). Recognition of Family Engagement in Young Children's Literacy Learning. *Australasian Journal of Early Childhood*, *41*(1), 73-81. https://doi.org/10.1177/183693911604100110

Nomaguchi, K., & Milkie, M. A. (2020). Parenthood and well-being: A decade in review. *Journal of Marriage and Family*, *82*(1), 198-223.

OECD (2013), *OECD Guidelines on Measuring Subjective Wellbeing*, OECD Publishing, Paris, https://doi.org/10.1787/9789264191655-en.

OECD (2016) *OECD Family Database*. Social Policy Division, Directorate of Employment, Labour and Social Affairs. Available at: www.oecd.org/els/family/database.htm

Oliva, E. & Villa, V.J. (2014) Hacia un concepto interdisciplinario de la familia en la globalización. *Justicia Juris. 10*(1), 11-20. http://www.scielo.org.co/scielo.php?script=sci_arttext&pid=S1692-85712014000100002&lng=en&tlng=es.

Ortiz-Ospina, E. (2024). Loneliness and Social Connections. *One World in Data*. https://ourworldindata.org/social-connections-and-loneliness#loneliness-solitude-and-social-isolation (retrieved: August 3rd, 2024).

Rojas, M. (2006) Communitarian versus individualistic arrangements in the family: What and whose income matters for happiness? In Estes., R. (ed.) *Advancing Quality of Life in a Turbulent World*, Springer, 153-167.

Rojas, M. (2018). Happiness in Latin America has social foundations. In J. Helliwell, R. Layard, & J. Sachs (eds.) *World Happiness Report 2018* (pp. 114–145). London: Springer.

Rojas, M. (2024) The joint enjoyment of life. explaining high happiness in Latin America. *Journal of Happiness Studies 25*, 100. https://doi.org/10.1007/s10902-024-00817-9

Rojas, M., Méndez, A., & Watkins, F. (2023) The Hierarchy of Needs. Empirical Examination of Maslow's Theory and Lessons for Development. *World Development*, 165, https://doi.org/10.1016/j.worlddev.2023.106185

Russo, M.T., Argandoña, A., & Peatfield, R. (eds) (2023) *Happiness and domestic life: the influence of the home on subjective and social well-being*. Routledge Advances in Sociology.

Thomas, P., H. Liu & Umberson, D. (2017) Family relationships and wellbeing. *Innovation in Aging. 1*(3), 1–11. doi:10.1093/geroni/igx025

Twenge, J. M., Campbell, W. K., & Foster, C. A. (2003). Parenthood and marital satisfaction: a meta-analytic review. *Journal of Marriage and Family*, *65*(3), 574-583.

United Nations (1993) Household. *Glossary of the 1993 Systems of National Accounts, Definition of terms*. http://data.un.org/Glossary.aspx?q=household

United Nations (2019) *Patterns and trends in household size and composition: Evidence from a United Nations dataset*. Department of Economic and Social Affairs Population Division. (ST/ESA/SER.A/433).

United Nations (2021) *World's Women 2020. Population and families*. Statistics. https://worlds-women-2020-data-undesa.hub.arcgis.com/apps/d5b5980632a8472a91211a7e94abcfb8/explore

US Census Bureau (2022) *Census Bureau Releases New Estimates on America's Families and Living Arrangements*. Press Release Number CB22-TPS.99. https://www.census.gov/newsroom/press-releases/2022/americas-families-and-living-arrangements.html

Velázquez, L. & Ramírez, M. (2012). La familia en Caldas: Características e importancia para el bienestar subjetivo. *RegionEs*, *7*(1), 7-42.

Velásquez, L. (2016). The Importance of Relational Goods for Happiness: Evidence from Manizales, Colombia. In M. Rojas (ed.) *Handbook of Happiness Research in Latin America* (91–112). Springer. DOI: 10.1007/978-94-017-7203-7_6

Wadsworth, T. (2016). Marriage and subjective wellbeing: How and why context matters. *Social Indicators Research*, *126*(3), 1025-1048.

Wall, K., & Gouveia, R. (2014). Changing meanings of family in personal relationships. *Current Sociology*, *62*(3), 352-373. https://doi.org/10.1177/0011392113518779

Watzlawick, P., Beavin Bavelas, J., & Jackson, D. D. (1967) *Pragmatics of human com-munication: A study of interactional patterns, pathologies, and paradoxes*. Norton.

You, W., Rühli, F. J., Henneberg, R. J., & Henneberg, M. (2018). Greater family size is associated with less cancer risk: An ecological analysis of 178 countries. *BMC Cancer*, *18*(1), 924. https://doi.org/10.1186/s12885-018-4837-0

Chapter 5

Connecting with others

How social connections improve the happiness of young adults

Rui Pei*
Department of Psychology, Stanford University

Jamil Zaki*
Department of Psychology, Stanford University

* Authors contributed equally

The authors would like to thank (1) the editors and other authors of the WHR, including Lara Aknin, John Helliwell, Sara Konrath, Richard Layard, and Jan-Emmanuel De Neve, for helpful comments, discussions, and feedback; (2) The Global Flourishing Study and Gallup World Poll for providing data used in Figures 5.1 to 5.7; and (3) the WHR production team, including Barry Grimes, Jonathan Whitney, and Leoni Boyle.

doi.org/10.18724/whr-zc5n-4f96

This version was last updated on 13 March 2025. Please see worldhappiness.report for the latest data.

The evidence we have reviewed so far provides strong support for the old proverb: “shared joy is a double joy; shared sorrow is half sorrow.”

5

Key Insights

Social connection is vital for the wellbeing of young adults: Social connection buffers people from the toxic effects of stress and significantly enhances subjective wellbeing during young adulthood.

Social disconnection is prevalent and increasing among young adults: In 2023, 19% of young adults across the world reported having no one that they could count on for social support, representing a 39% increase compared to 2006.

Early social ties during young adulthood have long-lasting effects: For university students, forming friendships in the first few weeks of college can increase the likelihood of flourishing and reduce the likelihood of developing depressive symptoms over the subsequent years.

Many young adults underestimate their peers' empathy, leading them to avoid connecting with others and miss out on opportunities for meaningful relationships.

Interventions can bridge this 'empathy perception gap': Field interventions that teach young adults about the empathy and care of their community can promote social connection. Undergraduate students exposed to these interventions see others as more empathic and are more likely to make new connections and build larger social networks.

Introduction

Young adults across the globe face increasing mental health challenges. Once considered one of the happiest phases of life, young adulthood has taken a troubling turn.[1] Young people in North America and Western Europe now report the lowest wellbeing among all age groups. In fact, *World Happiness Report 2024* found that the fall in the United States' happiness ranking was largely due to a precipitous decline in wellbeing among Americans under 30.[2]

This chapter centres on a key idea that illuminates the problem of low wellbeing among young adults and potential ways to reverse it: happiness is fundamentally social. Across cultures and generations, supportive relationships buoy mental health and happiness.[3] Social ties also buffer people from the toxic effects of stress,[4] reducing the risk that subclinical difficulties will escalate into mood disorders.[5]

But during the same period in which young adult wellbeing has declined, loneliness among this population has risen. A comprehensive analysis including 437 independent samples of young adults found that loneliness in this population has steadily increased over the past four decades.[6] This trend was exacerbated by the COVID-19 pandemic, with young adults reporting greater feelings of loneliness compared to other age groups during that time. Even after the pandemic, contrary to expectations, young adult loneliness did not return to pre-pandemic levels. The US National College Health Association's 2023 annual survey found that *half* of college undergraduates reported significant loneliness, representing a 4.7% increase compared to 2019.[7] Amidst busy campuses and despite a world saturated with instant communication, young people today report feeling increasingly distressed while lacking the connections that can help their psychological wellbeing.

In this chapter, we begin by presenting recent global patterns in social connection and wellbeing among young adults. Next, we review both classic and current research on community wellbeing, with a particular focus on young adults across the globe. We then zoom in on a large-scale, longitudinal project we have led, which explores social connection and wellbeing among multiple cohorts of one undergraduate student community across their four years at university. Data from this work advance the basic science of community wellbeing and provide avenues to improve it. We conclude by discussing open questions and how these research findings can inform policy to support the wellbeing of young adults worldwide.

Recent trends in wellbeing and social connection among young adults

In this chapter, we define young adults as individuals in the age range of 18 to 29, a period that marks the transition from late adolescence to adulthood. This life period is often accompanied by significant environmental changes as well as psychosocial developments.[8] During this time, many young adults leave home for education, work, romantic relationships, or personal growth. On the other hand, many young adults — especially in parts of Eastern Europe and East Asia[9] — continue living with their parents. This pattern has become increasingly common in other countries such as the United States, reflecting increasing economic challenges for the young generation.[10] Contemporary cohorts of young adults have also grown up alongside significant societal developments which have changed the nature of human relationships, such as changes in communication due to social media and, more recently, large language models such as ChatGPT.[11] These experiences may add to young adults' vulnerability to both loneliness and mental health difficulties.

As young adults strive for independence and transition to become less reliant on their family, they place greater emphasis on acquiring new friendships and expanding their social circles.

In addition to changes in their environment, young adulthood is accompanied by important developmental milestones, including the establishment of new personal and professional relationships.[12] As young adults strive for independence and transition to become less reliant on their family,[13] they place greater emphasis on acquiring new friendships and expanding their social circles.[14] Historically, young adulthood has been one of the most social periods of life, as young adults tend to form more friendships and spend more time socialising than people in other age groups. In addition to fulfilling social needs, young adult relationships lay the foundation for psychological and social growth in later life stages, providing a network of support that can sustain wellbeing and resilience in years to come.[15] However, as we will explore, young adults have also faced a disproportionate decline in social connection in recent years, potentially impacting their wellbeing

Defining social connection

Social connection is a multifaceted construct that captures different aspects of how we relate to others. As shown in Table 5.1, it includes three dimensions: quantity, quality, and structure. Each of these dimensions plays a distinct role in shaping our wellbeing, offering unique pathways for fostering connection, belongingness, and support.

The Global Flourishing Study (GFS) and Gallup World Poll (GWP) datasets

To explore wellbeing and social connection among young adults, we draw on Wave 1 of the Global Flourishing Study (GFS),[16] collected between April 2022 and December 2023. This dataset includes responses from over 200,000 participants from 22 countries and one territory, spanning six continents and representing a wide range of cultures and geographies. The GFS covers a robust set of measures on wellbeing, health, social, economic, political, religious, spiritual, psychological, and demographic variables.

Table 5.1: Dimensions of social connection

Dimension	Description	Implications for wellbeing
Quantity	Number of existing relationships an individual has. This ranges from complete social isolation (having no relationships) to being connected to many others.	A larger number of social relationships can provide access to a larger pool of interactions, support, and resources.
Quality	Actual or perceived support (e.g., emotional, social, or financial) provided by relationships.	Quality social relationships can provide emotional satisfaction and perceived reliability of support.
Structure	Structural makeup of one's social networks. One example is density, or how interconnected members of one's social network are with each other.	Shapes the diversity, cohesion, and accessibility of social resources available to an individual.

Notably, the GFS includes questions assessing the quantity and quality of respondents' social connections. For the quantity of social connection, participants were asked whether they had at least one special person in their life that they felt very close to. This measure, framed as a yes-or-no choice, does not capture the full range in the quantity of a person's connection, but it does help to identify people experiencing deep social isolation. With regard to the quality of social connection, participants rated the extent to which they could rely on other people in their lives for support when they needed help, using a scale from 0 (never) to 10 (always). The structure of individuals' social networks, such as density, was not assessed in the GFS.

We also draw on the Gallup World Poll (GWP) to explore the temporal trends of social connection among young adults. The GWP dataset offers valuable insight into the quality of social connection by asking respondents whether they can count on their relatives or friends for support when they are in trouble. Importantly, the GWP has been tracking respondents on this measure for over a decade, allowing us to characterise changes in young adult social connection globally. Here, we utilise GWP data from 2006 to 2023, including over 661,000 observations from 168 countries.

Regional patterns of social connection

We begin by examining the current state of the quantity and quality of social connection among young adults across countries, and then compare the state of social connection across age groups. Data from the GFS demonstrate that while most young adults report having at least one social connection, a significant number are socially isolated. Across the 22 countries and regions, 17% of the young adult population report not having anybody (including family and friends) that they feel close to (Figure 5.1A). Japan stands out starkly, with over 30% of the young adult population reporting social isolation. In contrast, in countries such as Nigeria, Egypt, and the Philippines, less than 10% of the young adult population report having no close relationships.

Countries also varied in the quality of social connection reported by young adults. Participants rated how often they could count on people in their lives, such as relatives or friends, to provide help whenever needed. Overall, about 76% of young adults in the GFS sample reported that they can often count on people in their life for social support (indicated by a rating of 5 or higher on a 0–10 scale). Israel ranks the highest in the quality of social connections, followed closely by Mexico and Argentina (Figure 5.1B), indicating that young adults in these countries generally feel confident about the availability of help. By contrast, young adults in Japan and Türkiye report the lowest levels of social support.

In Figure 5.2, we compare the social connection of young adults with other age groups. Consistent

Figure 5.1: The quantity and quality of social connection among young adults
Global Flourishing Study (2022–2023)

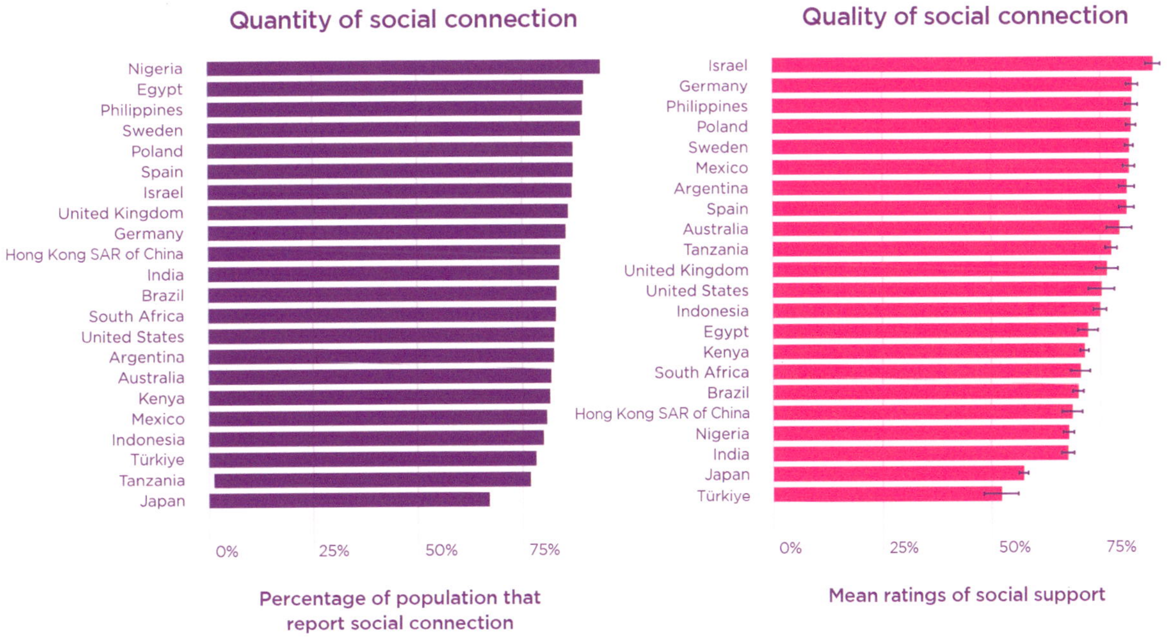

Note: Quantity of social connection refers to the percentage of the population that knows at least one person that they feel close to (includes both friends and families). Quality of social connection refers to answers to this question: "If you were in trouble, how often could you count on people in your life, like relatives or friends, to help you whenever you need them?", rated on a scale from 0 (never) to 10 (always).

Figure 5.2: Age differences in the quantity and quality of social connection
Global Flourishing Study (2022–2023)

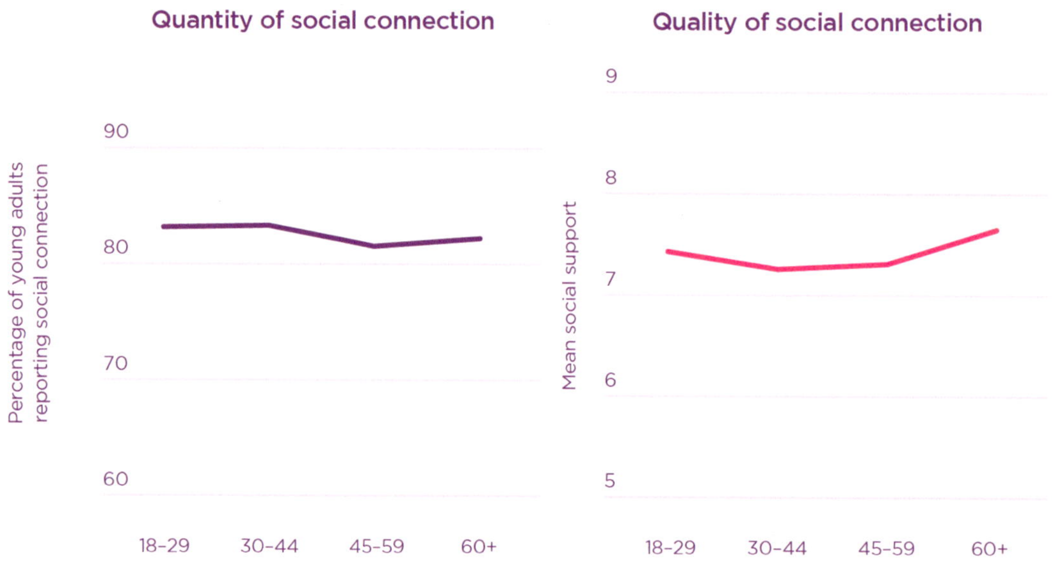

with previous observations,[17] adults older than 45 years report lower quantity of social connection compared to younger adults, representing higher levels of social isolation. For quality of social connection, the pattern follows a U-shaped curve, with both young adults (<30 years) and older adults (>60 years) reporting higher levels of social support. These patterns likely reflect the shifting priorities that come with age. Young adults often focus on expanding their social networks, while older adults may prioritise fewer but emotionally closer relationships, optimising their connections to benefit subjective wellbeing.[18]

Despite the overall trend that young adults report higher social connection than older adults, countries vary on the age-related differences in the quantity of social connection (Figure 5.3A). For example, this pattern is flipped in the United States, Japan, and Australia, where young adults report the lowest social connection among all age groups. In the United States, 18% of young adults (aged 18–29) reported not having anyone that they feel close to, whereas 15% of adults aged 30–44 reported no social connection.

Unlike other nations, young adults in the US also report lower quality of connection than other age groups (Figure 5.3B). Mirroring these patterns, *World Happiness Report 2024* also highlighted a decline in the US happiness ranking, largely driven by a drop in wellbeing in the young adult age group.[19] Although not definitive, this provides intriguing preliminary evidence that relatively low connection among young people might factor into low wellbeing among young Americans.

Figure 5.3A: Age differences in the quantity of social connection
Global Flourishing Study (2022–2023)

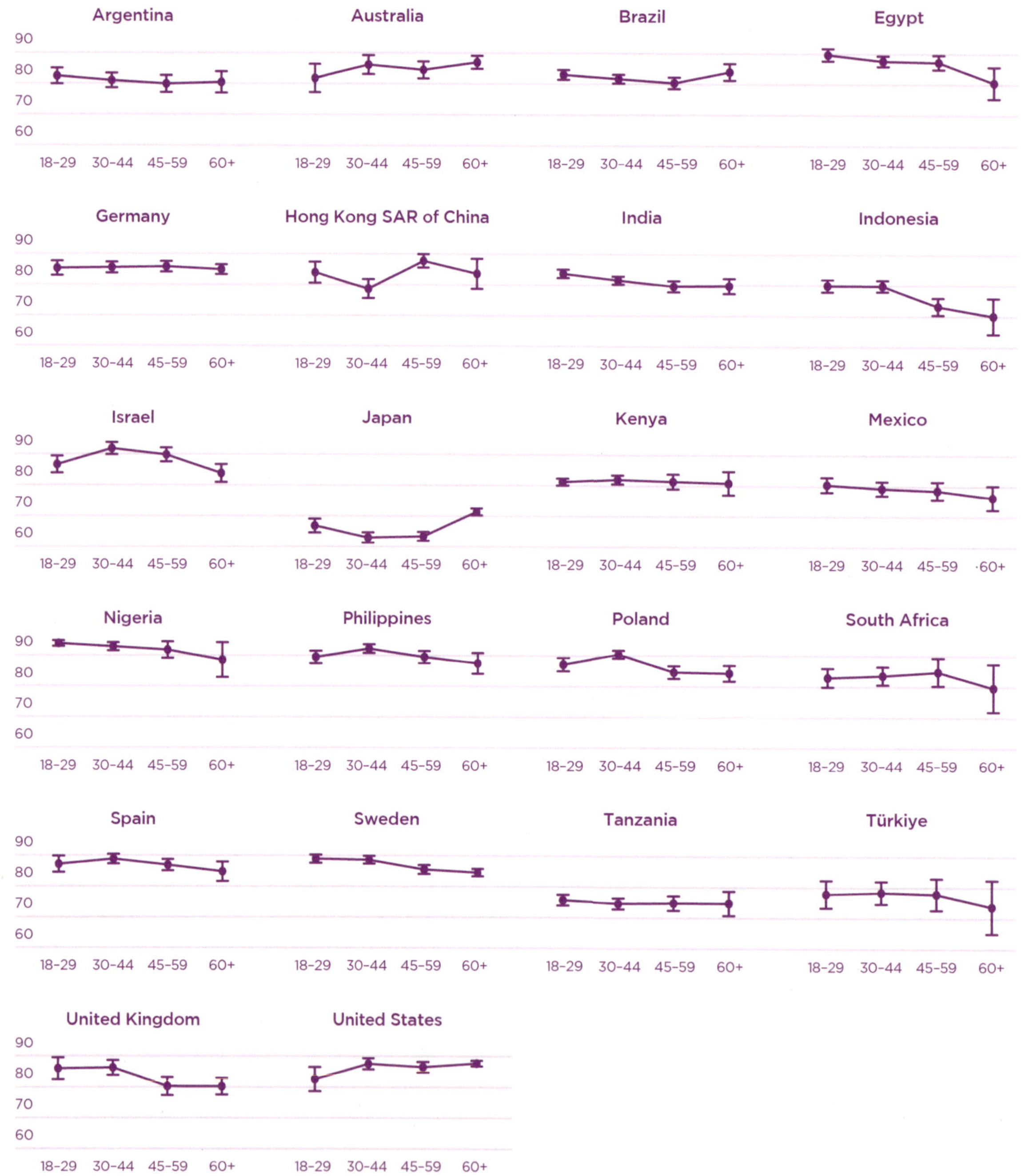

Note: The y-axis represents the percentage of each age group that reported having at least one individual that they feel close to.

Figure 5.3B: Age differences in the quality of social connection
Global Flourishing Study (2022–2023)

Note: The y-axis represents the mean rating of perceived social support on a 0–10 scale.

Temporal trends of social connection among young adults

Recent reports suggest that young adults are experiencing a decline in social connection and a rise in loneliness.[20] Yet, our analysis of the GFS dataset showed that young adults are more socially connected compared to older age groups. At first glance, these findings may seem contradictory, but examining the data over time provides helpful context. If young adults in the past were even more socially integrated than they are now, this age group could face increasing isolation while still remaining more connected than older adults. In this section, we explore this possibility using the Gallup World Poll (GWP) dataset, which includes data from young adults across 168 countries from 2006 to 2023.

First, we observed an overall decrease in the quality of social connection among young adults over time (Figure 5.4A). Each year, an additional 0.1% of young adults reported not having anyone that they could count on. This may seem negligible, but globally it represents 1.7 million more young adults reporting they have no one to count on each year.

Next, we explore these trends within the 22 countries in the GFS survey (Figure 5.4B). Some countries (especially Tanzania) demonstrated significant decreases, mirroring the global trend. Yet, three countries (Mexico, India, and Egypt) bucked this trend, showing significant increases in the quality of social connection among young adults during this period.

As described above, young adults could be losing social connection over time but still remain more connected than older adults, which would be reflected in a shrinking age gap in connectedness. Indeed, when comparing the difference in the quality of social connection between young adults (18–29) and older adults (60+), this gap has decreased over the last 17 years (Figure 5.5). In 2006, young adults were 6% more likely than older adults to report having someone to rely on. However, since 2020, the difference between the two groups has fallen to less than 1%. This indicates that the decrease in quality of social connection is specific to young adults, and not observed across age groups.

Figure 5.4: Trends in the quality of social connection among young adults
Gallup World Poll (2006–2023)

Note: The quality of social connection refers to the percentage of the population that reports having someone to count on when one is in trouble. Figure 5.4A: global trend of social connection quality (2006–2023), drawing from the Gallup World Poll dataset of 168 countries. Figure 5.4B: pink indicates significant negative trend, blue indicates significant positive trend, and purple indicates non-significant changes.

Figure 5.5: Trends in the quality of social connection by age group
Gallup World Poll (2006-2023)

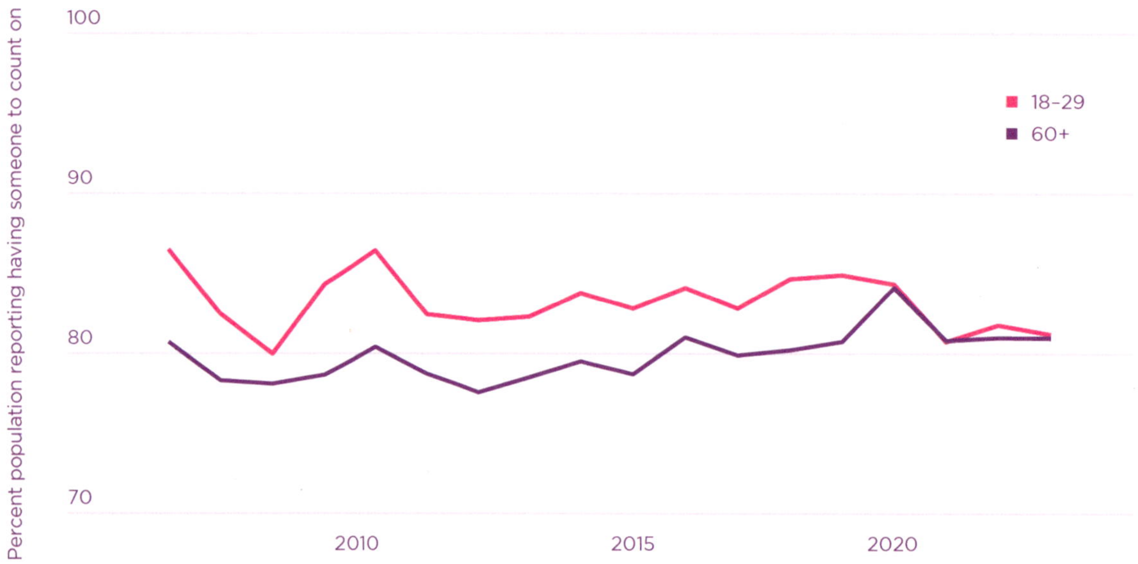

Social connection and wellbeing among young adults

So far, we explored temporal and regional patterns of social connection in young adults. Next, we examine the relationship between social connection and subjective wellbeing within the GFS dataset. In the GFS, subjective wellbeing was measured with the following life satisfaction question: "How satisfied are you with life as a whole these days?" Responses were rated on a scale from 0 (not at all satisfied) to 10 (completely satisfied). We find that countries where young adults report general higher levels of social connection and social support also report higher life satisfaction (Figure 5.6). This relationship was independently observed for both quantity and quality of social connection.

The link between social connection and wellbeing is not only observed at the national level but also for individuals. On average, young adults who report higher levels of both quantity and quality of social connection tend to feel more satisfied with their lives. Individuals who reported having at least one person they are close to are 16% more satisfied than individuals with no close contacts.[21]

In a few countries which scored highest on social connection (such as Nigeria and Egypt) we do not observe significant associations between social connection and wellbeing (Figure 5.7).

Figure 5.6: The relationship between social connection and subjective wellbeing
Global Flourishing Study (2022–2023)

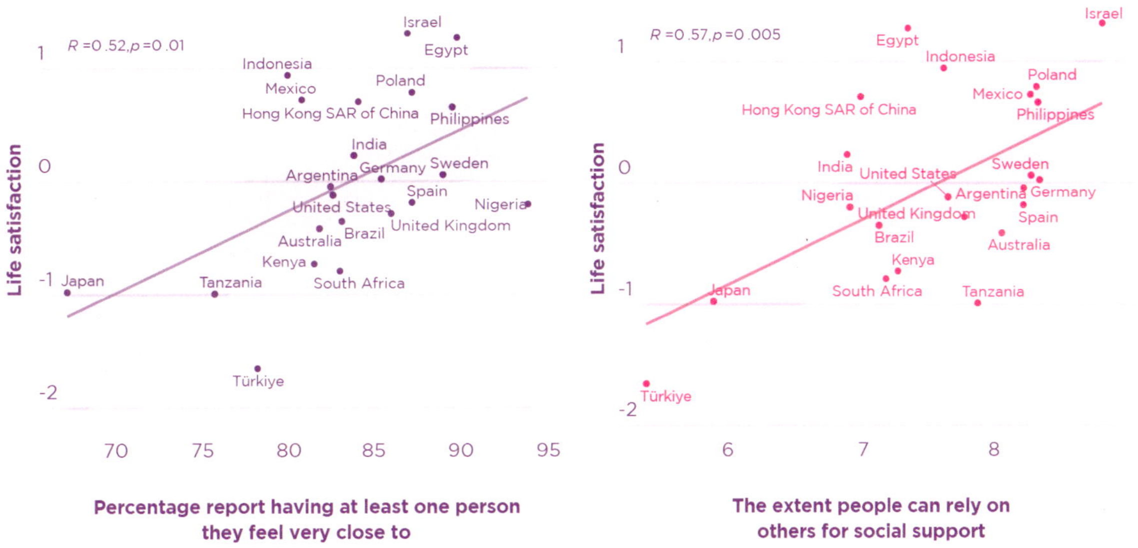

Note: Lines indicate the best-fitting regression model. Shaded areas indicate 95% confidence intervals.

For example, over 90% of young adults in Egypt reported having at least one person that they feel close to. Because only a small proportion of this group reports lacking social connection, it's difficult to relate social connection with wellbeing among this group.

Similarly, we observe a significant positive association between the quality of social connection and wellbeing. A 1-point increase in perceived social support is associated with a 0.29-point increase in life satisfaction.[22] Data from all 22 countries and regions in the GFS data showed a positive association, although the size of this positive association varied slightly across countries.

When these two factors of social connection are entered in the same model to predict life satisfaction, both the quantity and quality of social connection were significantly associated with life satisfaction, with comparable effect sizes.[23] This result indicates that the quantity and quality of social connection independently predict life satisfaction.

In the following sections, we present further evidence on the links between social connection and happiness, as well as the potential barriers preventing young people from fostering connections.

Figure 5.7: The relationship between social connection and subjective wellbeing by country
Global Flourishing Study (2022–2023)

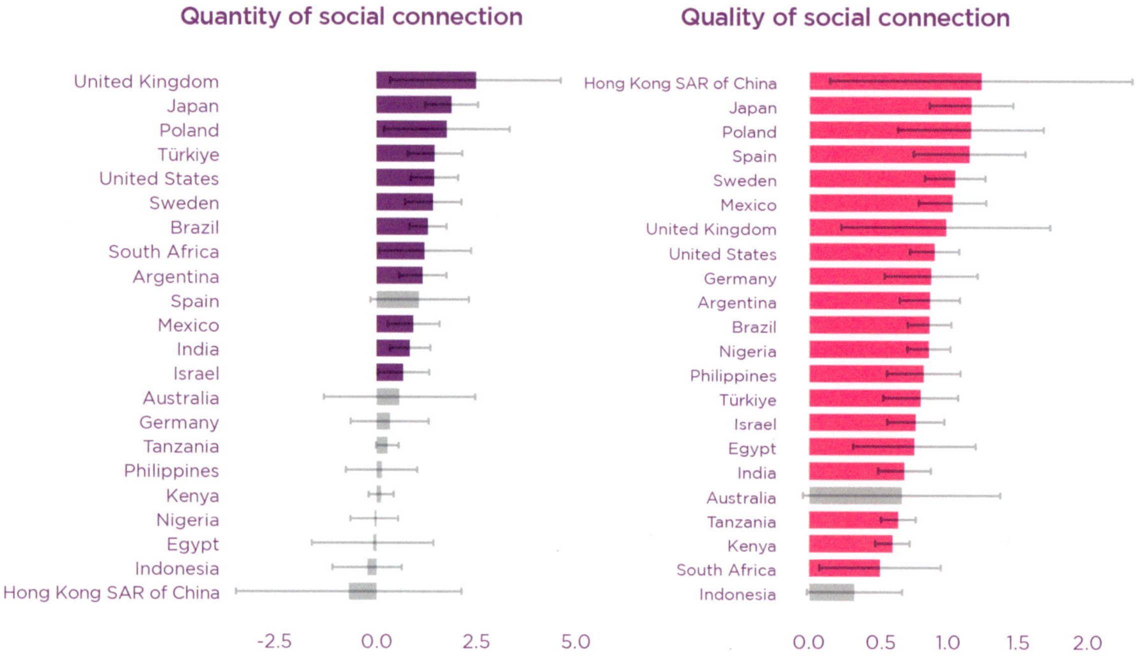

Note: Bars indicate standardised regression coefficients after controlling for age, gender, income, and marital status. Lines indicate 95% confidence intervals for the coefficients. Grey bars indicate nonsignificant model fits.

Literature review on social sources of wellbeing

Longitudinal tracking of young adult social connection

Social connection correlates with wellbeing both among individuals and across countries. These correlations raise an important question regarding causality: do healthy social relationships *lead* to greater wellbeing, or does feeling happy make people seek out social connections, or are these both true?[24] To characterise the direction of these relationships, it is important to go beyond measuring connection and wellbeing at a moment in time. For instance, researchers can track the same individuals over time to see if social connection predicts better wellbeing in the future, or if the reverse is true. Studies that take this approach indeed find that when people are socially connected, they are more likely to thrive in the future.

Consider the Harvard Adult Development Study, a long-running research project that investigates

Studies that take this approach indeed find that when people are socially connected, they are more likely to thrive in the future.

how individuals' health and wellbeing evolved as they grew up.[25] This program started in 1938 with 724 university students. Researchers continue to monitor the wellbeing of the original participants who are in their late 90s now. Over the years, the project expanded to include larger cohorts of participants. For over 80 years, researchers have tracked the participants' lives, collecting data on their health, relationships, and overall wellbeing through periodic interviews and medical checkups. One of the study's most significant findings is the importance of social relationships for long-term happiness and health. Researchers found that the people who stayed the healthiest and lived the longest tended to be those who had the strongest connections to others. For example, close relationships were found to delay mental and physical decline, and were better predictors of long and happy lives than social class, IQ, or even genetic factors.[26]

Other studies with wider population samples have also found a similar link. Social connection predicts later increases in life satisfaction and this holds after controlling for a wide range of demographic variables and stress.[27] In one study, using a large representative survey in Germany, researchers asked participants to report on their ideas for how they could improve their life satisfaction. The researchers then investigated which ideas predicted changes in life satisfaction one year later. The researchers found that those who had socially engaged goals (e.g., "I plan to spend more time with friends and family") often reported improvements in life satisfaction one year later. In contrast, those who had other goals (e.g., "I plan to find a better job") did not report increased life satisfaction.[28]

Social relationships are also significant predictors of wellbeing in longitudinal analyses of young adults specifically. In one study, researchers tracked 393 US students across their four years of university on a number of social factors as well as life satisfaction.[29] Students who were more extroverted reported higher life satisfaction four years later, in part because they formed stronger social connections in the university. In addition to social connection, several studies conducted across cultures including the US, Portugal,

You might expect that extroverts would derive greater joy from social interactions, but evidence suggests that both extroverts and introverts derive happiness from social interactions, but for different reasons.

Germany, Russia, and China found that students who report receiving more social support also reported higher wellbeing later in college.[30]

In addition to tracking individuals over long periods of time, recent studies have captured data over shorter but more intensive periods, such as days or weeks. This is done by briefly surveying participants multiple times per day (an approach called 'experience sampling') or passively collecting data from individuals (such as background sound from their mobile phone, an approach called 'passive sensing'). For example, researchers sometimes 'ping' participants throughout the day to assess their behaviour (e.g., whether the participant engaged in a social interaction) and whether they feel happy at that moment. These studies find that people generally feel happier after they engage in social interactions.

Experience sampling also allows researchers to examine *who* benefits most from social interactions. For example, you might expect that extroverts would derive greater joy from social interactions,[31] but evidence suggests that both extroverts and introverts derive happiness from social interactions, but for different reasons. Extroverts tend to experience a boost in mood after spending time with others. Introverts, on the other hand, tend to feel a stronger sense of connectedness after interactions, especially when those conversations are meaningful and deep.[32]

Experience sampling can also reveal *when* social interactions improve wellbeing. In one recent study, Krämer and colleagues tracked German-speaking participants with experience sampling and passive mobile sensing. Participants generally felt happier after social interactions, but this was

These studies provide robust evidence that people feel increased happiness after connecting and interacting with others.

only the case when the social interactions were aligned with their needs. For example, when people were socially interacting while desiring to be alone, they experienced decreases in happiness. On the other hand, when individuals deliberately engage in social interactions to seek comfort, celebrate, or commiserate, these connections tend to increase wellbeing.[33]

There is strong evidence that social connection is followed by greater happiness, but what about the reverse? Does being happy lead people to seek the company of other people? The relationship between happiness and social behaviour appears to be more nuanced. People do tend to report feeling more social when they are in a happier mood,[34] but they also tend to seek contact with others in times of distress.[35] This pattern suggests that we turn to others for different reasons depending on how we feel. When we are feeling down, we may seek out happiness-enhancing social relationships. On the other hand, when we are feeling good, we might be more willing to share this happiness with others or invest in less enjoyable social interactions – like resolving conflicts or developing new relationships – that could bring long-term benefits.[36]

Together, these studies provide robust evidence that people feel increased happiness after connecting and interacting with others. While these studies help us understand the correlation between social connection and happiness, it is important to note that they cannot establish social connection as the sole reason causing this positive boost in mood.

Causal links between social connection and wellbeing

Examining social interaction and happiness across people, or within people over time, provides intriguing links between these experiences. But to better understand if social interaction *causes* greater wellbeing, scientists need to conduct experiments that randomly assign participants to engage in social connection or not. One specific type of social behaviour, prosocial behaviour, has already been covered in Chapter 4 of *World Happiness Report 2023*.[37] But can other forms of social connection help protect against the harmful effects of stress and elevate happiness?

In difficult times, social connections act as a protective shield against the harmful effects of stress.[38] A series of experimental studies from the past two decades provide evidence that receiving social support (compared to no support) can buffer the negative impact of stressful events. These studies generally put participants under distress, such as applying a mild electric shock or giving a speech in front of others, and record participants' stress levels under different experimental circumstances.

In a series of classic experiments,[39] young adult participants were randomly assigned to deliver a speech either with or without access to social support. Participants in the social support condition exhibited lower blood pressure, a physiological indicator of stress, compared to those who did not have access to social support. This suggests that the presence of social support can mitigate physiological stress responses during stressful tasks.

More recent studies corroborate this effect with brain evidence. In one experiment, young adults received mild electric shocks while their brain activity was recorded in a fMRI scanner.[40] Each participant received these shocks either alone or when holding hands with an opposite-sex companion, such as a friend or romantic partner. Overall, when participants were holding hands, they rated the experience as less distressing and exhibited less activity in brain regions associated with the experience of threat.

Even when people cannot directly access their support systems, simply imagining the presence of a caring friend or loved one can soothe the distress of experimentally-induced pain.[41] In one experiment, participants viewed a picture of a loved one while experiencing mild pain from heat stimulation. Compared to viewing a picture of an object, seeing the photo of a loved one reduced their pain perception just as much as physically holding a partner's hand.

In addition to dampening stress, social interactions can be a powerful driver of happiness too.[42] When the adventurer Christopher McCandless faced his own eventual death after months alone in the Alaskan wilderness, one of his last reflections was "happiness is only real when shared".[43] While his experience may be extreme, people do tend to share their joyful moments with others to elevate happiness – a process called capitalisation.[44] For example, compared to those who wrote about a positive event privately, or shared this event with unresponsive peers, participants who shared this event with responsive others reported these events to be more positive and personally meaningful, demonstrating the power of social sharing in enhancing happiness and meaning.[45]

Happiness is only real when shared.

The benefits of social interaction go beyond sharing good news with friends and family. Even general interactions with strangers, though potentially nerve-wracking, can spark joy. In one study, some university students were instructed to either spend 30 minutes interacting with peers they did not know or to stop interacting whenever they wanted and to spend the remaining time sitting in solitude.[46] Students who were assigned to interact with strangers for the entire 30 minutes enjoyed the time more than participants who were allowed to spend some of the time in solitude. Thus, even brief interaction with strangers can elevate happiness.

30-minute conversations with strangers might be rare, but our day-to-day lives are filled with brief encounters: thanking the barista preparing our morning coffee, asking someone for directions, or exchanging a few words with a fellow commuter on a bus. Would something as simple as saying "hi" or "have a nice day" to these strangers contribute to happiness? Research on these "minimal social interactions" shows that even small exchanges with strangers can enhance happiness. In a study involving 265 Turkish university students and staff, the students who were asked to thank, greet, or express good wishes to their shuttle bus drivers experienced greater happiness than those who were asked to not speak with the drivers.[47] Similar effects were observed for bus and train commuters who were instructed to interact with a fellow commuter,[48] as well as customers at Starbucks who were instructed to have a brief social interaction with the barista, compared to those who were not asked to interact.[49] Together, these findings underscore that even brief interactions with strangers can buoy happiness.

Going beyond a single instance of socialising, a few studies have examined how being social for weeks can impact long-term happiness. For instance, one study investigated the benefits of acting more extroverted.[50] 114 Australian adults were randomly assigned either to act more extroverted (bold, outgoing, and talkative) or more introverted (quiet, sensitive, and calm) for a week. Those who acted extroverted experienced more positive moods compared to those who acted introverted. Even individuals who were more introverted by nature experienced more positive emotions when they acted more outgoing. Interestingly, this improvement in happiness was not due to more social interactions, but because of the way people behaved during them.

A network science approach to social connection

The evidence and data reviewed so far concern direct connections between people. Yet, our social world extends beyond our direct connections. A wealth of research reveals that the structure of a person's social network - the way that their relationships are organised - is also associated with wellbeing. These studies often employ a network science approach,[51] where individuals (or 'nodes') in the social network are connected by relationships (or 'ties') to form complex networks of social relationships. This type of diagram intuitively characterises how people connect and allows researchers to develop precise mathematical metrics that capture the structural composition of these networks. Such metrics can reveal important insights into someone's social world that would otherwise remain hidden, offering a more nuanced picture of how our social worlds may shape our capacity to thrive.

One network metric is *density*, the level of interconnectedness between nodes in a social network (Figure 5.8).[52] People who are situated in dense social networks tend to be less lonely and happier,[53] perhaps because these dense connections offer a sense of security, stability, and belonging.

In one recent study, researchers interviewed 2,485 individuals in Indiana, USA across three years during the COVID-19 pandemic.[54] In addition to wellbeing measures, they also assessed characteristics of the participants' social networks, such as size, closeness, and density. Overall, young adults were disproportionately affected by the pandemic, reporting larger drops in wellbeing compared to other age groups. Interestingly, young adults with more dense and interconnected social networks experienced smaller decreases in wellbeing compared to those with sparser networks. This buffering effect of social network density was stronger for young adults compared to other age groups.

Another aspect of social network characteristics that may contribute to mental health is *diversity*, the extent to which individuals connect with different groups of people. Diverse networks, which include a mix of close family ties and different types of peers (e.g., members of the

Interacting across group boundaries – such as differences in race or socioeconomic status – further amplifies these benefits.

Figure 5.8: Illustration of social network density

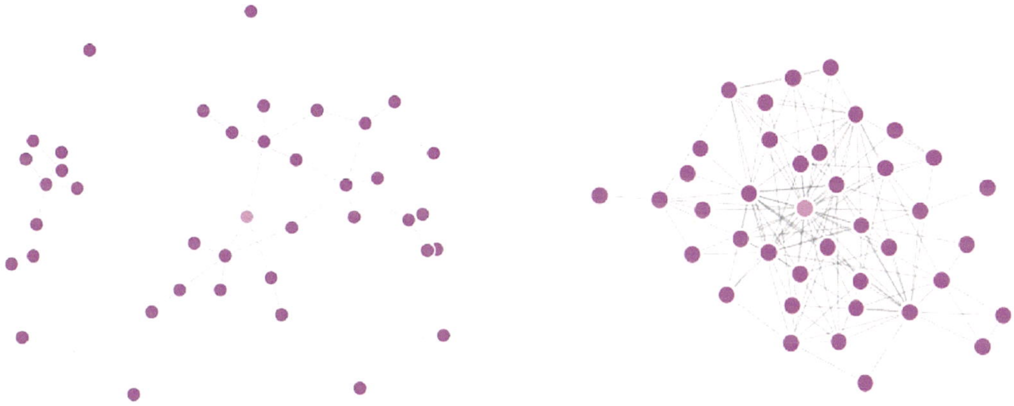

orchestra, track team, and culinary club), can protect against depression and other mental illnesses.[55] Interacting across group boundaries – such as differences in race or socioeconomic status – further amplifies these benefits. These connections foster a sense of belonging and can help reduce feelings of exclusion, especially for individuals from minority groups.[56]

A recent analysis including over 24,000 adults in England paints a nuanced relationship between social network diversity and subjective wellbeing.[57] While some level of network homophily (similarity between contacts) is linked with better wellbeing, social networks that are excessively homogeneous can undermine happiness. This finding underscores that a balance between homophily and diversity – combining in-group familiarity and out-group variety – may offer the greatest advantages and enhance subjective wellbeing.

In summary, work in this area so far demonstrates that the quantity and quality of social connections, along with the structure of social networks, shape wellbeing. Social networks that are dense and diverse can offer psychological security, belongingness, and opportunities for growth, all of which can elevate subjective wellbeing.

An in-depth study of one young adult community

The evidence we have reviewed so far provides strong support for the old proverb: "shared joy is a double joy; shared sorrow is half sorrow." Yet, recent studies and our new analysis paint a sobering picture: young adults globally are lonelier than before.[58]

While young adulthood is expected to be one of the happiest and most social life stages, young adults in the US reported the lowest happiness and social connection of all age groups. If social connection is so beneficial, why are young adults not connecting more? Answering this question will require more in-depth assessment of young adults' beliefs and attitudes towards their community. This section presents insights from our large-scale, longitudinal project, the Stanford Communities Project (SCP).

The goal of the SCP is to provide a detailed assessment of the social health of one young adult community. Since 2018, the project has assessed thousands of Stanford undergraduates, multiple times a year, to gather data on personality traits, wellbeing, social networks, and momentary assessments of social activity. The SCP provides a novel and comprehensive means to examine perceptions, social behaviour, and mental health in a young adult population. So far, the findings from this project underscore the profound impact of social connection on happiness and wellbeing, but they also highlight a critical gap: young people experience diminished connection when they perceive their peers as less empathic than their peers self-report. By examining this 'perception gap' and trying to reduce it, we can better understand how to foster meaningful connections and support the wellbeing of young adults.

Early social ties have long-lasting effects on wellbeing

As we reviewed above, scientists often track individuals over time to uncover the longitudinal link between social connection and wellbeing. The SCP did the same by following two cohorts of undergraduate students (N = 1,061) across their college years. Twice a year, we assessed changes in the students' friendships and wellbeing and found five distinct wellbeing trajectories that students followed during college. For example, some students experienced improving wellbeing ('getting better'), and others experienced worsening wellbeing ('getting worse'). Notably, 38% of students followed the 'getting worse' trajectory, where symptoms of depression intensified over the course of college.

We found that the number of social ties a student forms in their first few weeks of college predicts the long-term trajectory in the subsequent years. Every additional friendship was associated with a significant reduction in the likelihood of 'getting worse' compared to 'getting better'. This highlights the protective role of social connection during a critical time of transition.

The impact of friendships lasted well beyond those first few weeks. Across the college years, friendships change in interesting and meaningful ways, shaping the wellbeing trajectories that students may follow. Each new friendship increases the likelihood of 'getting better' by 17%. On the other hand, losing a friendship increases the chances of falling into a 'getting worse' trajectory by 19%. These findings add to existing evidence that stronger social connections are usually followed by better wellbeing down the road.

Cohesive 'social microclimates' support wellbeing

An individual's direct friendships contribute to their wellbeing. As we have seen, existing evidence suggests that tight-knit social circles can offer a sense of security and belonging, thus promoting mental health. But what happens when we zoom out to consider the larger social ecosystem that students inhabit? Each person resides in a unique 'social microclimate', characterised by the emotional traits of friends and community members, as well as the relationships among neighbours. Unlike direct friendships, a young adult's social microclimate is often beyond their control. Yet, various features of this microclimate can significantly affect their subjective wellbeing.

To test this hypothesis, our team leveraged an assignment process that many universities use for student housing.[59] At Stanford University, all first-year students were assigned to residential halls. This offers a unique opportunity to study how social microclimates shape wellbeing while

Unlike direct friendships, a young adult's social microclimate is often beyond their control. Yet, various features of this microclimate can significantly affect their subjective wellbeing.

Figure 5.9: The relationship between psychological distress and the density of social connections in residential halls
Courtney et al. (2024)

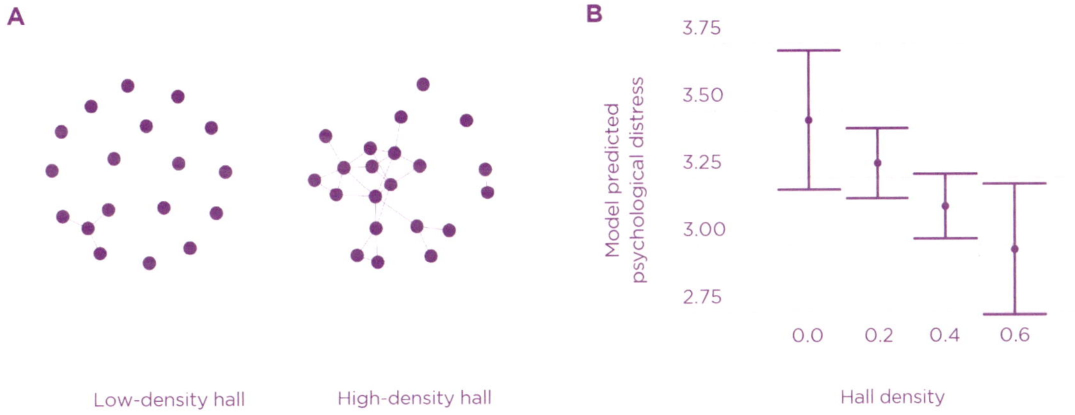

Note:
A: an illustration of social connections in a low-density hall vs. a high-density hall.
B: predicted psychological distress for a student living in halls of 0%, 20%, 40%, and 60% density.
Bars represent 95% confidence intervals. Figure adapted from Courtney et al. (2024).

controlling for the confounding factor of individuals selecting their own social groups. Since students did not choose who would live in the same hall with them, the researchers could isolate the effect of the broader social ecosystem from the effects of the direct, personal connections.

We collected data from 798 first-year students and assessed their personal traits, such as emotional stability and empathy, before the students arrived on campus. Midway through their first term, students were asked to report their subjective wellbeing and nominated their friends. This allowed us to capture two types of social factors. The first type consists of a student's direct social network: how many friends they had and how supportive and empathic those friends were. The second type concerns the social microclimate, which includes not only direct personal ties, but also the hallmates that the student is not directly friends with. For instance, we measured the overall density of social connections within each hall. A hall with high density would have many students nominating each other as friends, creating a more cohesive microclimate (Figure 5.9A).

The results showed that these social factors significantly influenced wellbeing. Consistent with the evidence reviewed above, attributes of a student's direct social network are significantly associated with subjective wellbeing. Moreover, the density of a student's social microclimate also plays a significant role in mental health. Students who lived in high-density halls, where community members are more interconnected, reported lower levels of psychological distress and higher life satisfaction (Figure 5.9B). These effects remained robust even after accounting for individual traits and characteristics of one's direct social network. Thus, the density and cohesion of a student's social microclimate may also shape subjective wellbeing, as well as their immediate social network.

Gaps in social perceptions hinder social connection and wellbeing

Close friendships and cohesive communities buoy happiness. Yet, even in supportive communities, people can still feel isolated and hesitate to reach out to others. As we described above, feelings of loneliness for young adults have increased by an average of 0.22% per year for the past four decades,[60] and the quality of social connection has decreased for young adults since 2006.

If social connection brings so many benefits, why do so many young people still feel lonely? Part of the answer may lie in their perceptions of others and their communities. Findings from new research reveal that inaccurate social perceptions can be a barrier for social connections. For instance, people tend to underestimate how fulfilled and happy they will feel after interacting with strangers,[61] having deep conversations with friends,[62] expressing gratitude,[63] giving compliments,[64] and asking others for help.[65]

People are not very accurate in forecasting how they will feel after engaging in social activities, leading them to miss out on opportunities to connect.

In short, people are not very accurate in forecasting how they will feel after engaging in social activities, leading them to miss out on opportunities to connect.

We hypothesise that there may be other factors at play in addition to inaccurate forecasting of future feelings. Perhaps people are not only misjudging their own emotional outcomes but also holding inaccurate beliefs about others and their communities. For example, students may underestimate the empathy and care in others, and this empathy perception gap might leave

Figure 5.10: The empathy perception gap

individuals socially risk averse and ultimately more isolated.

We tested this hypothesis using data from the SCP.[66] Drawing from two years of data involving over 5,000 undergraduate students, we assessed two types of data related to community empathy. First, we assessed 'empathy perception', where students estimated the empathy of their peers. We also asked about their own levels of empathy. By combining these two types of data, we can assess whether students' perceptions matched their peers' self-report.

Our results indicate a persistent empathy perception gap. Students tend to view other students as less empathic and caring than their peers see themselves. For instance, participants estimated that 87% of Stanford students would "act kindly by helping others who are feeling bad", whereas 96% of Stanford students responded positively to the same question, indicating a 9% empathy perception gap on this measure (Figure 5.10).

The consequences of this empathy perception gap were profound. Students who perceived their peers as less empathic and supportive were less likely to take social risks such as striking up conversations, sharing personal struggles, or reaching out for help; behaviours that are critical for building meaningful relationships. This social risk avoidance led to missed opportunities to connect and learn from others, perpetuating the misguided belief that those around them lack empathy and care. Over time, this empathy perception created a vicious cycle of misperception and social disconnection (Figure 5.11).

Figure 5.11: The vicious cycle of empathy misperception

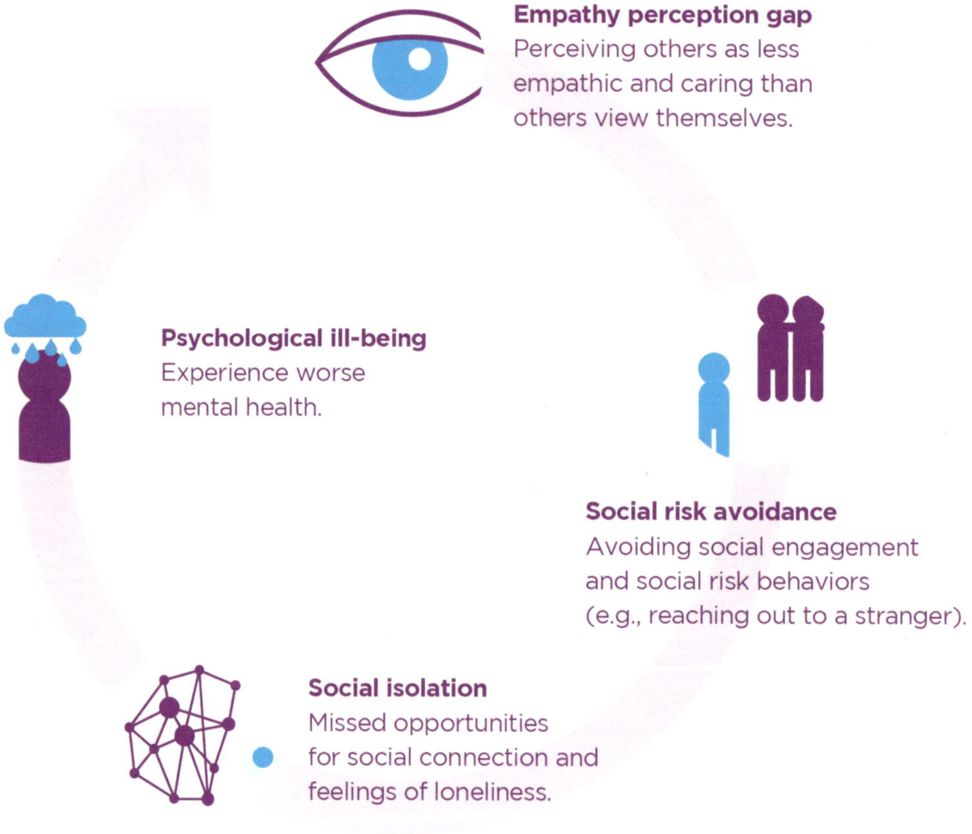

Campaigns to align perceptions can foster social connection

Empathy perception gaps help explain why young people are socially isolated and lonely. They also illuminate a potential opportunity to break the vicious cycle and promote social connection. When individuals view the people and community that surround them as supportive and caring, they are more likely to take the social risks of reaching out to strangers and seeking social support. Taking these social risks can help foster meaningful connections, expand social networks, and improve wellbeing.

Over the past two years, we have pioneered a new intervention to enhance social connection by addressing the perception gaps and providing opportunities for students to learn about the care and support in their community. We did this using two field experiments.

In the first field experiment, we presented students with data about their peers' high levels of empathy and interest in making friends. To do this, we put up posters around residential halls with statistics like "95% of Stanford students are likely to help others who are feeling down" (Figure 5.12). We paired the posters with a one-hour educational workshop designed to reinforce the message that their peers were more caring and supportive than they might think.

Students exposed to this data significantly shifted their perceptions. On average, participants in the control condition underestimated their peers' empathy by 0.4 points (on a 7-point scale). Participants in the experimental condition still underestimated their peer's empathy (by 0.1 points), but slightly less than participants in the control condition, representing a 75% reduction. Students also reported taking more social risks following our intervention – they reached out to classmates they didn't know, initiated conversations, and were more willing to share their vulnerabilities. On average, the frequency of these social risk behaviours increased by 11%.

In the second experiment, we expanded the intervention by adding behavioural nudges delivered directly to students' smartphones (Figure 5.13). These nudges encouraged students to engage in small, everyday acts of social risk-taking, such as complimenting a stranger or catching up with someone they hadn't spoken to in a while. Once again, our intervention reduced the empathy perception gap. On average, participants in the control condition underestimated their peers' empathy by 1.0 points on a 7-point scale. Those in the experimental condition also underestimated their peer's empathy, but to a lesser degree (0.9 points), representing a 10% reduction.

Similarly, our intervention also increased acts of social risk behaviour. In the week following the intervention, experimental condition participants were 89% more likely to report engaging in social risk-taking compared to those in the control group. We also found that these effects were long-lasting. Two months after the intervention, students in the experimental group were still twice as likely to sign up for a social event in

Figure 5.12: Intervention posters for shifting empathy perception

Figure 5.13: A behavioural nudge to promote social risk-taking

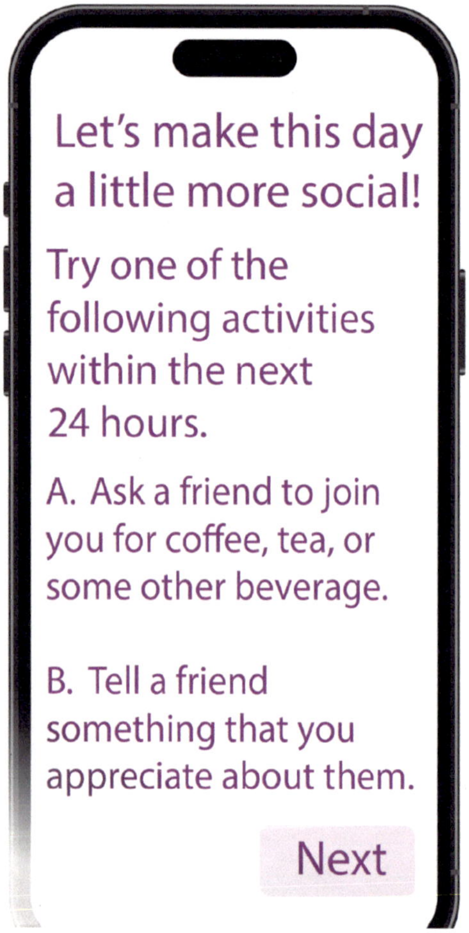

which they have an extended conversation with strangers compared to participants in the control group.

The intervention also strengthened students' social networks. We assessed students' friendship networks four months after the intervention and found that students in the experimental condition reported an average of 0.44 more close friends compared to those in the control group.

Together, results from these field experiments indicate that a community's empathy can be a powerful, yet underutilised, resource for mental health and happiness. We provide initial evidence that interventions highlighting a community's care and empathy, as well as behavioural nudges to encourage social risk-taking, can effectively shift people's perceptions and behaviours, as well as expand social networks. These findings point to the importance of creating caring social environments and helping individuals to recognise the empathy that surrounds them.

Open questions and future directions

In previous sections, we described the robust relationships between different aspects of social connection and overall happiness in young adults. Next, we discuss some open questions and opportunities for future exploration.

Towards a multifaceted measure of social connection

Social connection is a multifaceted construct that encompasses the quantity, quality, and structure of an individual's social network. Research so far has largely focused on the presence or absence of social ties, with an emphasis on social isolation and loneliness. This perspective overlooks the rich dimensions of social relationships. As the World Health Organization defines health as "a state of complete physical, mental, and social wellbeing, not merely the absence of disease or infirmity",[67] social health is more than the absence of social deficit, and thus encompasses the fulfilling, supportive, and flourishing aspects of social connection.

A first step towards a more nuanced understanding of social connection is the creation and utilisation of a common measure that (1) encompasses different aspects of social connection, and (2) can be applied to diverse cultural contexts. A standardised, multifaceted measure of social connection can serve as a powerful tool to assess the state of social health globally, and offer valuable new insights into the intricate ways different aspects of social connection shape wellbeing.

Furthermore, different factors of social connection could matter for the wellbeing of people that vary by cultural background, age group, and socio-

Social connection is a multifaceted construct that encompasses the quantity, quality, and structure of an individual's social network.

economic status. For example, existing research finds that the quantity and quality of social interactions may be valued differently depending on one's age. Individuals in their 20s may prefer quantity while those in their 30s may prefer quality.[68] Thus, a multifaceted measure of social connection assessing the global population can help researchers identify which aspects of social connection most strongly contribute to happiness across different groups.

An additional aspect that can be incorporated into existing measures of social connection is relationship diversity: the variety of relationship types within an individual's social network. Recent research has highlighted a robust correlation between happiness and having a variety of relationships, such as family, close friends, coworkers, and acquaintances.[69] For example, a study of over 50,000 people reveals that interacting with a more diverse set of relationship types predicts higher wellbeing. This effect, comparable with other established contributors of wellbeing such as marital status, held after controlling for total time spent socialising as well as the diversity of activities that people engaged in.[70]

We are now seeing efforts to establish a global indicator of social connection. In 2022, Gallup, Meta, and a group of academic advisors collaborated on the State of Social Connections study, a first-of-its-kind, in-depth look at people's social connections around the world. A second phase of the research, the State of Social Connections Gallup World Poll survey, expanded its global reach by running a select set of the State of Connections study questions on the Gallup World Poll, reaching over 140 countries and providing the ability to study overall life evaluations and the relative importance of the quantity, quality, and diversity of social connection.[71]

Prevention and intervention efforts to promote social connection

The trends of declining social connection among young adults, combined with the evidence on the associations between social connection and wellbeing, point to an urgent need to take action. Social connection generally occurs naturally among individuals and within communities. However, when it does not, intervention becomes necessary to reduce risk. We presented our own effort on intervention to enhance social connection among Stanford University students. Yet, much remains to be understood: which interventions work best, for whom, and under what circumstances? Below, we discuss a few directions for future research.

Technology-based interventions

Today's young adults are the first generation to have grown up completely immersed in technology-based communications. Also called 'digital natives', young adults today have had access to the internet and digital devices from a very young age. Researchers are starting to learn more about the role of social networking and messaging apps in social connection and loneliness.

With the development of artificial intelligence (AI) and large language models, research is needed to understand how best to use AI to enhance social connection and wellbeing. Preliminary research shows that AI-powered virtual companions — chatbots designed for conversation or emotional support — may offer short-term relief from loneliness. These tools simulate human interaction, providing immediate responses that mimic companionship. While their potential is exciting, robust research is essential to evaluate their long-term effectiveness and understand how best to integrate them into broader efforts to promote wellbeing.

Policy-based interventions

Social relationships are shaped, in large part, by systemic societal, economic, and technological factors. As such, there is growing interest in the role of policy-based interventions in fostering social connection and mitigating the growing trends in social isolation and loneliness.[72] These

interventions aim to create structural changes that promote prosocial behaviours and relationships at scale, moving beyond individual efforts to address the broader contexts in which relationships emerge. An example of such efforts is the introduction of social and emotional learning (SEL) curriculum in schools. By embedding these practices into educational systems, SEL programs create an environment where healthier relationships can flourish, helping students develop critical tools for connection.[73]

Policy-based interventions hold great potential for fostering social connections. When we examine other public health challenges such as smoking cessation, societal-level efforts such as taxation and public health campaigns generally outperform individual-level approaches.[74] Policy-based interventions that address social isolation and loneliness are still sparse, so future progress will require rigorous, evidence-based research to carefully guide policy design and implementation.

Together, future work is needed to identify how interventions can effectively promote social connection, particularly through the promising avenues of technology and public policy. As we discussed above, social connection is not a one-dimensional, catch-all concept. It encompasses the quantity, quality, and the structure of social relationships that individuals are embedded in. As such, we need integrated, multi-level strategies that account for the interplay of these factors.[75]

Equally important is understanding when these interventions may have no effect or backfire. Well-intentioned efforts could inadvertently deepen isolation or exacerbate disparities. Future interventions that simultaneously address individual, community, and societal levels in a systematic way are likely to be the most effective at promoting social connection.

Interventions that provide opportunities for students to learn about the empathy and care in their community can effectively shift empathy perceptions, encourage social risk-taking, and expand social networks.

Conclusion

This chapter has examined the critical role of social connection in the happiness and wellbeing of young adults.

First, drawing from the Global Flourishing Study and the Gallup World Poll, we showed that social disconnection is prevalent and growing in young adults, and that both the quantity and quality of social connection robustly map onto subjective wellbeing.

Second, we reviewed classic and contemporary studies that underscore the importance of social relationships for human flourishing. Evidence in this area points to a robust link between social connection and wellbeing, both across individuals (happier people tend to report better social connection) and within the same individuals over time (people report greater happiness when more socially engaged). Building on these correlational findings, there is growing evidence that credibly demonstrates a significant causal effect of social connection on improved mental health. Individuals who are randomly assigned with social engagement tend to report lower stress when exposed to distressing stimuli, regulate their emotions better, and report more positive affect.

Third, we zoomed in on the Stanford Community Project; a large-scale, longitudinal project that focuses on one undergraduate student community. Data from this work have produced several discoveries that advance the basic science of community wellbeing and provide avenues through which to improve it: (1) friendships formed in the first few weeks of college significantly shape the long-term mental health trajectories of students; (2) both direct friendships and the broader 'social microclimate' can significantly contribute to wellbeing; (3) an 'empathy perception gap' — the tendency for young people to underestimate the empathy of their peers — leads to missed opportunities for connection; and (4) interventions that provide opportunities for students to learn about the empathy and care in their community can effectively shift empathy perceptions, encourage social risk-taking, and expand social networks. These findings point to novel, promising ways of bolstering connection and happiness among this age group.

In summary, this chapter highlights the multi-faceted ways in which social connection influences the wellbeing of young adults. Our evidence points to practical opportunities to leverage social connection to enhance happiness. By targeting both individual relationships and the broader social environment, these strategies offer promising avenues for improving the happiness and wellbeing of young adults.

Endnotes

1 Blanchflower et al. (2024).

2 Helliwell et al. (2024).

3 Cohen and Wills (1985); Wickramaratne et al. (2022); Holt-Lunstad (2024).

4 Choi et al. (2023); Coan et al. (2006).

5 Stein and Smith (2015); Williams et al. (2018).

6 Buecker et al. (2021).

7 American College Health Association (2023).

8 Crone and Dahl (2012); Pei et al. (2019).

9 Sompolska-Rzechuła and Kurdyś-Kujawska (2022).

10 Fry et al. (2020).

11 Larson (2000).

12 Arnett (2000).

13 LaFreniere (2024).

14 Boneva et al. (2006).

15 Noble and McGrath (2011).

16 Johnson and VanderWeele (2022).

17 Blieszner and Roberto (2003).

18 Carstensen et al. (2006).

19 Helliwell et al. (2024).

20 Buecker et al. (2021).

21 $b = 0.92$, 95% CI = [0.85, 0.99], $p < 0.001$.

22 $b = 0.29$, 95% CI = [0.28, 0.30], $p < 0.001$.

23 Quantity: standardized $b = 0.77$, 95% CI = [0.70 - 0.85], $p < 0.01$; Quality: standardized $b = 0.84$, 95% CI = [0.82-0.87], $p < 0.01$.

24 We focus on subjective wellbeing in this chapter. Evidence also points to social connection as a critical factor for physical health. Interested readers can see Holt-Lunstad (2024) for a recent review on this topic.

25 Waldinger and Schulz (2023).

26 Vaillant (2008).

27 Cacioppo et al. (2010).

28 Rohrer et al. (2018).

29 Harris et al. (2017).

30 Cai et al. (2017); Figueira et al. (2017);Hu et al. (2020); Margraf et al. (2020); Velten et al. (2018); Ye et al. (2019).

31 Srivastava et al. (2008).

32 Sun et al. (2020).

33 Quoidbach et al. (2019); Rimé (2007).

34 Diener et al. (2015); Watson et al. (1995).

35 Thayer et al. (1994).

36 Quoidbach et al. (2019).

37 Rhoads and Marsh (2023).

38 Choi et al. (2023); Coan et al. (2006); Cohen & Wills (1985).

39 Lepore et al.; Uchino and Garvey (1997).

40 Coan et al. (2017).

41 Master et al. (2009); Pei et al. (2023).

42 Jacques-Hamilton et al. (2019); Schroeder et al. (2022).

43 Krakauer (1996).

44 Langston (1994); Peters et al. (2018).

45 Reis et al. (2010).

46 See experiment 5 in Kardas et al. (2022) They also find a dose/response relationship, with longer conversations showing progressively larger increases in enjoyment. (Kardas et al. 2022, Figure 7).

47 Gunaydin et al. (2021).

48 Epley and Schroeder (2014).

49 Sandstrom and Dunn (2014).

50 Jacques-Hamilton et al. (2019). Note that this study included young adults as well as older adults as participants. The experimental findings did not depend on the age of the participants.

51 Westaby et al. (2014).

52 Marsden (1993).

53 Granovetter (1985); Granovetter (1992); Stokes (1985); Zou et al. (2015).

54 Perry et al. (2024).

55 Collins et al. (2022).

56 Bowman and Park (2015).

57 Ramos et al. (2024). Note that this study includes adults of all age groups, though further analysis indicated that results held for both younger and older adults.

58 Buecker et al. (2021).

59 Courtney et al. (2024).

60 Buecker et al. (2021).

61 Epley and Schroeder (2014).

62 Kardas et al. (2022).

63 Zhao and Epley (2021a).

64 Zhao and Epley (2021b).

65 Zhao and Epley (2022).

66 Pei et al. (in prep).

67 World Health Organization (2006).

68 Carstensen et al. (1999).

69 Fingerman et al. (2020).

70 Collins et al. (2022).

71 https://www.gallup.com/analytics/509675/state-of-social-connections.aspx.

72 Office of the Surgeon General (2023).

73 Corcoran et al. (2018).

74 Lemmens et al. (2008).

75 Glasgow et al. (1999).

References

American College Health Association. (2023). American College Health Association-National College Health Assessment III: Undergraduate Student Reference Group Executive Summary Spring 2023. *Silver Spring, MD: American College Health Association.*

Arnett, J. J. (2000). Emerging adulthood. A theory of development from the late teens through the twenties. *The American Psychologist*, *55*(5), 469–480.

Blanchflower, D. G., Bryson, A., & Xu, X. (2024). *The Declining Mental Health of the Young and the Global Disappearance of the Hump Shape in Age in Unhappiness.* Retrieved from https://papers.ssrn.com/abstract=4794387

Blieszner, R., & Roberto, K. A. (2003). Friendship across the life span: Reciprocity in individual and relationship development. In *Growing Together* (pp. 159–182). Cambridge: Cambridge University Press. https://doi.org/10.1017/cbo9780511499852.007

Bowman, N. A., & Park, J. J. (2015). Not all diversity interactions are created equal: Cross-racial interaction, close interracial friendship, and college student outcomes. *Research in Higher Education*, *56*(6), 601–621. https://doi.org/10.1007/s11162-015-9365-z

Buecker, S., Mund, M., Chwastek, S., Sostmann, M., & Luhmann, M. (2021). Is loneliness in emerging adults increasing over time? A preregistered cross-temporal meta-analysis and systematic review. *Psychological Bulletin*, *147*(8), 787–805. https://doi.org/10.1037/bul0000332

Cacioppo, J. T., Hawkley, L. C., & Thisted, R. A. (2010). Perceived social isolation makes me sad: 5-year cross-lagged analyses of loneliness and depressive symptomatology in the Chicago Health, Aging, and Social Relations Study. *Psychology and Aging*, *25*(2), 453–463. https://doi.org/10.1037/a0017216

Cai, D., Zhu, M., Lin, M., Zhang, X. C., & Margraf, J. (2017). The bidirectional relationship between positive mental health and social rhythm in college students: A three-year longitudinal study. *Frontiers in Psychology*, *8*, 1119. https://doi.org/10.3389/fpsyg.2017.01119

Carstensen, L. L., Isaacowitz, D. M., & Charles, S. T. (1999). Taking time seriously. A theory of socioemotional selectivity. *The American Psychologist*, *54*(3), 165–181. https://doi.org/10.1037//0003-066x.54.3.165

Carstensen, L. L., Mikels, J. A., & Mather, M. (2006). Aging and the intersection of cognition, motivation, and emotion. In *Handbook of the Psychology of Aging* (pp. 343–362). Elsevier. https://doi.org/10.1016/b978-012101264-9/50018-5

Choi, K. W., Lee, Y. H., Liu, Z., Fatori, D., Bauermeister, J. R., Luh, R. A., ... Smoller, J. W. (2023). Social support and depression during a global crisis. *Nature Mental Health*, *1*(6), 428–435. https://doi.org/10.1038/s44220-023-00078-0

Coan, J. A., Beckes, L., Gonzalez, M. Z., Maresh, E. L., Brown, C. L., & Hasselmo, K. (2017). Relationship status and perceived support in the social regulation of neural responses to threat. *Social Cognitive and Affective Neuroscience*, *12*(10), 1574–1583. https://doi.org/10.1093/scan/nsx091

Coan, J. A., Schaefer, H. S., & Davidson, R. J. (2006). Lending a hand: social regulation of the neural response to threat. *Psychological Science*, *17*(12), 1032–1039. https://doi.org/10.1111/j.1467-9280.2006.01832.x

Cohen, S., & Wills, T. A. (1985). Stress, social support, and the buffering hypothesis. *Psychological Bulletin*, *98*(2), 310–357. Retrieved from https://www.ncbi.nlm.nih.gov/pubmed/3901065

Collins, H. K., Hagerty, S. F., Quoidbach, J., Norton, M. I., & Brooks, A. W. (2022). Relational diversity in social portfolios predicts well-being. *Proceedings of the National Academy of Sciences of the United States of America*, *119*(43), e2120668119. https://doi.org/10.1073/pnas.2120668119

Courtney, A. L., Baltiansky, D., Fang, W. M., Roshanaei, M., Aybas, Y. C., Samuels, N. A., ... Zaki, J. (2024). Social microclimates and well-being. *Emotion*, *24*(3), 836–846. https://doi.org/10.1037/emo0001277

Crone, E. A., & Dahl, R. E. (2012). Understanding adolescence as a period of social-affective engagement and goal flexibility. *Nature Reviews. Neuroscience*, *13*(9), 636–650. https://doi.org/10.1038/nrn3313

Diener, E., Kanazawa, S., Suh, E. M., & Oishi, S. (2015). Why people are in a generally good mood. *Personality and Social Psychology Review: An Official Journal of the Society for Personality and Social Psychology, Inc*, *19*(3), 235–256. https://doi.org/10.1177/1088868314544467

Epley, N., & Schroeder, J. (2014). Mistakenly seeking solitude. *Journal of Experimental Psychology. General*, *143*(5), 1980–1999. https://doi.org/10.1037/a0037323

Figueira, C. P., Marques-Pinto, A., Pereira, C. R., & Roberto, M. S. (2017). How can academic context variables contribute to the personal well-being of higher education students? *The Spanish Journal of Psychology*, *20*, E43. https://doi.org/10.1017/sjp.2017.46

Fingerman, K. L., Huo, M., Charles, S. T., & Umberson, D. J. (2020). Variety is the spice of late life: Social integration and daily activity. *The Journals of Gerontology. Series B, Psychological Sciences and Social Sciences*, *75*(2), 377–388. https://doi.org/10.1093/geronb/gbz007

Fry, R., Passel, J. S., & Cohn, D. 'vera. (2020, September 4). A majority of young adults in the U.S. live with their parents for the first time since the Great Depression. Retrieved October 27, 2024, from Pew Research Center website: https://www.pewresearch.org/short-reads/2020/09/04/a-majority-of-young-adults-in-the-u-s-live-with-their-parents-for-the-first-time-since-the-great-depression/

Glasgow, R. E., Vogt, T. M., & Boles, S. M. (1999). Evaluating the public health impact of health promotion interventions: the RE-AIM framework. *American Journal of Public Health*, *89*(9), 1322–1327. https://doi.org/10.2105/ajph.89.9.1322

Granovetter, M. (1985). Economic action and social structure: The problem of embeddedness. *American Journal of Sociology*, *91*(3), 481–510. https://doi.org/10.1086/228311

Granovetter, M. (1992). Economic institutions as social constructions: A framework for analysis. *Acta Sociologica*, *35*(1), 3–11. https://doi.org/10.1177/000169939203500101

Gunaydin, G., Oztekin, H., Karabulut, D. H., & Salman-Engin, S. (2021). Minimal social interactions with strangers predict greater subjective well-being. *Journal of Happiness Studies*, *22*(4), 1839–1853. https://doi.org/10.1007/s10902-020-00298-6

Harris, K., English, T., Harms, P. D., Gross, J. J., & Jackson, J. J. (2017). Why are extraverts more satisfied? Personality, social experiences, and subjective well-being in college: Extraverts and social experience. *European Journal of Personality*, *31*(2), 170–186. https://doi.org/10.1002/per.2101

Helliwell, Huang, H., Shiplett, H., Wang, S., & World Happiness Report. (2024). *Happiness of the younger, the older, and those in between*. University of Oxford. https://doi.org/10.18724/WHR-F1P2-QJ33

Holt-Lunstad, J. (2024). Social connection as a critical factor for mental and physical health: evidence, trends, challenges, and future implications. *World Psychiatry: Official Journal of the World Psychiatric Association (WPA)*, *23*(3), 312–332. https://doi.org/10.1002/wps.21224

Hu, S., Cai, D., Zhang, X. C., & Margraf, J. (2020). Relationship between social support and positive mental health: A three-wave longitudinal study on college students. *Current Psychology (New Brunswick, N.J.)*, *41*(10), 6712–6721. https://doi.org/10.1007/s12144-020-01175-4

Jacques-Hamilton, R., Sun, J., & Smillie, L. D. (2019). Costs and benefits of acting extraverted: A randomized controlled trial. *Journal of Experimental Psychology. General*, *148*(9), 1538–1556. https://doi.org/10.1037/xge0000516

Johnson, B. R., & VanderWeele, T. J. (2022). The Global Flourishing Study: A New Era for the Study of Well-Being. *International Bulletin of Mission Research*, *46*(2), 272–275. https://doi.org/10.1177/23969393211068096

Kardas, M., Schroeder, J., & O'Brien, E. (2022). Keep talking: (Mis)understanding the hedonic trajectory of conversation. *Journal of Personality and Social Psychology*, *123*(4), 717–740. https://doi.org/10.1037/pspi0000379

Krakauer, J. (1996). *Into the Wild*. Anchor Books.

LaFreniere, J. R. (2024). Defining and discussing independence in emerging adult college students. *Journal of Adult Development*, *31*(4), 293–303. https://doi.org/10.1007/s10804-023-09472-5

Langston, C. A. (1994). Capitalizing on and coping with daily-life events: Expressive responses to positive events. *Journal of Personality and Social Psychology*, *67*(6), 1112–1125. https://doi.org/10.1037/0022-3514.67.6.1112

Larson, R. W. (2000). Toward a psychology of positive youth development. *The American Psychologist*, *55*(1), 170–183. https://doi.org/10.1037/0003-066X.55.1.170

Lemmens, V., Oenema, A., Knut, I. K., & Brug, J. (2008). Effectiveness of smoking cessation interventions among adults: a systematic review of reviews. *European Journal of Cancer Prevention: The Official Journal of the European Cancer Prevention Organisation (ECP)*, *17*(6), 535–544. https://doi.org/10.1097/CEJ.0b013e3282f75e48

Lepore, S. J., Allen, K. A., & Evans, G. W. (1993). Social support lowers cardiovascular reactivity to an acute stressor. *Psychosomatic Medicine*, *55*(6), 518–524. https://doi.org/10.1097/00006842-199311000-00007

Margraf, J., Zhang, X. C., Lavallee, K. L., & Schneider, S. (2020). Longitudinal prediction of positive and negative mental health in Germany, Russia, and China. *PloS One*, *15*(6), e0234997. https://doi.org/10.1371/journal.pone.0234997

Marsden, P. V. (1993). The reliability of network density and composition measures. *Social Networks*, *15*(4), 399–421. https://doi.org/10.1016/0378-8733(93)90014-c

Master, S. L., Eisenberger, N. I., Taylor, S. E., Naliboff, B. D., Shirinyan, D., & Lieberman, M. D. (2009). A picture's worth: partner photographs reduce experimentally induced pain. *Psychological Science*, *20*(11), 1316–1318. https://doi.org/10.1111/j.1467-9280.2009.02444.x

Noble, T., & McGrath, H. (2011). Wellbeing and resilience in young people and the role of positive relationships. In *Positive Relationships* (pp. 17–33). Dordrecht: Springer Netherlands. https://doi.org/10.1007/978-94-007-2147-0_2

Office of the Surgeon General (OSG). (2023). *Our epidemic of loneliness and isolation: The U.S. surgeon general's advisory on the healing effects of social connection and community*. Washington (DC): US Department of Health and Human Services. Retrieved from https://pubmed.ncbi.nlm.nih.gov/37792968/

Pei, R., Courtney, A. L., Ferguson, I., Brennan, C., & Zaki, J. (2023). A neural signature of social support mitigates negative emotion. *Scientific Reports*, *13*(1), 17293. https://doi.org/10.1038/s41598-023-43273-w

Pei, R., Grayson, S. J., Appel, R. E., Bouwer, A., Huang, E., Jackson, M. O., ... Zaki, J. (in prep). *Bridging the Gap: Enhancing Empathy Perceptions Fosters Social Connection*.

Pei, R., Kranzler, E., Suleiman, A. B., & Falk, E. B. (2019). Promoting adolescent health: Insights from developmental and communication neuroscience. *Behavioural Public Policy*, *3*(1), 47–71. https://doi.org/10.1017/bpp.2018.30

Perry, B. L., Smith, N. C., Coleman, M. E., & Pescosolido, B. A. (2024). Social networks, the COVID-19 pandemic, and emerging adults' mental health: Resiliency through social bonding and cohesion. *American Journal of Public Health*, *114*(S3), S258–S267. https://doi.org/10.2105/AJPH.2023.307426

Peters, B. J., Reis, H. T., & Gable, S. L. (2018). Making the good even better: A review and theoretical model of interpersonal capitalization. *Social and Personality Psychology Compass*, *12*(7), e12407. https://doi.org/10.1111/spc3.12407

Quoidbach, J., Taquet, M., Desseilles, M., de Montjoye, Y.-A., & Gross, J. J. (2019). Happiness and social behavior. *Psychological Science*, *30*(8), 1111–1122. https://doi.org/10.1177/0956797619849666

Ramos, M. R., Li, D., Bennett, M. R., Mogra, U., Massey, D. S., & Hewstone, M. (2024). Variety is the spice of life: Diverse social networks are associated with social cohesion and well-being. *Psychological Science*, *35*(6), 665–680. https://doi.org/10.1177/09567976241243370

Reis, H. T., Smith, S. M., Carmichael, C. L., Caprariello, P. A., Tsai, F.-F., Rodrigues, A., & Maniaci, M. R. (2010). Are you happy for me? How sharing positive events with others provides personal and interpersonal benefits. *Journal of Personality and Social Psychology*, *99*(2), 311–329. https://doi.org/10.1037/a0018344

Rhoads, S. A., & Marsh, A. A. (2023). Doing Good and Feeling Good: Relationships Between Altruism and Well-being for Altruists, Beneficiaries, and Observers. Retrieved October 27, 2024, from https://worldhappiness.report/ed/2023/doing-good-and-feeling-good-relationships-between-altruism-and-well-being-for-altruists-beneficiaries-and-observers/

Rimé, B. (2007). The social sharing of emotion as an interface between individual and collective processes in the construction of emotional climates. *The Journal of Social Issues*, *63*(2), 307–322. https://doi.org/10.1111/j.1540-4560.2007.00510.x

Rohrer, J. M., Richter, D., Brümmer, M., Wagner, G. G., & Schmukle, S. C. (2018). Successfully striving for happiness: Socially engaged pursuits predict increases in life satisfaction. *Psychological Science*, *29*(8), 1291–1298. https://doi.org/10.1177/0956797618761660

Sandstrom, G. M., & Dunn, E. W. (2014). Social interactions and well-being: The surprising power of weak ties. *Personality & Social Psychology Bulletin*, *40*(7), 910–922. https://doi.org/10.1177/0146167214529799

Schroeder, J., Lyons, D., & Epley, N. (2022). Hello, stranger? Pleasant conversations are preceded by concerns about starting one. *Journal of Experimental Psychology. General*, *151*(5), 1141–1153. https://doi.org/10.1037/xge0001118

Sompolska-Rzechuła, A., & Kurdyś-Kujawska, A. (2022). Generation of young adults living with their parents in European Union countries. *Sustainability*, *14*(7), 4272. https://doi.org/10.3390/su14074272

Srivastava, S., Angelo, K. M., & Vallereux, S. R. (2008). Extraversion and positive affect: A day reconstruction study of person-environment transactions. *Journal of Research in Personality*, *42*(6), 1613–1618. https://doi.org/10.1016/j.jrp.2008.05.002

Stein, E. R., & Smith, B. W. (2015). Social support attenuates the harmful effects of stress in healthy adult women. *Social Science & Medicine*, *146*, 129–136. https://doi.org/10.1016/j.socscimed.2015.10.038

Stokes, J. P. (1985). The relation of social network and individual difference variables to loneliness. *Journal of Personality and Social Psychology*, *48*(4), 981–990. https://doi.org/10.1037/0022-3514.48.4.981

Sun, J., Harris, K., & Vazire, S. (2020). Is well-being associated with the quantity and quality of social interactions? *Journal of Personality and Social Psychology*, *119*(6), 1478–1496. https://doi.org/10.1037/pspp0000272

Boneva, B., Quinn, A., Kraut, R., Kiesler, S., & Shklovski, I. (2006). Teenage communication in the instant messaging era. In R. Kraut, M. Brynin, & S. Kiesler (Eds.), *Computers, phones, and the Internet: Domesticating information technology* (pp. 201–218). Oxford University Press.

Thayer, R. E., Newman, J. R., & McClain, T. M. (1994). Self-regulation of mood: Strategies for changing a bad mood, raising energy, and reducing tension. *Journal of Personality and Social Psychology*, *67*(5), 910–925. https://doi.org/10.1037/0022-3514.67.5.910

Uchino, B. N., & Garvey, T. S. (1997). The availability of social support reduces cardiovascular reactivity to acute psychological stress. *Journal of Behavioral Medicine*, *20*(1), 15–27. https://doi.org/10.1023/a:1025583012283

Vaillant, G. E. (2008). *Aging well: Surprising guideposts to a happier life from the landmark study of adult development.* Little, Brown.

Velten, J., Bieda, A., Scholten, S., Wannemüller, A., & Margraf, J. (2018). Lifestyle choices and mental health: a longitudinal survey with German and Chinese students. *BMC Public Health*, *18*(1), 632. https://doi.org/10.1186/s12889-018-5526-2

Waldinger, R., & Schulz, M. (2023). *The Good Life: Lessons from the World's Longest Scientific Study of Happiness*. Simon and Schuster.

Watson, D., Clark, L. A., Weber, K., Assenheimer, J. S., Strauss, M. E., & McCormick, R. A. (1995). Testing a tripartite model: II. Exploring the symptom structure of anxiety and depression in student, adult, and patient samples. *Journal of Abnormal Psychology*, *104*(1), 15–25. https://doi.org/10.1037/0021-843x.104.1.15

Westaby, J. D., Pfaff, D. L., & Redding, N. (2014). Psychology and social networks: a dynamic network theory perspective. *The American Psychologist*, *69*(3), 269–284. https://doi.org/10.1037/a0036106

Wickramaratne, P. J., Yangchen, T., Lepow, L., Patra, B. G., Glicksburg, B., Talati, A., ... Weissman, M. M. (2022). Social connectedness as a determinant of mental health: A scoping review. *PloS One*, *17*(10), e0275004. https://doi.org/10.1371/journal.pone.0275004

Williams, W. C., Morelli, S. A., Ong, D. C., & Zaki, J. (2018). Interpersonal emotion regulation: Implications for affiliation, perceived support, relationships, and well-being. *Journal of Personality and Social Psychology*, *115*(2), 224–254. https://doi.org/10.1037/pspi0000132

World Health Organization (2006). A state of complete physical mental and social well-being and not merely the absence of disease or infirmity. *Constitution of the World Health Organization Basic Documents*, *45*, 1–20.

Ye, J., Yeung, D. Y., Liu, E. S. C., & Rochelle, T. L. (2019). Sequential mediating effects of provided and received social support on trait emotional intelligence and subjective happiness: A longitudinal examination in Hong Kong Chinese university students. *International Journal of Psychology*, *54*(4), 478–486. https://doi.org/10.1002/ijop.12484

Zhao, X., & Epley, N. (2021a). Insufficiently complimentary?: Underestimating the positive impact of compliments creates a barrier to expressing them. *Journal of Personality and Social Psychology*, *121*(2), 239–256. https://doi.org/10.1037/pspa0000277

Zhao, X., & Epley, N. (2021b). Kind words do not become tired words: Undervaluing the positive impact of frequent compliments. *Self and Identity: The Journal of the International Society for Self and Identity*, *20*(1), 25–46. https://doi.org/10.1080/15298868.2020.1761438

Zhao, X., & Epley, N. (2022). Surprisingly Happy to Have Helped: Underestimating Prosociality Creates a Misplaced Barrier to Asking for Help. *Psychological Science*, *33*(10), 1708–1731. https://doi.org/10.1177/09567976221097615

Zou, X., Ingram, P., & Higgins, E. T. (2015). Social networks and life satisfaction: The interplay of network density and regulatory focus. *Motivation and Emotion*, *39*(5), 693–713. https://doi.org/10.1007/s11031-015-9490-1

Chapter 6

Supporting others

How prosocial behaviour reduces deaths of despair

Kelsey J. O'Connor
STATEC Research

Chiara Peroni
STATEC Research

Francesco Sarracino
STATEC Research

Giulia Slater
STATEC Research

Fengyu Wu
STATEC Research

The authors would like to thank Carol Graham for helpful comments and gratefully acknowledge the support of the Observatoire de la Compétitivité, Ministère de l'Economie, DG Compétitivité, Luxembourg, and STATEC, the National Statistical Office of Luxembourg. This project received funding from the European Union's Horizon 2022 research and innovation programme under grant agreement No 101094546. Views and opinions expressed in this article are those of the authors and do not reflect those of STATEC and funding partners.

doi.org/10.18724/whr-t5h9-dj68

This version was last updated on 13 March 2025. Please see worldhappiness.report for the latest data.

The evidence is consistent with the hypothesis that prosocial behaviour constitutes an informal safety net whose benefits go beyond the donors and recipients, and provides a buffer against adversities.

Key Insights

Increasing prosocial behaviour – donating, volunteering, and helping strangers – **is reliably connected to decreasing deaths of despair around the world.** Regression results indicate that a ten percentage-point increase in the share of people engaging in prosocial behaviour is associated with approximately one fewer death per year per 100,000. For a country like the United Kingdom, with an adult population of approximately 55 million, that is equivalent to about 550 preventable deaths per year.

On average, deaths of despair decreased around the world by nearly 5 deaths per 100,000 people over the period 2000–2019 in 59 middle- to high-income countries. This equates to approximately 2,750 people in a country like the UK. The largest declines occurred in northeastern European countries, such as Lithuania and Latvia, which tended to have very high initial levels.

Deaths of despair declined in three quarters of the 59 considered countries, but are still high and rising in a few cases, such as the United States and Republic of Korea. Slovenia had the highest level in 2019, with more than 50 deaths of despair per 100,000.

Deaths of despair are nearly four times higher among men than women, and more than double among those aged 60 and above compared to 15- to 29-year-olds. Three quarters of these deaths are due to suicide, followed by deaths due to alcohol and drug abuse.

Investing in the conditions that support prosocial behaviour could help create societies where people are more supportive, cooperative, and trusting, and where deaths of despair are lower.

Introduction

Deaths of despair are preventable deaths due to suicide, alcohol abuse, and drug overdose. Together, they represent a form of ill-being akin to extreme unhappiness. The term was introduced to describe troubling trends in the United States,[1] while little was known about global patterns[2]. However, rising feelings of distress around the world suggested this trend was not confined to the US.[3] Indeed, our data show that Slovenia, not the US, had the highest rate of deaths of despair in 2019, although the US did experience the greatest increase over the period 2000 to 2019.

In this chapter, we document the levels of deaths of despair around the world using the latest available data, show how these levels have changed over time, and provide an initial analysis of the factors that explain the rise or fall of deaths of despair within countries over time. We focus particularly on prosocial behaviour - donating, volunteering, and helping strangers - as a factor that could reduce deaths of despair.

Previous research has documented the rise in premature mortality due to suicides, opioid poisonings, and alcohol-related liver disease, particularly in North America, Australia, and parts of Europe.[4] The increase in these deaths is typically linked to factors such as extreme distress,[5] feelings of hopelessness,[6] social and economic marginalisation,[7] loss of social ties and decline in social capital,[8] pharmaceutical marketing of prescription opioids[9], rising occupational injuries and pain,[10] and stress stemming from perceived status loss, especially among white populations.[11] These deaths are more common among disadvantaged groups, leading to inequalities in mortality based on education,[12] income,[13] area-level deprivation,[14] and economic insecurity.[15] In addition, evidence from studies on suicide and wellbeing suggests that deaths of despair should be less prevalent in countries where people engage more in prosocial behaviour.[16]

Our research shows that the previously documented trends in deaths of despair do not extend around the world. In the United States, deaths of despair rapidly rose for the reasons mentioned above, especially among middle-aged men due to

In our sample of 59 countries, deaths of despair have declined on average from 2000 to 2019, due to declining suicide among older men.

increased drug abuse. Whereas in our sample of 59 countries, deaths of despair have declined on average from 2000 to 2019, due to declining suicide among older men. In any case, preventable deaths are concerning and it should be possible to reduce them further. Our regression results indicate that fostering prosocial behaviour should reduce deaths of despair.

Deaths of despair in 59 countries

Prior to the COVID-19 pandemic, the average number of deaths of despair per year, in our sample of 59 countries, was 23 per 100,000 persons aged 15 and above.[17] In a country like the United Kingdom, this equates to approximately 12,500 preventable deaths each year.[18] Grenada had the fewest deaths of despair with 4 per 100,000, while Slovenia had the greatest at 53 per 100,000.[19]

Figure 6.1 shows the ranking of deaths of despair in 2019,[20] grouped by region and broken down into their three components. The highest rates are recorded in Slovenia, followed by Guyana, the United States, and Lithuania. Grenada, and Antigua and Barbuda have the lowest rates. Most Northern European countries (such as Finland) and Central and Eastern European countries (such as Slovenia and Lithuania) have scores above average. The fact that Nordic countries top the European ranking of deaths of despair, as well as the global ranking of subjective wellbeing (see Chapter 2), poses a puzzle that we will discuss later. In the United States, Canada, and Australia, deaths of despair are also above average, comparable to Central and Eastern European countries. Among the six Asian countries in our sample, the Republic of Korea and Kazakhstan stand out as the two cases with the highest

Figure 6.1: Deaths of despair around the world by cause
World Health Organisation (2019)

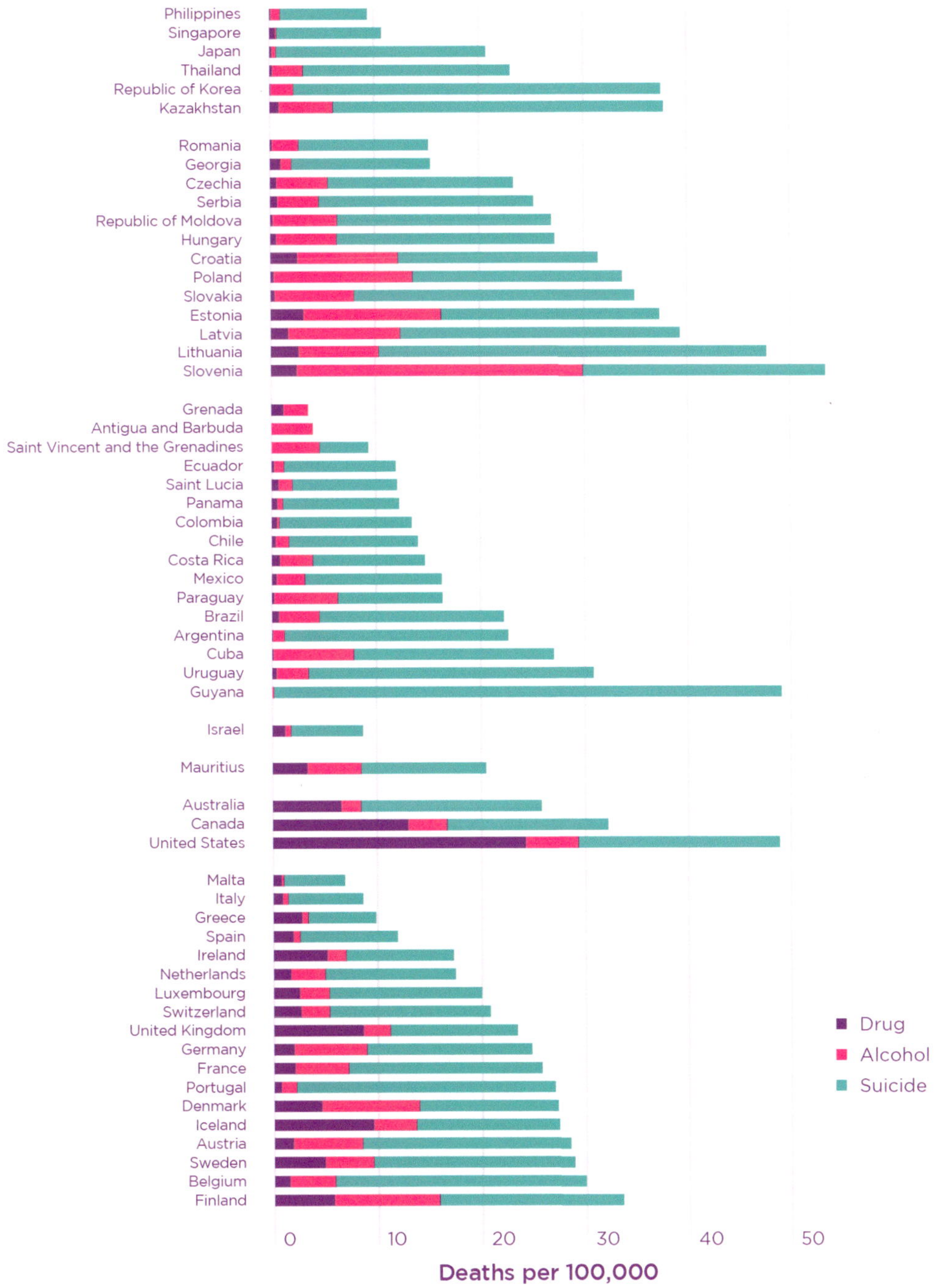

Note: Deaths are per 100,000 for the population aged 15 and above. Data for Republic of Moldova are for 2018. South Africa is excluded due to comparability, see endnote 19 for details.

scores, mainly due to high levels of suicides. Mediterranean and Latin American countries are well below the international average.

Our measure of deaths of despair includes deaths due to alcohol use disorder, drug use disorder, self-inflicted injuries, and ill-defined injuries/accidents from the WHO Mortality Database.[21] Self-inflicted and ill-defined injuries/accidents (of undetermined intent) are summed to obtain suicides, consistent with previous research[22] because ill-defined deaths could include suicide. Previous studies used different variants of deaths of despair, including slightly different causes of mortality depending on data source. Online Appendix A provides more details on how we define and operationalise deaths of despair.

While we used the best available data, caution should be exercised when interpreting differences in deaths across countries. Cultural and institutional differences can affect 'cause of death' reporting and limit the comparability of data. For more details on the quality of data and issues with the comparability of cause of death across countries please see Box 6.1.

Our sample is limited to 59 countries because many countries, especially low-income countries, do not provide the necessary data to meet the standards for international comparison.[23] We focus on the period 2000 to 2019 because it gives us the longest and broadest sample of countries possible. Fewer countries have data before 2000 and after 2019. We present mortality figures for populations aged 15 and older in order to match the surveyed populations used to obtain prosocial behaviour figures. Tables B1 to B3 in the online appendix provide sample details, listing the included countries and the reasons for excluding others.

Box 6.1: WHO mortality database and deaths of despair data quality

There are a few reasons why cross-country comparisons of deaths of despair pose an issue. These reasons include systematic differences in the process of assigning cause of death, the capacity for obtaining and storing such information, as well as legal or societal differences.

National statistics on mortality by cause are sourced from the WHO Mortality Database which collects and harmonises information from country civil registration and vital statistics systems. When a death occurs, it is registered at the local civil registry with information on the cause of death, which is typically filed by health professionals such as doctors or nurses. Health professionals document the injuries or diseases that led to the death of the person and list what they believe is the "underlying cause of death" on the death certificate. In some cases, the death registration process may be different, such as suicides and sudden deaths, when it is a coroner who determines and reports the cause of death. Subsequently, the cause of death is assigned an International Classification of Diseases (ICD) death code, registered in the national databases, and sent to the WHO by each country. National statistics offices, or the ministry of health or registrar-general's office, are responsible for compiling the cause of death data for submission to WHO every year.

The WHO Mortality Database does not include death statistics for all countries. Some do not report their mortality data to the WHO and some send data that are not in standard ICD or do not have ICD codes at all. In other countries, such as China and India, the total deaths reported to the WHO represent less than 5% of the population of the country, hence their data is deemed unreliable and not made available on the WHO Mortality Database. In many countries, cause of death information is difficult to obtain because the system for recording such information is not functioning or non-existent. In addition, there may be a lack of medical certifiers to complete the death certificates.[24] As a result, many deaths may go uncounted or be recorded without listing a cause. When the WHO receives countries' data, they assess their completeness and quality. They display the data on the portal if they are estimated to include at least 65% of all deaths occurring in a country, with the appropriate cause of death recorded.

The large variations in the systems and processes to define mortality causes imply there may be very different numbers of deaths that are registered with a specific cause. This creates a problem for cross-country comparisons of mortality by cause in general, and even more so for deaths of despair, and suicides in particular.

The person responsible for writing the cause of death on the death certificate may be different across countries. In some countries, the police are responsible, while in others a medical doctor, coroner, or judicial investigator takes on this role. Differences in doctors' training, access to medical records, and autopsy requirements contribute to these discrepancies. The legal or judicial systems that decide causes of death also vary. For instance, in some countries suicide is illegal and is not listed as a classifiable cause of death, leading to underreporting or misclassification of suicides as accidents, violence, or deaths of "undetermined intent."[25]

Data on suicides, even when reported, can be inaccurate due to social factors as well. In some countries, suicide might be taboo and highly stigmatised, so the families and friends of the person who committed suicide might decide to misreport or not disclose the mortality cause, causing underreporting of its incidence. In other societies, such as Northern Europe, there is less stigma attached to suicides, and alcohol and drug use.

Lastly, another layer of complexity with cross-country comparisons of deaths of despair is given by the differences in the ICD codes used to categorise deaths, as countries may adopt different versions at different times, complicating data harmonisation and comparison.

Although the WHO Mortality Database provides ICD-codes harmonised data per country, cross-country comparisons are still discouraged. As a result, we analyse variations of deaths of despair over time *within* countries, rather than focusing on cross-country comparisons.

Table 6.1: Deaths of despair summary data
World Health Organisation (2019)

Income Groups	Countries	Mean	Std. dev.	Min	Max
Lower Mid	11	22.95	11.97	9.40	49.10
Upper Mid	21	22.55	12.19	2.41	47.68
High	27	23.37	12.00	3.96	53.47
Total	**59**	**23.00**	**11.86**	**2.41**	**53.47**

Income Groups	Countries	DoD	Suicide	Alcohol	Drug
Lower Mid	11	22.95	19.33	3.22	0.40
Upper Mid	21	22.55	16.48	5.05	1.02
High	27	23.37	15.14	4.22	4.01
Total	**59**	**23.00**	**16.40**	**4.33**	**2.27**

Note: Deaths are per 100,000 for the population aged 15 and above. Income groups are defined based on 2005 values of GNI per capita in US dollars and World Bank (2024a).

On average, nearly 75% of deaths of despair are due to suicides (16.40 per 100,000), followed by alcohol abuse (4.33 per 100,000) and drug overdose (2.27 per 100,000) (see Table 6.1). Total deaths of despair represent 2.2% of all deaths – 23 out of 1,045 deaths per 100,00, per year, on average.[26]

Table 6.1 also shows that deaths of despair are fairly equally distributed across lower-middle (22.95), upper-middle (22.55), and high-income countries (23.37), with no accurate data available for low-income countries (and sufficient time coverage). It is worth emphasising that figures on deaths of despair in China and India are missing.

In Table 6.2, we break down the data by age and gender. Deaths of despair are nearly four times higher among men than among women. In the case of alcohol abuse, it is five times higher. Deaths of despair, independently from their cause, are more than double among men and women aged 60+ (30.45) compared to those aged 15 to 29 (13.45). However, mortality due to drug overdose is more frequent among working-age adults. The average mortality rate is 3.15 among people aged 30 to 59, and 1.47 among others.

Table 6.2: Deaths of despair by age and gender
World Health Organisation (2019)

	Full	Women	Men
Despair			
All Ages 15+	23.00	9.98	36.86
15-29	13.23	6.08	20.06
30-44	20.03	7.50	32.38
45-59	26.88	10.86	43.46
60+	30.45	13.73	51.87
Suicide			
All Ages 15+	16.40	7.29	26.10
15-29	11.20	4.96	17.18
30-44	14.23	5.13	23.21
45-59	17.88	7.41	28.77
60+	21.32	10.37	35.38
Alcohol			
All Ages 15+	4.33	1.45	7.41
15-29	0.20	0.05	0.33
30-44	2.18	0.73	3.58
45-59	6.32	1.93	10.84
60+	8.02	2.55	15.04
Drug			
All Ages 15+	2.27	1.24	3.35
15-29	1.83	1.07	2.55
30-44	3.63	1.63	5.59
45-59	2.67	1.51	3.85
60+	1.11	0.81	1.45

Note: Deaths are per 100,000 for the population aged 15 and above.

Since the year 2000, the number of deaths of despair has declined in nearly 75% of the considered countries (see Figure 6.2). The decline approaches -2 deaths per 100,000 persons per year in Lithuania, Latvia, and Estonia, followed by Kazakhstan, Finland, Switzerland, Luxembourg, and Denmark. The United States, on the other hand, has seen an average yearly increase of 1.3 deaths per 100,000. Hence, as previously documented, the United States is characterised by both high and increasing levels of deaths of despair. The Republic of Korea and Slovakia have the second and third highest yearly increase in deaths of despair, with average annual increases below 1. Among Western European countries, the yearly growth rates in the Netherlands, Portugal, Greece and the United Kingdom are positive, but very close to zero.

Table 6.3 indicates that the yearly decline of deaths of despair was -0.41 deaths per 100,000 in upper-middle-income countries, -0.28 in lower-middle-income countries, and nearly zero (-0.09) in high-income countries where only suicides declined, whereas both drug and alcohol abuse increased. These changes are driven largely by a general decline in suicide (-0.28), whereas drug overdoses are on the rise in all countries (0.05), especially among high-income countries (0.09). Hence, although the levels of deaths of despair appear unrelated to a country's income (Table 6.1), the yearly changes indicate that deaths of despair are more concerning in high-income countries.

Countries that had initially high levels of deaths of despair also tended to have large decreases over time. Among the 15 countries with the highest levels (Figure 6.1), nine experienced decreases (Figure 6.2), and at rates that were among the highest, e.g., the Baltic nations: Estonia, Latvia, and Lithuania. To assess whether this observation is statistically meaningful, we performed convergence tests of whether the initial level of deaths of despair influences its subsequent growth.[27]

Table 6.4 indicates that higher initial levels of deaths of despair generally correspond to negative changes over time (-0.025, significant at 1%). In other words, deaths of despair decreased more in countries where the incidence of deaths of despair was higher. This result is remarkably stable, both in magnitude and significance, for men and women and for people in various age groups. Investigating the causes of this relationship is beyond the scope of this analysis. However, it is possible that this

Figure 6.2: Average annual change in deaths of despair
World Health Organisation (2000–2019)

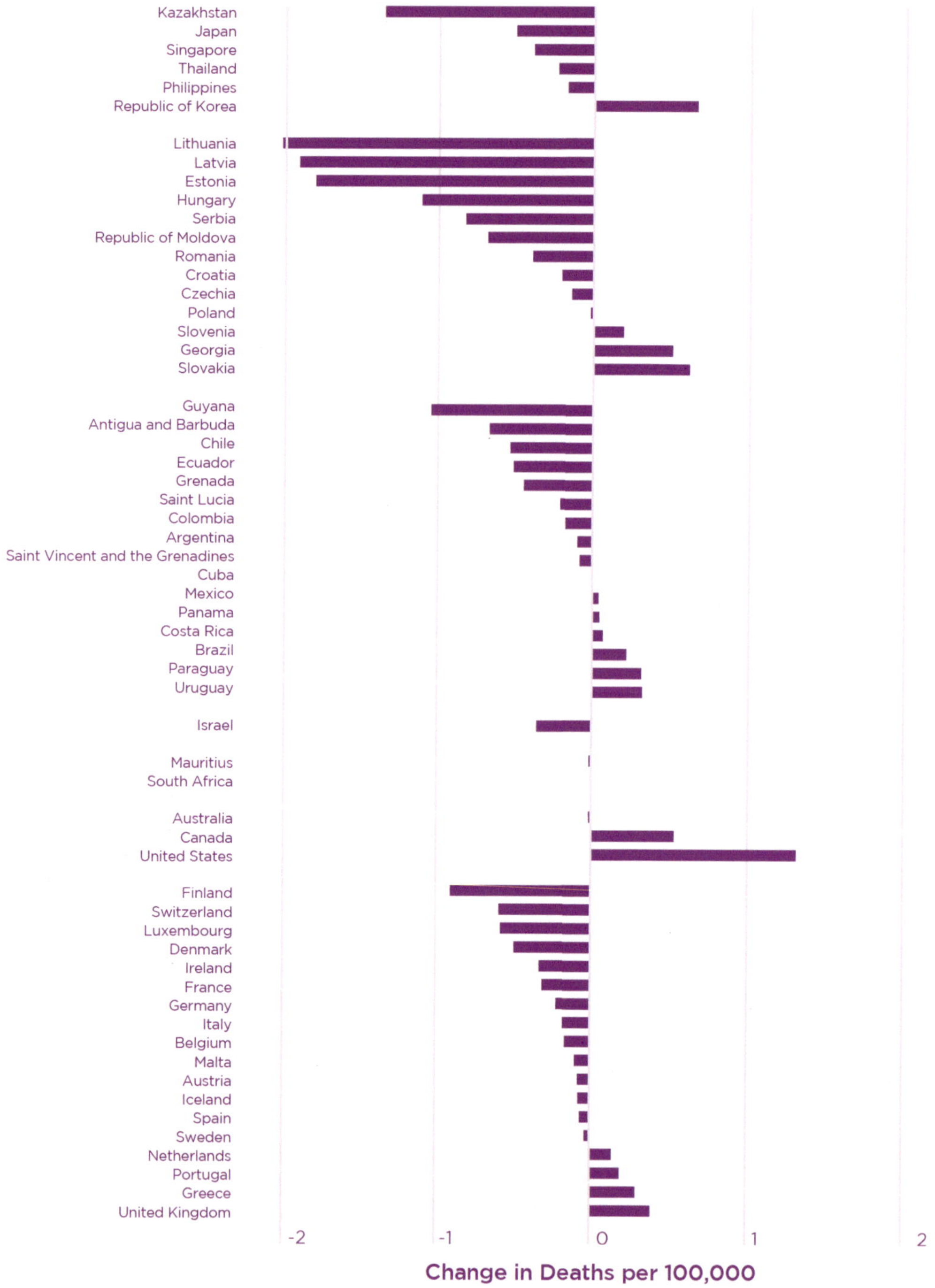

Note: Deaths are per 100,000 for the population aged 15 and above. Guyana's data begins in 1999. Republic of Moldova and South Africa's data ends in 2018.

Table 6.3: Average annual change in deaths of despair
World Health Organisation (2000–2019)

Income Groups	Countries	Mean	Std. dev.
Lower Mid	11	-0.28	0.57
Upper Mid	21	-0.41	0.73
High	27	-0.09	0.47
Total	**59**	**-0.24**	**0.60**

Income Groups	Countries	DoD	Suicide	Alcohol	Drug
Lower Mid	11	-0.28	-0.33	0.04	0.01
Upper Mid	21	-0.41	-0.38	-0.04	0.01
High	27	-0.09	-0.19	0.02	0.09
Total	**59**	**-0.24**	**-0.28**	**0.00**	**0.05**

Note: Deaths are per 100,000 for the population aged 15 and above. Income groups are defined based on 2005 values of GNI per capita in US dollars and World Bank (2024a). Guyana's data begins in 1999. Republic of Moldova and South Africa's data ends in 2018.

happened because societies experiencing high mortality rates adopted formal and/or informal strategies to prevent deaths of despair.

More generally, deaths of despair declined among both men and women independently of their age (see Table B4 in the online appendix). This was mainly due to decreasing suicides across age groups. Mortality due to alcohol abuse did not change much on average – it diminished among the working-age population but increased for those aged 60+. Among the three causes of mortality composing deaths of despair, only drug abuse shows positive growth rates on average (0.05). This trend concerns both men and women, especially in the age group 30 to 59.

Three notable patterns emerged among the countries that experienced increasing deaths of despair. The majority witnessed increasing deaths among older men due to alcohol abuse. In the US, Canada, and the UK, deaths of despair increased among men aged 30 to 59 primarily due to rising drug abuse. In the Republic of Korea, deaths of despair increased largely among men aged 60+ due to increasing suicide. These distinct patterns indicate that the mechanisms behind the rise in deaths of despair in the United States do not apply generally.

Table 6.4: Regressions determining convergence, average annual DoD change on initial level of DoD
World Health Organisation (2000–2019)

	(1) Full	(2) 15-29	(3) 30-44	(4) 45-59	(5) 60+
Initial DoD	-0.025***	-0.028***	-0.027***	-0.033***	-0.020***
	(0.003)	(0.002)	(0.003)	(0.002)	(0.005)
Observations	59	59	59	59	59
R-Squared	0.559	0.602	0.537	0.719	0.271
Initial DoD among women	-0.024***	-0.022***	-0.023***	-0.033***	-0.025***
	(0.003)	(0.003)	(0.005)	(0.004)	(0.003)
Observations	59	59	59	59	59
R-Squared	0.381	0.232	0.274	0.526	0.372
Initial DoD among men	-0.026***	-0.032***	-0.028***	-0.034***	-0.018***
	(0.003)	(0.002)	(0.004)	(0.002)	(0.005)
Observations	59	59	59	59	59
R-Squared	0.614	0.693	0.575	0.766	0.270

Note: Standard errors in parentheses (clustered by country); * $p<0.10$, ** $p<0.05$, *** $p<.01$

In sum, the available evidence from 59 countries around the world shows that deaths of despair have declined since 2000 in nearly 75% of countries. Countries like Lithuania, Latvia, and Estonia have seen the largest decreases, while deaths of despair increased in the United States, Republic of Korea, and Slovakia. Suicides are the most prevalent cause of despair-related mortality. On average, deaths of despair are nearly four times more prevalent among men than women and twice as prevalent among those aged 60+ compared to 15- to 29-year-olds. The only exception is drug overdose, which is more frequent for men and women of working age than for others.

The level of deaths of despair appears unrelated to the income level of a country. However, this is not the case for the changes over time. On average, deaths of despair declined faster in poorer countries than in rich ones. The number of suicides declined at a rate that more than offset the increases in drug- and alcohol-related deaths in lower-middle and upper-middle-income countries, but the suicide decline is lower in high-income countries. In the latter case, the average annual decrease of 0.19 deaths per 100,000 persons was almost offset by the increase in drug- and alcohol-related deaths.

Why do deaths of despair rise in some countries and fall in others? Differences in how prosocial behaviour evolved in different countries may offer an explanation. In the next section, we describe the reasons why prosocial behaviour could contribute to less deaths of despair.

Social relationships foster emotional support, provide a sense of belonging, and promote meaning, self-esteem and purpose in life which, in turn, support mental wellbeing and reduce the likelihood of engaging in harmful behaviours.

The case for prosocial behaviour

Prosocial behaviour – sometimes called acts of kindness or altruism – includes behaviours that are done for the benefit of others. It is a behavioural measure of social capital which includes trust, behaviours, norms and shared values enabling a society to cooperate to achieve common goals. For both prosocial behaviour and social capital, there are numerous studies from multiple disciplines that document their beneficial impacts on individuals and society.

Prosocial behaviour should contribute to limiting deaths of despair for numerous reasons. People who engage in prosocial behaviour are healthier[28] and happier,[29] and they experience a greater sense of purpose and meaning in life[30] as well as improved psychological flourishing.[31] Each of these aspects should reduce the risk of deaths of despair. Prosocial behaviour also strengthens and expands individuals' social networks, thereby increasing access to social support and information, which improves coping strategies that mitigate stress perception and physiological responses.[32] Importantly, prosocial behaviour may reduce deaths of despair by buffering individuals against the harmful effects of stressors and life challenges. In times of uncertainty and distress, these social networks provide practical assistance, emotional support, and advice which boost wellbeing[33] and reduce the likelihood of resorting to maladaptive coping mechanisms, such as substance abuse.[34] Finally, prosocial behaviour fosters self-esteem, which serves as a protective factor against life challenges.[35]

The benefits of prosocial behaviour extend beyond the direct effects on those engaging in them. Prosocial behaviour contributes to social capital by fostering trust in others, shared values and sense of responsibility, and cooperative norms both within civic networks and across diverse groups.[36] See Chapter 2 for additional details.

Social capital, more broadly, also affects health and wellbeing outcomes directly and indirectly, both through individual-level psychosocial mechanisms (e.g., emotional support, stress buffering, and behaviour modulation)[37] and community-level social cohesion (e.g., collective efficacy and social norms).

Social capital can directly affect individual health and wellbeing via cognitive, emotional, behavioural, and biological pathways.[38] Indeed, social relationships foster emotional support, provide a sense of belonging, and promote meaning, self-esteem and purpose in life[39] which, in turn, support mental wellbeing and reduce the likelihood of engaging in harmful behaviours.[40] In fact, the influence of social capital extends beyond individual psychosocial support as it reinforces positive community outcomes. Social norms within communities affect behavioural mechanisms by fostering accountability, encouraging health-promoting habits, and discouraging risky or harmful behaviours and the adoption of maladaptive coping mechanisms, such as substance abuse, and excessive alcohol consumption.[41] For example, previous research demonstrated that individuals embedded in strong social networks are more likely to seek help for mental health issues, which, in turn, lowers the chances of harmful behaviour.[42]

At the community level, social capital fosters life expectancy, longevity, and public health, and reduces all-cause mortality.[43] Community social capital is linked to lower death rates, including from heart disease,[44] and to lower mortality from cancer, cardiovascular disease, and suicide.[45] Furthermore, collective efficacy, i.e., perception of mutual trust and willingness to help each other, has been associated with positive societal outcomes including reduced rates of assaults, homicide, premature mortality, and asthma.[46]

Changes in prosocial behaviour

In 2019, on average, 31.1% of respondents engaged in prosocial behaviour, with 45.3% stating that they helped a stranger, 29.9% donated money, and 18.2% volunteered in groups or associations (see Table 6.5). These data are sourced from the Gallup World Poll (GWP) and refer to the population aged 15 and over. Our measure of prosocial behaviour is the average of the shares of respondents answering "yes" to each of the three components: donating money, volunteering, and helping strangers.[47] Our sample includes 50 countries and covers the period 2005–06 to 2019 when data on prosocial behaviour and deaths of despair are available. See Tables B1–B3 in the online appendix for details.

Prosocial behaviour is more frequent in high-income countries where 35.7% of the population

Table 6.5: Summary data for prosocial behaviour (PSB) in 50 countries
Gallup World Poll (2019)

Income Groups	Countries	Mean	Std. dev.	Min	Max
Lower Middle	10	31.41	5.24	25.02	40.39
Upper Middle	16	23.91	6.67	11.86	33.93
High	24	35.74	9.87	13.19	53.47
Total	**50**	**31.09**	**9.58**	**11.86**	**53.47**

Income Groups	Countries	PSB	Donation	Volunteer	Helped
Lower Middle	10	31.41	24.10	17.79	52.43
Upper Middle	16	23.91	19.21	12.23	40.40
High	24	35.74	39.38	22.24	45.58
Total	**50**	**31.09**	**29.87**	**18.15**	**45.29**

Note: Income groups are defined based on 2005 values of GNI per capita in US dollars and World Bank (2024a). Countries using 2018: Republic of Moldova and South Africa.

reported that they engaged in prosocial behaviour, mainly helping others (45.6%) or donating money (39.4%). Helping others is the main form of prosocial behaviour in lower-middle-income countries where more than 50% reported helping others.

Over the period 2005–6 to 2019, engagement in prosocial behaviour decreased in most countries. We exclude the years after 2019, which are marked by rising prosocial behaviours (see Chapter 2) due to the limited availability of mortality data and because the pandemic might have affected prosocial behaviours and deaths of despair in exceptional ways. Figure 6.3 presents the average yearly changes in prosocial behaviour by country and world region. Most changes are below 1 percentage point per year in absolute value. The countries where prosocial behaviour decreased at a faster rate are Czechia, Switzerland, Belgium, and Japan. Prosocial behaviour decreased in most Western European countries, in some Latin American countries, in Japan, the Philippines, and the Republic of Korea, as well as in North America and Australia.

Figure 6.3: Annual average changes in prosocial behaviour
Gallup World Poll (2005-06 to 2019)

Note: Countries using 2018: Republic of Moldova and South Africa.

The average change is -0.23 percentage points per year (see Table 6.6). However, the decrease was steeper in high-income countries, where engagement decreased by -0.45 percentage points per year, compared to upper-middle-income countries, where the rate of decrease was -0.14. In both cases, the decrease was prevalently associated with a decrease in donating money which, in high-income countries, proceeded at a rate of -0.99%, followed by decreasing engagement in volunteering activities and helping others. This general decrease in prosocial behaviour is consistent with findings from earlier studies that documented a long-term decrease of social capital, prevalently in industrialised countries.[48]

Prosocial behaviour increased by 0.13 percentage points per year in lower-middle-income countries. However, this increase masks two contrasting trends: on one hand, an increase in the share of people helping others (0.56); on the other hand, a decrease in volunteering (-0.15).

The decrease in prosocial behaviour is about two times larger for women than for men, but this difference occurs to varying degrees across age groups and behaviours (see Table B5 in the online appendix). The most striking gender difference concerns people helping others. The modest increase in the share of people helping others (0.05) is driven entirely by men. Among women, helping others declined slightly on average, especially among women aged 60+.

Table 6.6: Average annual change in prosocial behaviour (PSB)
Gallup World Poll (2005–06 to 2019)

Income Groups	Countries	Mean	Std. dev.	Min	Max
Lower Mid	10	0.13	0.55	-0.75	0.87
Upper Mid	16	-0.14	0.64	-1.55	0.78
High	24	-0.45	0.55	-1.42	0.75
Total	**50**	**-0.23**	**0.61**	**-1.55**	**0.87**

Income Groups	Countries	PSB	Donation	Volunteer	Helped
Lower Mid	10	0.13	-0.01	-0.15	0.56
Upper Mid	16	-0.14	-0.30	-0.14	-0.02
High	24	-0.45	-0.99	-0.27	-0.11
Total	**50**	**-0.23**	**-0.58**	**-0.21**	**0.05**

Note: Income groups are defined based on 2005 values of GNI per capita in US dollars and World Bank (2024a). Countries using 2018: Republic of Moldova and South Africa.

In sum, engagement in prosocial behaviour decreased in 30 out of 50 countries over the period 2005–06 to 2019. This decrease was observed for both men and women of all ages and is the result of a general decrease in donations and volunteering, especially among upper-middle and high-income countries. Do these changes help us understand how deaths of despair have changed over time? In particular, does prosocial behaviour buffer against despair leading to death?

Evidence from three countries (the United States, Republic of Korea, and Finland) provides a starting point to address these questions.

The US and Korea stand out for their high and rapidly increasing deaths of despair, while Finland is notable for high but decreasing deaths of despair. Indeed, the pace of increasing deaths of despair in the US and Korea is approximately the opposite of the decrease in Finland (see Figure 6.2). One potential explanation for this contrast is differing prosocial behaviour trends. In both the US and Korea, prosocial behaviour declined over time, whereas in Finland, it increased (see Figure 6.3). This pattern strengthens the hypothesis that prosocial behaviour plays a role in reducing deaths of despair, which is the subject of the next section.

Explaining deaths of despair over time: the role of prosocial behaviour

In our sample of 50 countries, regression analysis confirms that increasing prosocial behaviour correlates with decreasing deaths of despair over time. Donations are the most important component of prosocial behaviour and they have a larger and more precisely estimated impact on men compared to women, and on older compared to younger people. The regression analysis allows us to account for a host of relevant control variables as alternative explanations, such as religiosity and poverty. Prosocial behaviour maintains a similar relationship in all instances.

Estimation technique and control variables

We estimate the relationship between prosocial behaviour and deaths of despair using a standard regression framework which simultaneously considers the influence of multiple variables including prosocial behaviour. This approach allows us to isolate the relationship between prosocial behaviour and deaths of despair from the influence of multiple explanatory factors. For instance, deaths of despair could be influenced by economic, demographic, governmental, and societal characteristics. To this end, we consider as many variables as possible that might confound the relationship. We begin with the full set of variables with adequate data and then reduce the set using a standard variable selection technique.[49]

We considered the following economic factors: GDP per capita, GDP per capita growth rate, inflation rate, unemployment rate, and labour force participation rate (separately defined by gender). During periods of recessions, with declining purchasing power and rising unemployment, people experience more financial pressure and suicides tend to rise. According to Galbraith (2009), during the early years of the Great Depression (1929-1932), suicides increased from 17.0 to 21.3 per 100,000 in the United States. The labour force participation rate is the population share that is either working or seeking work. Decreases for men correspond, in part, with people giving up on seeking work due to factors such as bad health,[50] discouragement, and desperation.

We included the population age structure, measured using population shares for ages 30–44, 45–59, 60–64, and 65+[51]; the female population share; divorce rate; and degree of urbanisation. The population structure accounts for the varying age and gender compositions of countries and the varying degrees to which they engage in prosocial behaviour or die from deaths of despair. For instance, we observe that those aged 60–64 are much more likely to engage in prosocial behaviour, especially donating, than those aged 65+, which is why we distinguish them in the analysis.[52] The female population share, divorce rate, and degree of urbanisation are included as deaths of despair are greater among men (see Table 6.2), in communities with greater divorce rates,[53] and in rural areas.[54]

We also considered religious importance, quality of governance, health expenditures, and the three variables from the *World Happiness Report* that are used to explain differences in life evaluations around the world, specifically: support, freedom, and perceptions of corruption. Religiosity is an important control variable because, in many cases, religions forbid suicides and discourage the use of drugs. Religiosity is also often associated with volunteering, donating money, and a high propensity to help strangers. Quality of governance could affect the resources available and opportunities for societal improvement. Religiosity and quality of governance have also previously been found to be associated with suicide.[55] Health expenditures capture the quality of health care available to individuals. Feelings of support, having someone to count on, should alleviate feelings of despair; freedom could help one climb out of a bad place. Likewise, a lack of freedom corresponds to a lack of agency and despair. Perceived corruption operates somewhat similarly to freedom and the opposite of quality of governance. If the system is perceived to be corrupt, then individuals may believe they have less ability to affect their community and own lives.

Lastly, we included indicator variables for each country and year, so-called country and year fixed effects. Country fixed effects account for all fixed characteristics of a country including, for instance, latitude, which is predictive of climate and suicide[56]. When including country fixed effects, the estimated relations correspond to changes within countries over time, not differences across countries. Year fixed effects account for any common trends over time that exist across countries. See Online Appendix C for a list of data sources.

Increasing prosocial behaviour is associated with decreasing deaths of despair

The results of rigorous analyses – in which we isolate the relation between changing prosocial behaviour and deaths of despair from fixed country characteristics and the contemporaneous effects of numerous control variables – make us confident that prosocial behaviour does indeed contribute to decreasing deaths of despair. Our results are presented in Table 6.7.[57] The negative coefficient means that when prosocial behaviour increases within a country by one percentage point, we can expect a decrease in deaths of despair in the same country by 0.096 deaths per 100,000. In the case of the United States, with a population aged 15+ of nearly 270 million, this amounts to about 260 persons for each percentage point.

Columns 2 to 4 refine the analysis by checking how each of the three prosocial behaviour components explains deaths of despair. Donating money, volunteering, and helping a stranger are all associated with lower levels of deaths of despair, but only donating money is statistically different from zero. This could be because donating money is the only component of prosocial behaviour that contributes to decreasing deaths of despair, or because the data do not provide enough statistical power to identify significant effects for all three variables. Either way, considering the beneficial role played by prosocial behaviour in general, and by donating money in particular, decreasing engagement observed in many high-income countries until 2019 is a cause of concern.

The variable selection process, discussed in footnote 48, reduced our set of explanatory variables (including prosocial behaviour) from the 20 discussed above to the 7 (excluding fixed effects) presented in Table 6.7. Perhaps surprisingly, variables such as religiosity, GDP, and unemployment were not among the most important variables. This may be because country fixed effects account for all country characteristics that do not change over time, and even some factors that do not vary much or frequently. Thus, the impact of religiosity on deaths of despair is accounted for in our model, to the extent that religiosity remains stable. GDP and unemployment vary over time more than religiosity but also vary along with the other variables, especially the economic ones that were retained, i.e., inflation and labour force participation. The remaining variables, together with the fixed effects, account for the influences of those that were dropped.[58] In any case, the analysis is not intended to uncover the causal impacts of the control variables (e.g., inflation and support) and, for this reason, we do not recommend interpreting them causally.

In addition to the 20 previously mentioned variables, we tested five more variables that were left out of the previous analysis because they have reduced data coverage, specifically: the poverty rate, income inequality, public social expenditures, population share living alone, and the population share with at least secondary education (separately by gender). The poverty rate – measured as the population share that earns less than $3.65 a day (adjusted for purchasing power) – is the population share considered to earn too little to meet basic material needs in lower-middle-income countries. Income inequality – measured using the Gini coefficient – is often connected with a sense of unfairness and distrust, which in turn reduces one's sense of control over their life. Public social expenditures represent the social safety net, e.g., expenditures on health, unemployment benefits, and child and elderly care. The population share living alone reflects one form of social isolation, which could be related to loneliness and despair (see Chapter 4). The population share with at least upper-secondary education could be important as deaths of despair tend to be greater among less educated groups in the United States.[59] Tables B7 and B8 in the online appendix show respectively the results of regressions with the full set of control

Table 6.7: Fixed effects regressions predicting deaths of despair using prosocial behaviour, underlying components, and control variables

	(1)	(2)	(3)	(4)
Prosocial	-0.096*			
	(0.050)			
Donations		-0.093**		
		(0.041)		
Volunteering			-0.040	
			(0.048)	
Helped Stranger				-0.033
				(0.029)
Labor Force Part. Male	-0.727***	-0.680***	-0.732***	-0.744***
	(0.222)	(0.210)	(0.226)	(0.229)
Inflation Rate	0.078	0.077	0.083	0.076
	(0.067)	(0.066)	(0.066)	(0.067)
Pop. Share 65+	-1.773***	-1.792***	-1.724***	-1.734***
	(0.631)	(0.625)	(0.629)	(0.625)
Pop. Share 60-64	-0.944	-0.831	-1.019	-0.988
	(0.925)	(0.894)	(0.918)	(0.911)
Support	-0.153*	-0.148*	-0.163*	-0.155*
	(0.078)	(0.076)	(0.083)	(0.081)
Corruption	0.130**	0.125**	0.125**	0.129**
	(0.050)	(0.049)	(0.050)	(0.051)
Constant	103.316***	99.881***	102.330***	102.563***
	(17.267)	(15.937)	(17.334)	(17.107)
Observations	620	620	620	620
# of Countries	50	50	50	50
R-Squared	0.291	0.300	0.283	0.284

Note: Regressions of deaths of despair in levels on indicated variables and fixed effects for year and country. Standard errors in parentheses (clustered by country); * $p<0.10$, ** $p<0.05$, *** $p<0.01$

Table 6.8: Fixed effects regressions of deaths of despair in different population subgroups on prosocial behaviour and control variables

	Full	15-29	30-44	45-59	60+	Full	15-29	30-44	45-59	60+
Prosocial	-0.096*	-0.025	-0.066	-0.175**	-0.208**					
	(0.050)	(0.038)	(0.054)	(0.081)	(0.103)					
Donations						-0.093**	-0.044	-0.064	-0.164**	-0.178***
						(0.041)	(0.034)	(0.043)	(0.074)	(0.062)
R-Squared	0.291	0.196	0.266	0.384	0.189	0.300	0.202	0.270	0.392	0.196
Women	**Full**	**15-29**	**30-44**	**45-59**	**60+**	**Full**	**15-29**	**30-44**	**45-59**	**60+**
Prosocial	-0.032	-0.004	-0.008	-0.043	-0.101					
	(0.026)	(0.017)	(0.023)	(0.033)	(0.082)					
Donations						-0.034*	-0.008	-0.012	-0.056*	-0.095*
						(0.017)	(0.012)	(0.017)	(0.033)	(0.051)
R-Squared	0.239	0.092	0.163	0.322	0.200	0.244	0.092	0.164	0.330	0.205
Men	**Full**	**15-29**	**30-44**	**45-59**	**60+**	**Full**	**15-29**	**30-44**	**45-59**	**60+**
Prosocial	-0.165*	-0.044	-0.125	-0.321**	-0.330**					
	(0.085)	(0.065)	(0.094)	(0.142)	(0.145)					
Donations						-0.161**	-0.079	-0.123	-0.290**	-0.286***
						(0.074)	(0.063)	(0.077)	(0.128)	(0.086)
R-Squared	0.289	0.207	0.278	0.377	0.185	0.298	0.214	0.283	0.385	0.194
Observations	620	620	620	620	620	620	620	620	620	620
# of Countries	50	50	50	50	50	50	50	50	50	50

Note: Fixed effect regressions of deaths of despair on prosocial behaviour and: male labour force participation rate, inflation rate, population share 60-64, and population share 65+, social support, perceptions of corruption, and fixed effects for year and country. Time series sample, 620 observations for 50 countries. Standard errors in parentheses (clustered by country); *p<0.10 **p<0.05 ***p<0.01

variables and with the additional five ones just described. Across nearly all of the regressions, prosocial behaviour maintains a statistically significant and negative relation with a broadly similar magnitude.

Additional analysis reveals differences between the average relationship for the full population and for population subgroups. Columns 1–5 in Table 6.8 summarise the results of the same model presented in column 1 of Table 6.7 but this time for men and women of various age groups. As in the previous results, greater prosocial behaviour is associated with fewer deaths of despair, but the relation is only statistically significant for men in later stages of life. In these cases, the coefficient magnitudes are larger than for the full population. For instance, increasing prosocial behaviour by 1 percentage point would reduce deaths of despair among people aged 60+ by 0.21 people per 100,000 (compared with 0.096 for the full population); and the relation for men is more than three times larger than for women. We also replicate the analysis using donations, from column 2 of Table 6.7. The results, presented in columns 6–10 of Table 6.8, indicate donations are likewise negatively related to deaths of despair, and statistically significant for both men and women in older ages.

Social capital and deaths of despair

In this section, we complement the previous analysis using social trust and group membership.[60] These are two widely used measures of social capital and we expect that their increase should be associated with lower deaths of despair within countries for the reasons discussed earlier.[61] Stated briefly, societies with higher trust in others or more participation in groups and associations should be more cohesive than others and therefore better equipped to look after those who fall behind. Group membership also provides a sense of belonging and community to those who participate in it, and in many cases, a service to the broader society.

Trust and group membership data come from the integrated World Values Survey and European Values Study dataset, which allows us to consider a longer time period. We include all the countries with at least three waves of observations between waves 2 and 7 (1991–2019) and the necessary mortality data. Table B6 in the online appendix provides summary statistics.

We use the same analytical technique as the previous section and the same reduced set of control variables as presented in Table 6.7. However, we exclude the controls sourced from the Gallup World Poll (social support and perceptions of corruption) which are unavailable before 2005.

The results are broadly consistent with the previous ones. Both trust and membership are negatively related to deaths of despair, although the coefficients are not statistically different from zero (see Tables 6.9 and 6.10). This could be because there are fewer observations and countries, which reduces estimation precision. Indeed, the estimates are fairly similar across social capital measures. An increase in prosocial behaviour is associated with 0.096 fewer deaths per 100,000 (Table 6.7), while increases in trust and membership are associated with 0.095 (Table 6.9) and 0.074 (Table 6.10) fewer deaths per 100,000.

While no statistically significant relationships are observed in the full population, distinct patterns emerge within specific gender and age groups. For social trust, the magnitudes are larger for men compared to women and for the upper-middle-aged compared to other age groups. Indeed, for upper-middle-aged individuals, increasing social trust is significantly associated with a decrease in deaths of despair for both men and women. An increase in membership correlates with fewer deaths of despair in the full sample of women on average. For men, an increase in membership is significantly linked to a reduction in deaths of despair within the youngest age group.

Table 6.9: Relation between deaths of despair and trust, dependent variable: deaths of despair
WVS-EVS (1991–2019)

	(1) All ages	(2) 15-29	(3) 30-44	(4) 45-59	(5) 60+
Female & Male					
Social trust	-0.095	-0.025	-0.104	-0.316*	0.019
	(0.114)	(0.103)	(0.135)	(0.16)	(0.202)
R-squared	0.205	0.144	0.321	0.313	0.09
Female					
Social trust	-0.071	-0.033	-0.055	-0.148*	-0.077
	(0.053)	(0.05)	(0.05)	(0.078)	(0.152)
R-squared	0.181	0.053	0.23	0.255	0.127
Male					
Social trust	-0.142	-0.029	-0.175	-0.520*	0.085
	(0.194)	(0.183)	(0.235)	(0.28)	(0.299)
R-squared	0.225	0.183	0.339	0.318	0.11
Observations	154	154	154	154	154
# of countries	40	40	40	40	40

Note: Fixed effects regressions of deaths of despair on population share 60–64, population share 65+, male labour force participation, inflation rate, and wave dummies. Standard errors in parentheses (clustered by country); * $p<0.10$, ** $p<0.05$, *** $p<0.01$

How do deaths of despair relate to (un)happiness?

Although Northern European countries typically top the rankings in life evaluations, many of them (e.g., Finland) also experience relatively high rates of deaths of despair (see Figure 6.1). This seeming contradiction has been referred to as a puzzle in the past[62] and requires an explanation. In this section, we find that such countries are exceptions and we discuss the differences between deaths of despair and life evaluations that could give rise to these exceptional cases.

Figure 6.4 presents the relationship between annual, within-country changes in life evaluations and deaths of despair over time. The trend line illustrates increasing life evaluations are associated with decreasing deaths of despair. Generally, high life evaluations do *not* go with high deaths of despair. At the same time, the dispersion in dots indicates that the variables are not simply the opposite of each other. In fact, changing life evaluations explain less than 10% of changing deaths of despair. This inexact match is what empirically allows for countries like Finland to be high on both.

Table 6.10: Relation between deaths of despair and membership, dependent variable: deaths of despair
WVS-EVS (1991–2019)

	(1) All ages	(2) 15-29	(3) 30-44	(4) 45-59	(5) 60+
Female & Male					
Any membership	-0.074	-0.064	-0.088	-0.074	-0.097
	(0.058)	(0.04)	(0.058)	(0.083)	(0.104)
R-squared	0.213	0.163	0.329	0.296	0.099
Female					
Any membership	-0.053*	-0.015	-0.032	-0.052	-0.125
	(0.03)	(0.018)	(0.025)	(0.037)	(0.089)
R-squared	0.194	0.052	0.232	0.247	0.15
Male					
Any membership	-0.097	-0.115*	-0.152	-0.108	-0.058
	(0.096)	(0.068)	(0.097)	(0.143)	(0.135)
R-squared	0.229	0.204	0.348	0.3	0.111
Observations	154	154	154	154	154
# of countries	40	40	40	40	40

Note: Fixed effects regressions of deaths of despair on population share 60-64, population share 65+, male labour force participation, inflation rate, and wave dummies. Standard errors in parentheses (clustered by country); * p<0.10, ** p<0.05, *** p<0.01

Although conceptually related, deaths of despair are distinct from life evaluations. Deaths of despair, as suggested by the name, result from negative feelings of despair, whereas life evaluations represent a cognitive evaluation of how one feels about their life. Negative feelings and cognitive evaluations correlate but are not simply the opposite of each other.[63] Furthermore, despair is more forward-looking than current negative feelings. Consequently, it is even less related to life evaluations conceptually than many other negative feelings.

We use the same technique presented in the previous section to analyse the empirical relation between deaths of despair and life evaluations. The presence of distinct patterns in their associations with our explanatory variables will help us pinpoint where deaths of despair and life evaluations differ. For this purpose, we expand the list of explanatory variables from the model presented in Table 6.7 by adding back GDP per capita and satisfaction with freedom, which were dropped in our variable selection process for deaths of despair, but are important in the

Figure 6.4: Within-country changes in deaths of despair and life evaluations over time

World Health Organization & Gallup World Poll (2006–2019)

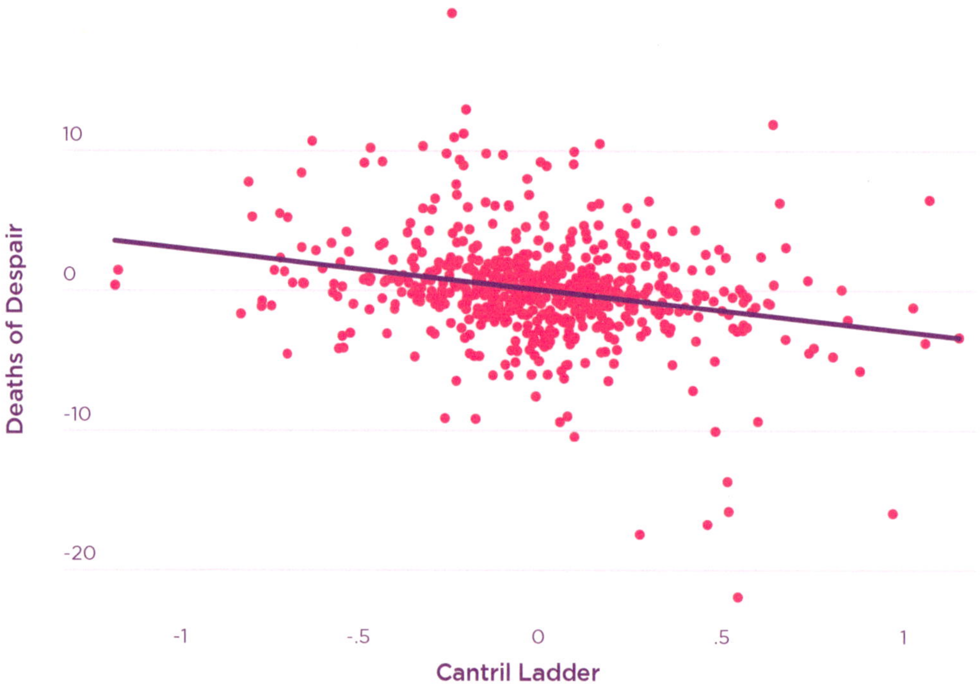

Note: The values represent annual deviations from the within-country mean-level of each variable. As such, the trend line illustrates the correlation between deaths of despair and life evaluations when controlling for country fixed effects.

World Happiness Report for life evaluations. We first assess prosocial behaviour as an explanatory variable and then one of its components, donations, because the *World Happiness Report* typically explains life evaluations using, among others, donations (labelled as generosity).

To allow comparison across variables, we report the results in Table 6.11 using standardised variables, each adjusted to a common scale in line with standard practice. This is why the coefficient of prosocial behaviour in the first column of Table 6.11 is different from the coefficient in Table 6.7. The two coefficients are from the same model, except that the one in Table 6.11 comes from a standardised variable. Table 6.11 shows that, in standardised terms, donations have equivalent relations with both deaths of despair and life evaluations.[64]

The first thing to note is that the variables explain life evaluations better than deaths of despair. The R-squared, a measure of how well the model fits the data, is 0.40 for the Cantril Ladder and 0.29 for deaths of despair (see the bottom of columns 1 and 2).

Table 6.11: Fixed effects regressions of deaths of despair and life evaluations using prosocial behaviour and control variables

	(1) DoD	(2) Cantril Ladder	(3) DoD	(3) Cantril Ladder
Prosocial	-0.076**	0.038		
	(0.048)	(0.369)		
Donations			-0.129**	0.122**
			(0.019)	(0.017)
Labor Force Part. Male	-0.340***	0.242**	-0.317***	0.220*
	(0.003)	(0.042)	(0.003)	(0.063)
Inflation Rate	0.032	-0.016	0.031	-0.015
	(0.235)	(0.559)	(0.235)	(0.571)
Pop. Share 60-64	-0.091	-0.091	-0.079	-0.105
	(0.344)	(0.229)	(0.392)	(0.167)
Pop. Share 65+	-0.714***	-0.014	-0.723***	0.004
	(0.006)	(0.954)	(0.005)	(0.989)
Support	-0.071*	0.213***	-0.067*	0.210***
	(0.088)	(0.007)	(0.091)	(0.005)
Corruption	0.202***	-0.186**	0.194**	-0.184**
	(0.009)	(0.015)	(0.011)	(0.011)
ln(GDP pc)	-0.050	0.938***	-0.046	0.953***
	(0.831)	(0.004)	(0.843)	(0.003)
Freedom	-0.031	0.124**	-0.035	0.118**
	(0.488)	(0.020)	(0.442)	(0.023)
Observations	620	620	620	620
# of Countries	50	50	50	50
R-Squared	0.293	0.400	0.302	0.409

Note: Regressions of deaths of despair and Cantril Ladder in levels on indicated variables and fixed effects for year and country. Standard errors in parentheses (clustered by country); * $p<0.10$, ** $p<0.05$, *** $p<0.01$

The results also reveal deaths of despair are more sensitive to prosocial behaviour and the population structure, especially the population share aged 65+. Life evaluations are more sensitive to GDP per capita, freedom, and social support. In common, they have the male labour force participation rate (although of somewhat different magnitudes) and perceptions of corruption. The inflation rate and population share aged 60–64 are not statistically significant for either outcome.

In sum, while it is possible for countries to have high deaths of despair and high life evaluations, these are exceptions. Conceptually, deaths of despair are related to life evaluations, but also quite distinct. The former is a behaviour connected to hopelessness, while life evaluations are cognitive evaluations of one's current life as a whole. Empirically, their determinants are similar but not the same. In any case, increasing prosocial behaviours, specifically donations, improves both outcomes, reducing deaths of despair and increasing life evaluations.

While it is possible for countries to have high deaths of despair and high life evaluations, these are exceptions.

Conclusion

Deaths of despair - due to suicide, alcohol abuse, and drug overdose - generally declined from 2000 to 2019 in our sample of 59 countries, based on data from the World Health Organization. However, the number is still high and has risen in some cases, notably in the United States and Republic of Korea. On average, 23 per 100,000 people died from deaths of despair in 2019. One factor that can contribute to reducing deaths of despair is prosocial behaviour. Our analysis indicates that a ten percentage-point increase in the share of people engaging in prosocial behaviour is associated with approximately 1 fewer death per 100,000 people per year. For a country like the United Kingdom, which has a 15-and-older population of approximately 55 million, that is equivalent to about 550 people per year.[65]

Deaths of despair are not equally distributed across and within countries. In our sample, deaths in 2019 were fewest in Grenada and greatest in Slovenia. However, differences in cultural and institutional characteristics call for caution when comparing deaths of despair across countries. Within countries, deaths of despair were nearly four times higher among men than women; more than double among those aged 60+ compared to 15- to 29-year-olds; and were primarily due to suicide, which accounted for three out of every four deaths.

Prosocial behaviour generally decreased between 2006 and 2019. According to Gallup World Poll data, volunteering and donating money decreased, especially in high-income countries. Only the share of people helping others showed signs of growth, but this was limited to lower-middle-income countries. The general decrease in prosocial behaviours concerned men and women of all age groups in a similar manner. However, as shown in Chapter 2, there are signs of increasing prosocial behaviours since 2019, after the period of our analysis. Since data on deaths of despair end in 2019, we do not know whether these increases are associated with declining deaths of despair.

We extended the regression analysis to two other common measures of social capital: social trust and group participation. Sourced from the integrated European Values Study and World Values Survey dataset, these variables are available for a longer period of time, but for a smaller sample of countries compared to our analysis of prosocial behaviours. Results indicate similar negative within-country correlations between the two measures of social capital and deaths of despair but, in general, they were not statistically significant.

There are two aspects of our analysis worth emphasising. The first is that the technique we used evaluates the association between variables by matching their within-country changes over time. This limits concerns of international comparability which are a major problem in the analysis of mortality by cause. The second aspect is that the negative relationship between prosocial behaviour and deaths of despair holds after considering the contemporaneous changes of other variables, such as economic growth, unemployment rate, labour force participation, religiosity, and the share of elderly people, among others. This evidence is consistent with the hypothesis that prosocial behaviour constitutes an informal safety net whose benefits go beyond the donors and recipients,[66] and provides a buffer against adversities.[67]

We also investigated how a country could have both high deaths of despair and high life evaluations. First, deaths of despair and life evaluations reflect two related but distinct concepts. Second, our empirical results indicate that different factors contribute to each variable. Deaths of despair are more sensitive to the population structure, while life evaluations are more sensitive to GDP per capita, freedom, and social support. However, both variables are related to components of prosocial behaviour. In particular, a rise in donations similarly increases life evaluations as it decreases deaths of despair.

Although much of the previous research on deaths of despair has focussed on the United States, the trends documented in this chapter illustrate that the US is not representative of global trends. In the US and, to a lesser extent, Canada and the United Kingdom, deaths of despair rapidly increased from 2000 to 2019, especially among men aged 30 to 59, due to increasing drug abuse. However, most countries experienced decreasing deaths of despair over this period. Even among the countries that did experience rising deaths of despair, the pattern generally differed from the US. In these countries, the rise was due largely to increasing alcohol abuse among men aged 45+. The Republic of Korea exhibited yet another trend, experiencing a rise in suicides among older men aged 60+. The existence of such unique trends motivates additional research on deaths of despair around the world.

It is well established that prosocial behaviour contributes to individual wellbeing and to societies in which people are more supportive, cooperative, and trusting. This chapter further demonstrates that increasing prosocial behaviour is reliably connected to decreasing deaths of despair. Societies could benefit from investing in the conditions supporting prosocial behaviour.

Endnote

1 Case and Deaton (2015, 2017).

2 While preparing this chapter, a new article came out documenting global trends of deaths of despair. Shirzad et al. (2024) obtain some similar results describing the trends, but their aim and analyses differ.

3 Daly and Macchia (2023).

4 Allik et al. (2020); Karanges et al. (2016); Peacock et al. (2018); Roxburgh et al. (2017).

5 Blanchflower and Oswald (2020).

6 Graham and Pinto (2019, 2021); O'Connor and Graham (2019).

7 Case and Deaton (2017).

8 Case and Deaton (2020a).

9 Friedman et al. (2020).

10 Friedman et al. (2020).

11 Diez Roux (2017); Siddiqi et al. (2019).

12 Case and Deaton (2017).

13 Zeglin et al. (2019).

14 Allik et al. (2020); Brown et al. (2019).

15 Knapp et al. (2019).

16 Helliwell (2006, 2007).

17 We present rates of deaths of despair as the number of deaths per 100,000 for the population aged 15 and above because this allows us to compare countries of different population sizes. Throughout the chapter, we will use this normalisation. Any reference to the number of deaths of despair is per 100,000.

18 23 ÷ 100,000 x 55,000,000 ≅ 12,500. The population aged 15 and above in the UK is approximately 55 million.

19 South Africa had fewer deaths of despair than Grenada. However, the level of deaths of despair in South Africa were not very comparable, because we had to adjust the figures for South Africa. Our measure of suicides generally includes deaths due to undetermined intent, in accordance with the literature as described in Online Appendix A. However, due to a break in the series for South Africa, we dropped deaths due to undetermined intent for South Africa, which mechanically reduces their levels of deaths. This adjustment should not greatly affect the changes in deaths of despair over time for South Africa, and has thus been left in the rest of the sample.

20 Data are from the year 2019 for most countries. However, Republic of Moldova uses the most recent available year, 2018. We focus on the pre-COVID-19 period due to data availability and because the pandemic might have affected prosocial behaviours and deaths of despair in exceptional ways.

21 The WHO Mortality Database includes annual mortality statistics by sex, age, and harmonised causes of death. Only countries that record at least 65% of deaths with a specific cause, classified according to International Classification of Diseases (ICD) were included by the WHO. The alternative WHO Health Inequality Data Repository (https://whoequity.shinyapps.io/heat/) includes similar cause of death data by sex, in five year intervals from 2001 to 2021, and with more countries than the WHO Mortality Database. We use the WHO Mortality Database for two reasons. First, the statistics from both databases stem from the same source, i.e., the WHO Global Health Estimates. However, the WHO Mortality Database only retains the countries with reliable data (WHO 2020). Second, the WHO Health Inequality Data Repository does not disaggregate by age nor include annual data.

22 For instance, Camacho et al. (2024).

23 See footnote 21.

24 Mikkelsen et al. (2015).

25 Snowdon and Choi (2020).

26 1,045 deaths per 100,000 is the average value of deaths from all causes across countries in 2019, using the same sample of countries and data from WHO Mortality Database on all cause mortality.

27 In economics, convergence refers to the observation that countries with low initial Gross Domestic Product (GDP) exhibit faster rates of GDP growth. This concept has been recently extended to changes of social capital over time (Sarracino and Mikucka, 2017).

28 Kumar et al. (2012).

29 Aknin et al. (2012); Dunn et al. (2008); Curry et al. (2018); Rowland and Curry (2019); Hui et al. (2020).

30 Thoits and Hewitt (2001); Musick and Wilson (2003).

31 Nelson et al. (2016).

32 Cohen and Wills (1985).

33 Helliwell et al. (2014).

34 Thoits (1986).

35 Thoits and Hewitt (2001).

36 Putnam (2000).

37 Berkman (1995); Cohen and Wills (1985); Kawachi and Berkman (2001); Cohen (2004); Thoits (2011a); Kawachi et al. (2008); Berkman et al. (2000); Helliwell and Putnam (2004); Helliwell (2007); Helliwell et al. (2014).

38 Cohen and Wills (1985).

39 Thoits (1985); Thoits (2011a); Berkman (1995).

40 Kawachi and Berkman (2001).

41 Thoits (2011b); Umberson et al. (2010).

42 Pescosolido (1992).

43 Cohen and Wills (1985); House et al. (1988); Chuang et al. (2015); Ronnerstrand (2014); Lynch et al. (2000); Nyqvist et al. (2014); Holt-Lunstad et al. (2010); Kim et al. (2011); Kennedy et al. (1998).

44 Lochner et al. (2003).

45 Islam et al. (2008).

46 Cohen et al. (2008).

47 Each component is assessed through responses to the question: "In the past month, have you done any of the following? A. Donated money to a charity; B. Volunteered your time to an organisation; C. Helped a stranger or someone you didn't know who needed help." For a discussion of prosocial behaviour more generally, see Chapter 2.

48 Putnam (2000); Costa and Kahn (2001); Bartolini et al. (2013).

49 We have identified too many relevant variables with overlapping concepts. As opposed to subjectively choosing a subset of variables, we perform a data-driven variable selection procedure. We begin our analysis including each variable, then we sequentially drop the least statistically significant variable one at a time and rerun the regression. We continue this procedure until all of the retained variables reach a specified level of statistical significance (with t-statistics above 1.00). Through this procedure, we reduced our initial set of 20 explanatory variables (including prosocial behaviour) to a set of 7 without a significant drop in explanatory power. This analysis is presented in Table B7 in the online appendix. For robustness, we also used the alternative lasso approach for variable selection and retained a similar set of variables.

50 Graham and Pinto (2021).

51 We omit the population shares 0–14 and 15–29 because it is necessary to omit one group to estimate the model and our full sample is based on ages 15+.

52 These groups also tend to have different labour market statuses, which is important to account for in order for labour force participation rate to capture the mechanism that we intend it to. Decreases in the participation rate conditional on the population structure are more likely to capture discouraged workers than students or retirees.

53 Helliwell (2006).

54 Graham and Pinto (2019).

55 Helliwell (2007).

56 For a discussion of latitude and suicide, see Helliwell (2007).

57 The full set of controls before variable selection are in Table B7 in the online appendix.

58 As evidenced by the adjusted R-squareds presented in Table B7 in the online appendix. They are nearly the same in the regression with 20 variables as in the regression with 7.

59 Case and Deaton (2022).

60 Social trust is the share of people (0–100) who respond "most people can be trusted" to the following question: "Generally speaking, would you say that most people can be trusted or that you need to be very careful in dealing with people?" Group membership is the share of people (0-100) who are a member of any of the following types of organisation: (1) church or religious organisation, (2) sport or recreational organisation, (3) art, music or educational organisation, (4) environmental organisation.

61 As was found in earlier studies of suicides using earlier samples of WVS data, and reported in Helliwell (2006, 2007) and Helliwell and Wang (2011).

62 Helliwell (2007).

63 OECD (2013); Kapteyn et al. (2015).

64 This replicates the results using earlier WVS data, where both memberships (and also social trust) were found to have the same standardised effects in equations for life satisfaction and suicide (Helliwell 2007, Figure 5).

65 Additional factors may simultaneously explain deaths of despair in certain countries, such as those discussed in the introduction (e.g., social and economic marginalisation). The focus in this chapter has been on prosocial behaviour, the relation of which we isolated from plausible alternative explanations (including numerous economic and social variables) to provide a more reliable estimate of its impact. We encourage researchers to explore additional explanations in future analysis.

66 Frey and Meier (2004); Shang and Croson (2009).

67 Raposa et al. (2016); Sin et al. (2021).

References

Aknin, L. B., Dunn, E. W., & Norton, M. I. (2012). Happiness runs in a circular motion: Evidence for a positive feedback loop between prosocial spending and happiness. *Journal of Happiness Studies*, *13*, 347-355.

Allik, M., Brown, D., Dundas, R., & Leyland, A.H. (2020). Deaths of despair: cause-specific mortality and socioeconomic inequalities in cause-specific mortality among young men in Scotland. *International Journal for Equity in Health*, *19*, 215

Azevedo, J.P. (2011) "wbopendata: Stata module to access World Bank databases," Statistical Software Components S457234, Boston College Department of Economics. http://ideas.repec.org/c/boc/bocode/s457234.html

Bartolini, S., Bilancini, E., & Pugno, M. (2013). Did the decline in social connections depress Americans' happiness?. *Social Indicators Research*, *110*, 1033-1059.

Berkman, L. F., Glass, T., Brissette, I., & Seeman, T. E. (2000). From social integration to health: Durkheim in the new millennium. *Social Science & Medicine*, *51*(6), 843–857.

Berkman, L. F. (1995). The role of social relations in health promotion. *Psychosomatic Medicine*, *57*(3), 245-254.

Blanchflower, D. G., & Oswald, A. J. (2020). Trends in extreme distress in the United States, 1993–2019. *American Journal of Public Health*, *110*(10), 1538-1544.

Camacho, C., Webb, R. T., Bower, P., & Munford, L. (2024). Risk factors for deaths of despair in England: An ecological study of local authority mortality data. *Social Science & Medicine*, *342*, 116560.

Case, A., & Deaton, A. (2015). Rising morbidity and mortality in midlife among white non-Hispanic Americans in the 21st century. *Proceedings of the National Academy of Sciences of the United States of America*, *112*(49), 15078–15083.

Case, A., & Deaton, A. (2017). Mortality and morbidity in the 21st century. *Brookings Papers on Economic Activity 2017*(1), 397–476.

Case, A., & Deaton, A. (2020a). *Deaths of Despair and the Future of Capitalism*. Princeton University Press.

Case, A., & Deaton, A. (2020b). The epidemic of despair: Will America's mortality crisis spread to the rest of the world?. *Foreign Affairs*, *99*(2), 92–102.

Case, A., & Deaton, A. (2022). The great divide: education, despair, and death. *Annual Review of Economics*, *14*(1), 1-21.

Chuang, Y. C., Huang, Y. L., Tseng, K. C., Yen, C. H., & Yang, L. H. (2015). Social capital and health-protective behavior intentions in an influenza pandemic. *PloS one*, *10*(4), e0122970.

Cohen, S., & Wills, T. A. (1985). Stress, social support, and the buffering hypothesis. *Psychological Bulletin*, *98*(2), 310.

Cohen, S. (2004). Social Relationships and Health. *American Psychologist*, *59*(8), 676-684.

Cohen, D. A., Inagami, S., & Finch, B. (2008). The built environment and collective efficacy. *Health & Place*, *14*(2), 198-208.

Costa, D., & Kahn, M. E. (2001). Understanding the decline in social capital, 1952-1998. *NBER Working Papers 8295*

Curry, O. S., Rowland, L. A., Van Lissa, C. J., Zlotowitz, S., McAlaney, J., & Whitehouse, H. (2018). Happy to help? A systematic review and meta-analysis of the effects of performing acts of kindness on the well-being of the actor. *Journal of Experimental Social Psychology*, *76*, 320-329.

Daly, M., & Macchia, L. (2023). Global trends in emotional distress. *Proceedings of the National Academy of Sciences*, *120*(14), e2216207120.

Diez Roux, A. V. (2017). Despair as a cause of death: more complex than it first appears. *American Journal of Public Health*, *107*, 1566–1567.

Dunn, E. W., Aknin, L. B., & Norton, M. I. (2008). Spending money on others promotes happiness. *Science*, *319*(5870), 1687–1688.

EVS (2021). *EVS Trend File 1981–2017*. GESIS Data Archive, Cologne. [ZA7503 Data file Version 3.0.0]. https://doi.org/10.4232/1.14021.

Frey, B. S., & Meier, S. (2004). Social comparisons and pro-social behavior: Testing "conditional cooperation" in a field experiment. *American Economic Review*, *94*(5), 1717-1722.

Friedman, S.R., Krawczyk, N., Perlman, D.C., Mateu-Gelabert, P., Ompad, D.C., Hamilton, L., Nikolopoulos, G., Guarino, H., & Cerd´a, M., (2020). The opioid/overdose crisis as a dialectics of pain, despair, and one-sided struggle. *Frontiers in Public Health*, *8*, 540423.

Galbraith, J. K. (2009). *The Great Crash 1929*. Houghton Mifflin Harcourt.

Gallup (2024). Gallup World Poll. Version 031524, Washington DC: Gallup.

Graham, C., & Pinto, S. (2019). Unequal hopes and lives in the USA: Optimism, race, place, and premature mortality. *Journal of Population Economics*, *32*(2), 665-733.

Graham, C., & Pinto, S. (2021). The geography of desperation in America: Labor force participation, mobility, place, and well-being. *Social Science & Medicine*, *270*, 113612.

Haerpfer, C., Inglehart, R., Moreno, A., Welzel, C., Kizilova, K., Diez-Medrano, J., Lagos, M., Norris, P., Ponarin, E., & Puranen, B., et al. (Eds.). (2022). *World Values Survey Trend File (1981-2022) Cross-National Data-Set*. Madrid, Spain & Vienna, Austria: JD Systems Institute & WVSA Secretariat. [Data File Version 2.0.0]. https://doi.org/10.14281/18241.27

Helliwell, J. F. (2006). Well-being, social capital and public policy: what's new?. *The Economic Journal*, *116*(510), C34-C45.

Helliwell, J. F. (2007). Well-being and social capital: Does suicide pose a puzzle?. *Social indicators Research*, *81*, 455-496.

Helliwell, J. F., & Huang, H. (2008). How's your government? International evidence linking good government and well-being. *British Journal of Political Science*, *38*(4), 595-619.

Helliwell, J. F., Huang, H., & Wang, S. (2014). Social capital and well-being in times of crisis. *Journal of Happiness Studies*, *15*, 145-162.

Helliwell, J. F., & Putnam, R. D. (2004). The social context of well-being. *Philosophical Transactions of the Royal Society of London. Series B: Biological Sciences*, *359*(1449), 1435-1446.

Helliwell, J. F., & Wang, S. (2011). Trust and wellbeing. *International Journal of Wellbeing*, *1*(1), 42–78.

House, J. S., Landis, K. R., & Umberson, D. (1988). Social relationships and health. *Science*, *241*(4865), 540-545.

Holt-Lunstad, J., Smith, T. B., & Layton, J. B. (2010). Social relationships and mortality risk: a meta-analytic review. *PLoS medicine*, *7*(7), e1000316.

Hui, B. P., Ng, J. C., Berzaghi, E., Cunningham-Amos, L. A., & Kogan, A. (2020). Rewards of kindness? A meta-analysis of the link between prosociality and well-being. *Psychological Bulletin*, *146*(12), 1084.

ILO (2014). Table B.16. Public social protection expenditure, 1995 to latest available year (percentage of GDP). Available at http://www.social-protection.org/gimi/gess/RessourceDownload.action?ressource.ressourceId=54614.

ILO (2024). Annex 6 General government actual expenditure on social protection including and excluding health care, latest available year (percentage of GDP). In World Social Protection Report 2024–26: Universal social protection for climate action and a just transition, ILO Geneva. Accessed October 2024. https://doi.org/10.54394/ZMDK5543

IMF (2024). Government Finance Statistics (GFS), Expenditure by Function of Government, updated June 27, 2020. Accessed: June 2024. Available at https://data.imf.org/?sk=388dfa60-1d26-4ade-b505-a05a558d9a42&sid=1479329334655

Islam, M. K., Gerdtham, U. G., Gullberg, B., Lindström, M., & Merlo, J. (2008). Social capital externalities and mortality in Sweden. *Economics & Human Biology*, *6*(1), 19-42.

Karanges, E.A., Blanch, B., Buckley, N.A., Pearson, S.A., (2016). Twenty-five years of prescription opioid use in Australia: a whole-of-population analysis using pharmaceutical claims. *British Journal of Clinical Pharmacology*, *82*, 255–267.

Kapteyn, A., Lee, J., Tassot, C., Vonkova, H., & Zamarro, G. (2015). Dimensions of subjective well-being. *Social Indicators Research*, *123*, 625-660.

Kawachi, I., & Berkman, L. F. (2001). Social ties and mental health. *Journal of Urban Health: Bulletin of the New York Academy of Medicine*, *78*, 458–467.

Kawachi, I., Subramanian, S. V., & Kim, D. (2008). Social capital and health: A decade of progress and beyond. In I. Kawachi, S. Subramanian, & D. Kim (Eds.), *Social Capital and Health* (pp. 1-26). Springer New York.

Kennedy, B. P., Kawachi, I., & Brainerd, E. (1998). The role of social capital in the Russian mortality crisis. *World Development*, *26*(11), 2029-2043.

Kim, D., Baum, C. F., Ganz, M. L., Subramanian, S. V., & Kawachi, I. (2011). The contextual effects of social capital on health: a cross-national instrumental variable analysis. *Social Science & Medicine*, *73*(12), 1689-1697.

Knapp, E.A., Bilal, U., Dean, L.T., Lazo, M., Celentano, D.D., (2019). Economic insecurity and deaths of despair in US counties. *American Journal of Epidemiology*, *188*, 2131–2139.

Kumar, S., Calvo, R., Avendano, M., Sivaramakrishnan, K., & Berkman, L. F. (2012). Social support, volunteering and health around the world: Cross-national evidence from 139 countries. *Social Science & Medicine*, *74*(5), 696-706.

Lochner, K. A., Kawachi, I., Brennan, R. T., & Buka, S. L. (2003). Social capital and neighborhood mortality rates in Chicago. *Social Science & Medicine*, *56*(8), 1797-1805.

Lynch, J., Due, P., Muntaner, C., & Smith, G. D. (2000). Social capital—Is it a good investment strategy for public health?. *Journal of Epidemiology & Community Health*, *54*(6), 404-408.

Mikkelsen, L., Phillips, D. E., AbouZahr, C., Setel, P. W., De Savigny, D., Lozano, R., & Lopez, A. D. (2015). A global assessment of civil registration and vital statistics systems: monitoring data quality and progress. *The Lancet*, *386*(10001), 1395-1406.

Musick, M. A., & Wilson, J. (2003). Volunteering and depression: The role of psychological and social resources in different age groups. *Social Science & Medicine*, *56*(2), 259-269.

Nelson, S. K., Layous, K., Cole, S. W., & Lyubomirsky, S. (2016). Do unto others or treat yourself? The effects of prosocial and self-focused behavior on psychological flourishing. *Emotion*, *16*(6), 850–861.

Nyqvist, F., Pape, B., Pellfolk, T., Forsman, A. K., & Wahlbeck, K. (2014). Structural and cognitive aspects of social capital and all-cause mortality: a meta-analysis of cohort studies. *Social Indicators Research*, *116*, 545-566.

O'Connor, K. J., & Graham, C. (2019). Longer, more optimistic, lives: Historic optimism and life expectancy in the United States. *Journal of Economic Behavior & Organization*, *168*, 374-392.

OECD (2024). Social Expenditure - Aggregated data. OECD, Accessed May 2024. Available at https://stats.oecd.org/Index.aspx?DataSetCode=SOCX_AGG#.

OECD (2013). OECD Guidelines on Measuring Subjective Well-being, OECD Publishing. http://dx.doi.org/10.1787/9789264191655-en.

Peacock, A., Leung, J., Larney, S., Colledge, S., Hickman, M., Rehm, J., Giovino, G.A., West, R., Hall, W., et al., (2018). Global statistics on alcohol, tobacco and illicit drug use: 2017 status report. *Addiction*, *113*, 1905–1926.

Pescosolido, B. A. (1992). Beyond rational choice: The social dynamics of how people seek help. *American Journal of Sociology*, *97*(4), 1096-1138.

Putnam, R. D. (2000). *Bowling alone: The collapse and revival of American community*. Simon Schuster.

Raposa, E. B., Laws, H. B., & Ansell, E. B. (2016). Prosocial behavior mitigates the negative effects of stress in everyday life. *Clinical Psychological Science*, *4*(4), 691-698.

Rowland, L., & Curry, O. S. (2019). A range of kindness activities boost happiness. *The Journal of Social Psychology*, *159*(3), 340-343.

Roxburgh, A., Hall, W.D., Dobbins, T., Gisev, N., Burns, L., Pearson, S., Degenhardt, L., (2017). Trends in heroin and pharmaceutical opioid overdose deaths in Australia. *Drug and Alcohol Dependence*. *179*, 291–298.

Rönnerstrand, B. (2014). Social capital and immunization against the 2009 A (H1N1) pandemic in the American States. *Public Health*, *128*(8), 709-715.

Sarracino, F., & Mikucka, M. (2017). Social capital in Europe from 1990 to 2012: Trends and convergence. *Social Indicators Research*, *131*(1), 407-432.

Shang, J., & Croson, R. (2009). A field experiment in charitable contribution: The impact of social information on the voluntary provision of public goods. *The Economic Journal*, *119*(540), 1422-1439.

Shirzad, M., Yenokyan, G., Marcell, A. V., & Kaufman, M. R. (2024). Deaths of despair-associated mortality rates globally: a 2000–2019 sex-specific disparities analysis. *Public Health*, *236*, 35-42.

Siddiqi, A., Sod-Erdene, O., Hamilton, D., Cottom, T.M., & Darity, W., (2019). Growing sense of social status threat and concomitant deaths of despair among whites. *SSM - Population Health*, *9*, 100449.

Sin, N. L., Klaiber, P., Wen, J. H., & DeLongis, A. (2021). Helping amid the pandemic: Daily affective and social implications of COVID-19-related prosocial activities. *The Gerontologist*, *61*(1), 59-70.

Snowdon, J., & Choi, N. G. (2020). Undercounting of Suicides: Where Suicide Data Lie Hidden. *Global Public Health*, *15*(12), 1894–1901.

Solt, F. (2019). The Standardized World Income Inequality Database, Versions 8-9, https://doi.org/10.7910/DVN/LM4OWF, Harvard Dataverse, V11. swiid9_6_summary.csv

Thoits, P. A. (1985). Social support and psychological well-being: Theoretical possibilities. In I.G. Sarason, I.G. & B.R. Sarason (Eds.), *Social support: Theory, research and applications* (pp. 51–72). Springer Netherlands.

Thoits, P. A. (1986). Social support as coping assistance. *Journal of Consulting and Clinical Psychology*, *54*, 416– 423.

Thoits, P. A. (2011a). Mechanisms Linking Social Ties and Support to Physical and Mental Health. *Journal of Health and Social Behavior*, *52*(2), 145-161.

Thoits, P. A. (2011b). Perceived Social Support and the Voluntary, Mixed, or Pressured Use of Mental Health Services. *Society and Mental Health*, *1*(1), 4-19.

Thoits, P. A., & Hewitt, L. N. (2001). Volunteer work and well-being. *Journal of Health and Social Behavior*, *42*(2), 115–131.

Umberson, D., Crosnoe, R., & Reczek, C. (2010). Social relationships and health behavior across the life course. *Annual Review of Sociology*, *36*(1), 139-157.

WHO (2020) WHO methods and data sources for country-level causes of death 2000-2019, Global Health Estimates Technical Paper WHO/DDI/DNA/GHE/2020.2, WHO, Geneva.

WHO (2024a) WHO Mortality Database: Interactive platform visualizing mortality data, Deaths by sex and age group for a selected country or area and year, World Health Organization, Accessed: 10 April 2024: Alcohol Use Disorders; All Causes; Drug Use Disorders; Self-Inflicted Injuries, and Accessed: 20 August 2024: Ill-Defined Injuries.

WHO (2024b) WHO Mortality Database: Interactive platform visualizing mortality data, Data usability: completeness and proportion of ill-defined or non-specific causes, World Health Organization, Accessed: 25 September 2024.

World Bank (2024a) World Bank Country and Lending Groups, Accessed 30 July 2024, https://datahelpdesk.worldbank.org/knowledgebase/articles/906519-world-bank-country-and-lending-groups.

World Bank (2024b) World Development Indicators, World Bank, Accessed: 27 November 2024, using Azevedo, J.P. (2011).

World Bank (2024c) Worldwide Governance Indicators 2024 Update, World Bank, Accessed 30 October 2024. https://www.worldbank.org/en/publication/worldwide-governance-indicators.

Zeglin, R. J., Niemela, D. R. M., & Baynard, C. W., (2019). Deaths of despair in Florida: assessing the role of social determinants of health. *Health Education & Behaviour*, *46*, 329–339.

Chapter 7

Trusting others

How unhappiness and social distrust explain populism

Yann Algan
HEC, Cepremap

Corin Blanc
Cepremap, EconomiX, Université Paris-Nanterre

Claudia Senik
Sorbonne University, PSE, Cepremap

We are grateful to Eugenie de Laubier for excellent research assistance, and to Elizabeth Beasley who contributed to an earlier companion paper.

doi.org/10.18724/whr-hhje-y762

This version was last updated on 13 March 2025. Please see worldhappiness.report for the latest data.

In the context of post-industrial societies that have become increasingly individualistic, subjective attitudes play a much greater role in shaping values and voting behaviour than traditional ideologies or class struggle.

Key Insights

The decline in happiness and social trust in Europe and the US explains a large share of the rise in political polarisation and anti-system votes.

Subjective attitudes such as life satisfaction and trust play a much greater role in shaping values and voting behaviour than traditional ideologies or class struggle.

Unhappy people are attracted by the extremes of the political spectrum. Low-trust people are found more often on the far right, whereas high-trust people are more inclined to vote for the far left.

We highlight the political implications of both the presence and absence of caring and sharing practices.

Introduction

During the last decade, Western countries have seen a cascade of anti-system political victories, from Brexit in 2015 to the election of Donald Trump in 2024. During this period, resentment of 'the system' has grown in most European countries, particularly in Austria, Denmark, Finland, France, Germany, Hungary, Italy, Poland, Sweden, and Switzerland. Figure 7.1 illustrates the global rise in anti-system populist parties over the 20th century, both on the far left and the far right, with a sharp increase since the 1980s.

Voters question immigration and globalisation, protectionism is on the rise, and attacks on experts and mainstream media are increasingly common. At first glance, these patterns can be surprising, given the historically unprecedented levels of prosperity and security after World War II and the fall of the Berlin Wall in 1989.

A number of studies suggest that this anti-system wave originates in a loss of confidence in traditional political parties, both right and left, and a more general loss of confidence in elites, giving rise to "anti-party parties".[1] This loss of confidence in the system may be attributed to economic factors - such as rising economic insecurity and the economic consequences of globalisation, trade, and automation[2] - or cultural factors leading to a backlash against modernity and growing hostility towards immigrants.[3]

These explanations rarely address the important question of why some anti-system voters respond to these pressures by moving to the far left, while others move to the far right.[4] Equally puzzling are

Figure 7.1: The rise in anti-system votes
Funke et al. (2023)

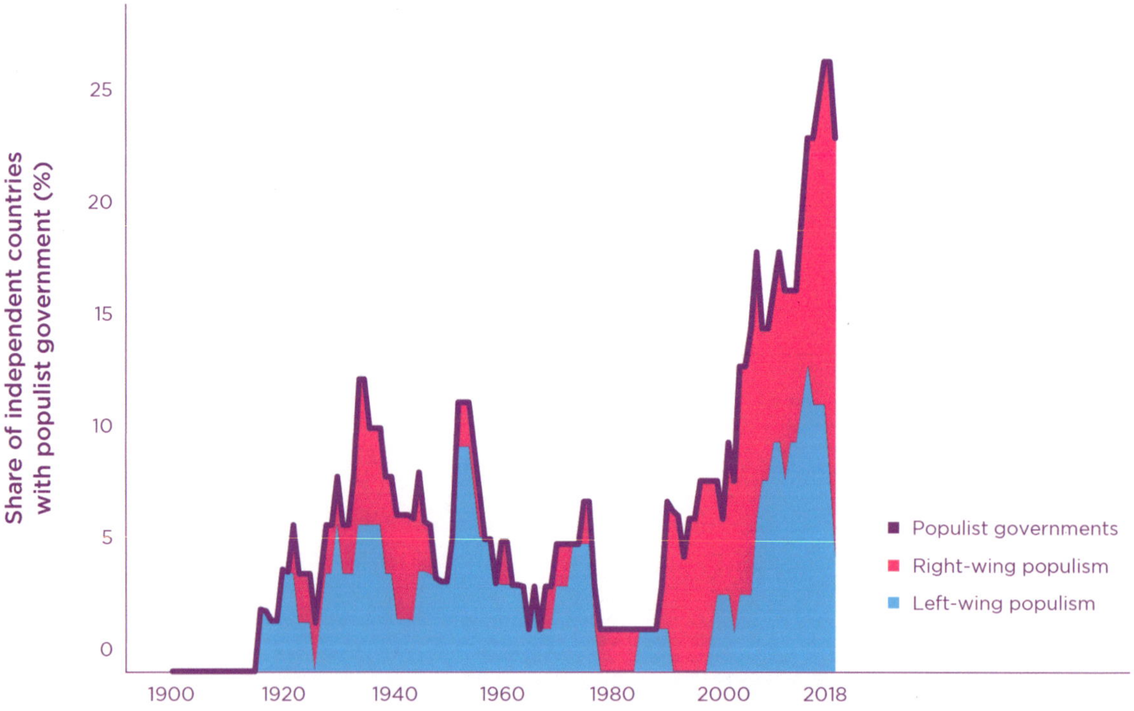

the different agendas of the populist right and the extreme left in terms of income inequality and the persistent paradox that poor people, who would benefit from income redistribution, often vote for parties that oppose it.[5]

The recent victory of Donald Trump in 2024 is a perfect illustration of this puzzle. Likewise, in the French context, during the first round of the 2022 presidential election, anti-system parties, both from the far left (La France Insoumise) and the far right (Rassemblement National) gathered around 20% of the votes, often coming from the same social background. Le Pen, the populist right candidate, and Mélenchon, the extreme left candidate, are both able to attract the votes of similar anti-system, blue-collar workers despite proposing very different policy programs on immigration and redistribution.

To explain these patterns of political behaviour, one must acknowledge the decline of the traditional class-based divide in voting behaviour and search for a new framework that can explain the rise in anti-system votes. In particular, why they flow either to the populist right (with anti-redistribution, anti-immigrant, and more parochial ideology) or the far left (pro-redistribution, pro-immigrant, and more universalist values).

In this chapter, we propose such a framework, where subjective attitudes such as life satisfaction and interpersonal trust play a crucial role. In the context of post-industrial societies that have become increasingly individualistic, subjective attitudes play a much greater role in shaping values and voting behaviour than traditional ideologies or class struggle. In line with previous findings,[6] we show that (low) life satisfaction is highly related to distrust in institutions and voting preferences for anti-system candidates, both in the United States (US) and in Europe, using various international databases.

Then we highlight a key new element: the role of social trust in explaining how these anti-system forces are oriented to the right or left of the political spectrum. Far-left voters have a higher level of social trust, while right-wing populists have a very low level of social trust. For the populist right, this low trust is not limited to strangers, but also extends to others in general, from homosexuals to their own neighbours. The xenophobic inclination of the populist right, well-documented worldwide,[7] seems to be a particular case of a broader distrust towards the rest of society. Right-wing populists throughout the world share xenophobic and anti-immigration inclinations. The Sweden Democrats, the Danish People's Party, the Finns Party, the Freedom Party of Austria, Greece's Golden Dawn, the Northern League and Fratelli in Italy, the National Rally in France, and a fraction of the Republican Party in the US are all built on strong anti-immigration foundations.

Far-left voters have a higher level of social trust, while right-wing populists have a very low level of social trust.

To encompass these patterns in a unifying framework, we construct a matrix of interpersonal trust and life satisfaction and show how voters in the United States and Western Europe are positioned in the matrix over the last decade. We test the relevance of this framework by combining various databases: the Gallup World Poll (2004–2024), the European Social Survey (2002–2023), the General Social Survey (1970–2023), and the World Values Survey (1981–2023). We show how our new paradigm offers additional insight for a more general analysis of political preferences. In particular, how social trust and life satisfaction combine to explain otherwise puzzling ideological attitudes. For example, why populist right-wing voters, while poorer than the average, tend not to support income redistribution policies.

An additional key element is to explain the rise of anti-system parties over time. If happiness and social trust can explain anti-system votes and ideologies not only in cross-section but also over time, we should see both attitudes moving in opposite directions over the last decade, especially among particular groups of people. We show, with the new release of the Gallup data, an acceleration in the decrease of life satisfaction in the US and, to a lesser extent, in Western European countries, as already documented in *World Happiness Report 2024*.[8]

We also provide new evidence for a strong decline in social trust. As an illustration, the share of American people who trust others has almost halved since the 1970s, dropping from 50% to 30%. These new findings resonate with Bob Putnam's seminal book, *Bowling Alone*, and with the original data discussed in Chapter 3 of this report,[9] that documents a sharp increase in the number of Americans who dine alone every day of the week. We show that the country-wide evolution of happiness and trust is highly associated with the rise in the likelihood of voting for anti-system parties in Western Europe and the United States, and discuss the groups of people whose life satisfaction and social trust have declined the most during this period, pushing them to anti-system and populist parties.

A new paradigm to explain anti-system votes

A growing literature has established the importance of looking at subjective attitudes, such as life satisfaction, for explaining voting behaviour. In particular, it has already been shown that unhappy people have less faith in political parties and the political system, are more likely to agree with authoritarian ideas such as having a strong leader to rule the country, and are more likely to reject incumbents, both in the European and Brexit contexts.[10] In the US context, low life satisfaction was highly predictive of Trump's election victory in 2016.[11] In addition, negative emotions measured by international surveys and social media, were also highly predictive of populist votes in the US and Western Europe.[12]

This evidence adds to the extensive literature documenting the political consequences of economic shocks and risks such as the financial crisis, globalisation, and the rise in income inequality. It makes a crucial addition by showing how measures of life satisfaction capture the impact of these shocks on life experience much more accurately than simple socio-demographic measures. It also shows how these negative

Our contribution here is to show the importance of adding social trust as a second dimension – in addition to life satisfaction – to understand the partitioning of unsatisfied citizens into the far left or far right.

subjective experiences translate into the blaming of institutions and elites for offering insufficient protection from these shocks.

Yet, this literature cannot explain why unhappy voters turn either to the left or to the right along the spectrum of anti-system parties, nor their different ideologies (i.e., economic and cultural values). Our contribution here is to show the importance of adding social trust as a second dimension – in addition to life satisfaction – to understand the partitioning of unsatisfied citizens into the far left or far right. We also pay attention to abstainers, whose lack of social inclusion is associated with a withdrawal from the political game and the refusal to vote. This is particularly revealing in the case of the United States, where this behaviour represents an addition to the limited options offered by the bi-partisan system.

This partitioning resonates with previous work showing the role of universalist versus parochial moral values in the rise of the pro-Trump vote in the US context.[13] It also talks to other studies that have identified cultural causes as the root of the upswing in populism. For instance, Inglehart and Norris claim that populist voting is mainly driven by the generation born between the two World Wars, driven by a rejection of cultural modernity, diversity, and the emancipation of women and sexual minorities.[14] However, these traditional explanations fall short of explaining why, in some countries, populist platforms attract at least as much support from the youth as from the older generation. The rise in populist votes is ubiquitous, cutting across generations and categories. Cultural backlash is equally unable to explain the dramatic drop in trust in government observed throughout the Western world, across all cohorts. Economic factors are also unable to explain why there is a strong xenophobic undercurrent in some anti-system parties but not others, or why many populist parties and their poor constituencies are hostile to income redistribution policies.

Instead, we build on abundant social science research showing how social trust shapes ideologies. This literature shows how an important part of economic values - in particular citizens' predisposition to pay taxes, finance public goods, or favour more or less redistributive policies - is related to the level of social trust. For example, the willingness to pay taxes depends on the belief that others are also contributing their share, a finding confirmed by laboratory experiments of public good games on the funding of collective services.[15] People who are most in favour of redistribution and, more generally, countries where the welfare state system is more generous, have higher levels of trust. By contrast, distrust undermines support for income redistribution. In the French context, trust explains a large share of economic and cultural values, especially attitudes towards immigration or homosexuality.[16]

The two diagrams below illustrate our proposed paradigm. Figure 7.2A shows how life satisfaction and social trust explain ideologies (both cultural and economic attitudes) and political attitudes and votes. Figure 7.2B shows how life satisfaction and social trust are both necessary to fully understand anti-system movements.

Figure 7.2A: The structure of relations between subjective attitudes, values, and votes

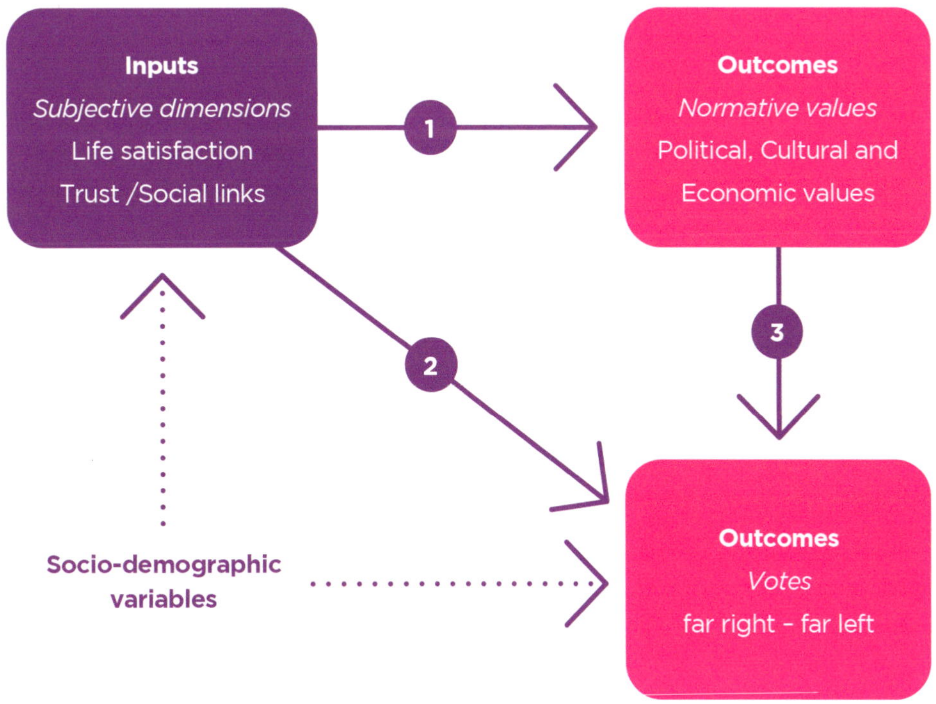

Figure 7.2B: The partition of values and votes depending on life satisfaction and trust

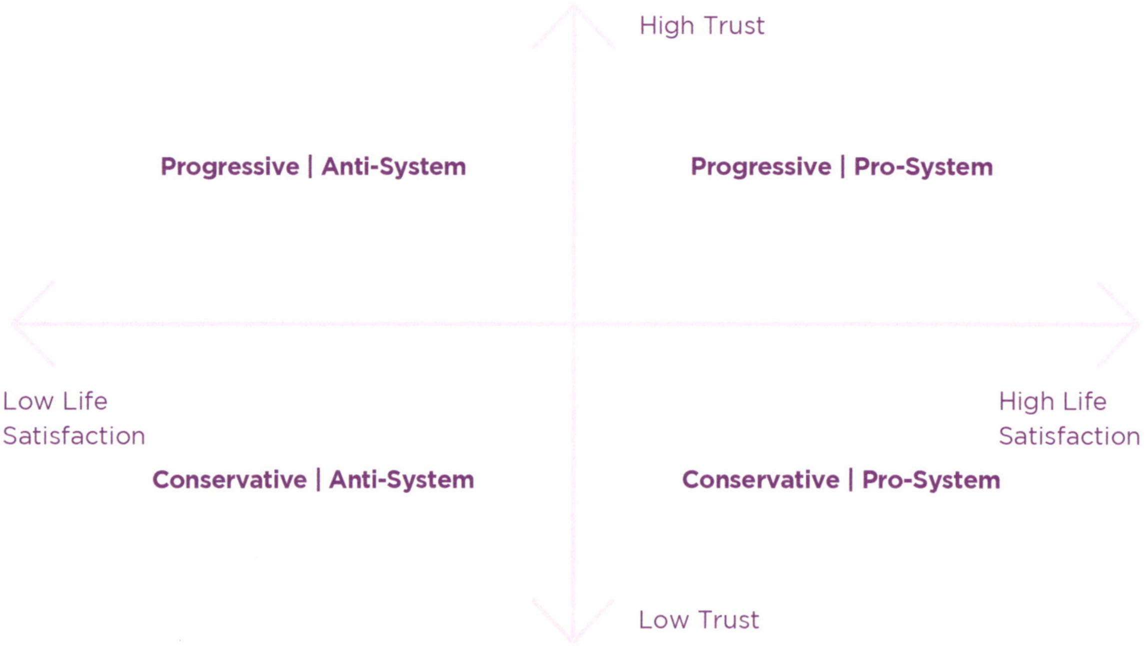

In the following sections, we apply this paradigm using several individual-level, cross-national surveys that contain information on subjective life evaluation, trust in others, and values in Western Europe and the United States.[17] In the next section, we show the relationship between these variables and political, economic, and cultural values (corresponding to arrow 1 of Figure 7.2A). In the following section, we illustrate the political consequences of these attitudes on votes (arrows 2 and 3 of Figure 7.2A and the partitioning in Figure 7.2B).

The influence of life satisfaction and trust on ideology

We start by illustrating the relationship between life satisfaction and social trust, on the one hand, and political, cultural, and economic values on the other hand.

Life satisfaction is measured with the question, "Are you satisfied with your life in general?" with responses on a scale from 0 (very dissatisfied) to 10 (very satisfied). Social trust is measured by interpersonal trust, which is the trust that people have in other people. Importantly, it is different from institutional trust, which is the trust that people have in institutions (government, parliament, legal, etc.). To measure interpersonal trust, we use the question, "In general, would you say that most people can be trusted or that one can never be too careful when dealing with others?"

This question has become a standard reference in international surveys for estimating the social fabric in a society, and the ability to cooperate outside the family and the private circle. We also propose additional measures of social trust based on concrete measures of social relationships.

We implement our framework and look at the interplay between life satisfaction and social trust. For simplicity, we split the population of each country into high trust versus low trust groups and into high life satisfaction versus low life satisfaction (as compared to the median level). We then represent the average values of the four groups defined by this partition.

Political values

Figure 7.3 shows how satisfaction with democracy is associated with life satisfaction and interpersonal trust using individual estimates in Europe and the US and controlling for income, education, gender, and age. Results for trust in the parliament, the legal system, politicians, and support of European integration are reported in the online appendix.

The green dots at the right of the figure represent the attitudes of individuals who declare a high level of life satisfaction and trust. Compared with this benchmark, the other coloured dots correspond to the three other groups: high life satisfaction and low trust (orange), low life satisfaction and low trust (pink), and low life satisfaction and high trust (blue).

As can be seen, dissatisfaction with democracy is particularly high when people express a low level of life satisfaction (pink and blue), especially in Europe. This is true whether they have a high (blue) or low (pink) level of social trust. In Europe, people with low life satisfaction and low social trust (pink) are 1.8 points less satisfied with democracy compared to the baseline, on a 0–10 scale. For the US, it is 0.8 points. People with high social trust but low life satisfaction (blue) also tend to be highly dissatisfied with democracy compared to their counterparts with high life satisfaction (green). We find similar results for distrust in legal institutions.

The main takeaway of this first finding is the anti-system political attitudes of citizens with the lowest level of life satisfaction. Those who are dissatisfied with their lives may feel that the system has failed them and consider that the democratic system and legal institutions have not protected them against life risks.

Figure 7.3: Satisfaction with democracy in the United States and Western Europe
ANES (2020), ESS (2018–2023)

Note: Controls: gender, age, income, education, year and country fixed effects (for Europe). Standard errors clustered at the country level (for Europe). Variables are rescaled to vary between 0 and 10.

Cultural values

The group of citizens with the lowest level of satisfaction, the anti-system group, is divided in terms of cultural values and what should replace 'the system', depending on their level of social trust. Figure 7.4 shows that social trust is the main predictor of the cultural divide in terms of attitudes towards homosexuality and immigration. We measure positive attitudes towards homosexuality with questions on whether gay and lesbian people should be able to adopt children (in the US) or are free to live their lives as they wish (in Europe). We measure positive attitudes towards immigrants with questions on whether immigrants are good for the country's economy or could take the jobs of natives.

The crucial role of social trust in shaping cultural values is powerfully illustrated by the contrast between the two groups of citizens with low life satisfaction. The attitudes of those with low trust (pink) are almost twice as negative as those with high trust (blue). These less satisfied but trusting citizens (blue) are more tolerant towards homosexuality and immigration than highly satisfied but distrusting people (orange). Thus, unlike political ideology, which maps onto the distribution of life satisfaction, the distribution of cultural ideology is aligned along the axis of interpersonal trust, as shown earlier in Figure 7.2B.

Economic values

The same patterns hold for the relationships between life satisfaction, social trust, and economic values. Social trust appears as the main predictor of the ideological divide in terms

Figure 7.4: Attitudes towards homosexuality and immigration in the United States and Western Europe
ANES (2020), ESS (2018–2023)

Note: Controls: gender, age, income, education, year and country fixed effects (for Europe). Standard errors clustered at the country level (for Europe). Variables are rescaled to vary between 0 and 10.

of income redistribution. We measure attitudes towards redistribution using a question that asks whether "Government should reduce differences in income levels" (1–10 scale) for Europe and "There should be greater incentives for individual effort" versus "Incomes should be made more equal" (1–10 scale) for the United States.

Figure 7.5 shows that support for income redistribution is generally associated with higher social trust and lower life satisfaction (blue). Those who are highly satisfied but less trusting (orange) are the ones who most often oppose income redistribution. The contrast is even more salient within the group of citizens with the lowest life satisfaction. They are less supportive of income redistribution if they have a low level of interpersonal trust (pink) and more in favour if they trust others (blue).

Figure 7.5: Attitudes towards income redistribution in the US and Western Europe
WVS (2017), ESS (2018–2023)

Note: Controls: Gender, age, income, education, year and country fixed effects (for Europe). Standard errors clustered at the country level (for Europe).

Citizens with low interpersonal trust may be sceptical of the social contract and doubt reciprocity on behalf of other people. Indeed, the economic literature has shown that the level of interpersonal trust explains an important part of citizens' predisposition to finance public goods, pay taxes, or support redistributive policies.[18] This sheds light on the key puzzle of why a large part of the working class, who are the least satisfied with life in general, might vote for parties that oppose redistributive policies.

The influence of life satisfaction and trust on voting behaviour

We now turn to voting behaviour. First, we study the connection between life satisfaction, interpersonal trust, and votes (the second arrow of Figure 7.2A) and then the relationship between economic and cultural values and votes (the third arrow of Figure 7.2A).

Western Europe

In Western Europe, Table 7.1 shows that high levels of life satisfaction are negatively associated with votes for far-left and far-right parties at the last national election. By contrast, voters for centre-left and centre-right parties are more satisfied with their lives on average. However, a low level of trust is only associated with an attraction to the far right, and to a lesser extent to the right, but not to the left and the far left.

Figure 7.6 shows that, in Western Europe, far-left and far-right voters both declare low levels of life

Table 7.1: Life satisfaction, trust, and voting behaviour in Western Europe
ESS (2018–2023)

	Dependent variable				
	Far left (1)	Left (2)	Center (3)	Right (4)	Far right (5)
Life satisfaction	-0.010***	−0.005***	0.005***	0.015***	−0.005***
	(0.001)	(0.001)	(0.001)	(0.001)	(0.001)
Trust in others	0.004***	0.014***	0.003***	−0.006***	−0.016***
	(0.001)	(0.001)	(0.001)	(0.001)	(0.001)
Observations	32,450	32,450	32,450	32,450	32,450
R^2	0.068	0.083	0.088	0.073	0.083
Adjusted R^2	0.067	0.082	0.087	0.072	0.082
Residual Std. Error (df=32414)	0.289	0.487	0.332	0.443	0.334

Note: Weighted least squares. Estimates of the likelihood to vote for either party at the last national election. Additional controls: gender, income, age, education level, country and year fixed effects. Standard errors clustered at the country level. Abstainers are excluded from the sample.

*p<0.1; **p<0.05; ***p<0.01

satisfaction but are symmetrically positioned on high trust (far left) and low trust (far right) levels. This is consistent with our previous findings on political, economic, and cultural values. Anti-system ideology and institutional distrust are related to low life satisfaction but low life satisfaction alone is not sufficient to generate a populist far-right movement. The populist right requires the dimension of low interpersonal trust which is associated with anti-immigration and anti-redistribution attitudes.

Figure 7.7 highlights these results for a selection of European countries from the south (Spain), the centre (France and Germany), and the north (Sweden). In all countries, far-right voters stand much lower on social trust measures than electors for any other political party. In general, voters for centre-right, or centre-left parties have above-average life satisfaction and social trust.

The French elections are a textbook illustration of our paradigm. There, both the far left (Mélenchon) and the far right (Le Pen) have prominent electorates (18–25% in the 2017 and 2022 presidential elections). Citizens with high interpersonal trust are more likely to vote for the left and those with lower trust are more likely to

Figure 7.6: Life satisfaction, trust, and voting behaviour in Western Europe
ESS (2018–2023)

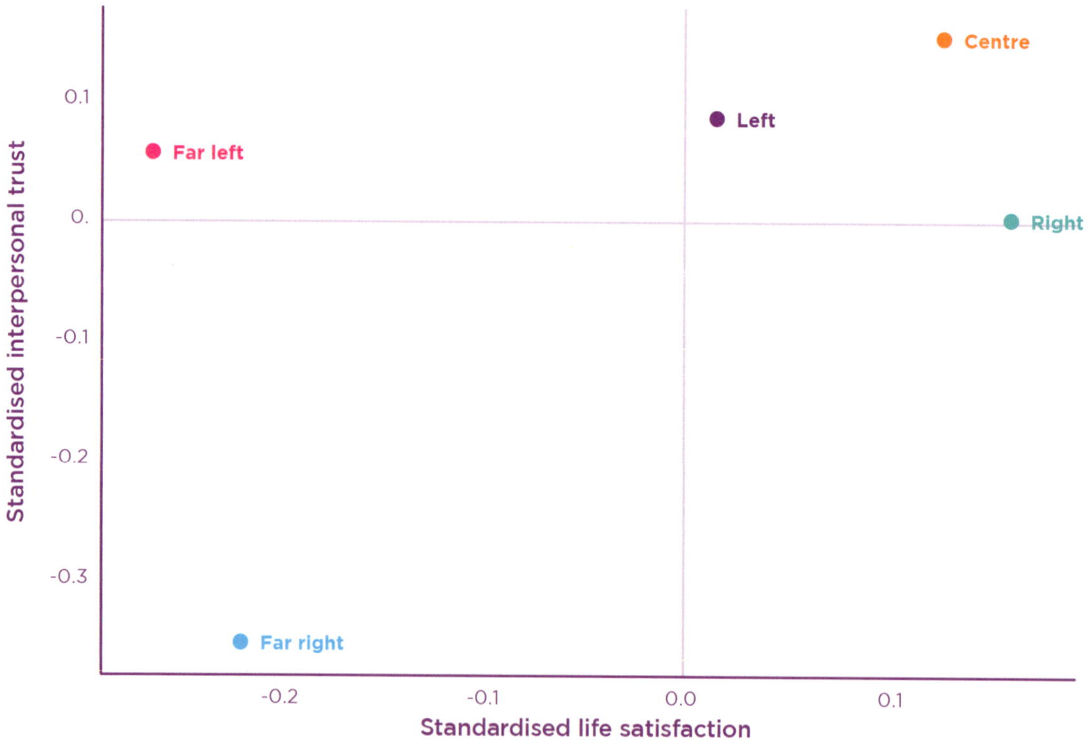

Note: Countries: Austria, Finland, France, Germany, Italy, Netherlands, Norway, Spain, Sweden, United Kingdom. Weighted mean and standardized levels of life satisfaction and interpersonal trust by votes in the last national elections. Life satisfaction and interpersonal trust are standardized with a mean of 0 and a standard deviation of 1.

Figure 7.7: Life satisfaction, trust, and voting behaviour in France, Germany, Spain, and Sweden
ESS (2018–2023)

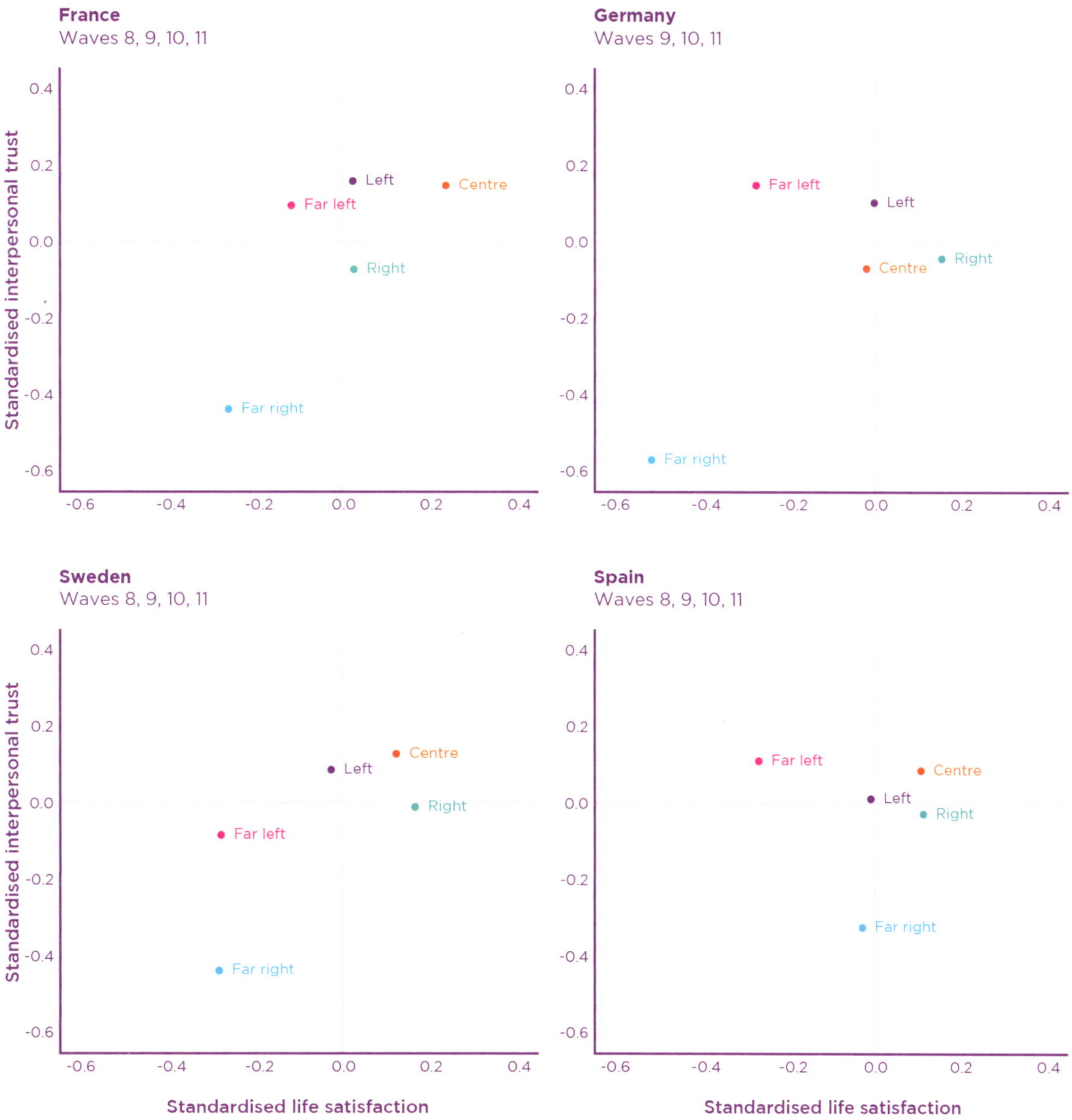

Note: Weighted mean and standardised levels of life satisfaction and interpersonal trust by votes in the last national elections. Life satisfaction and interpersonal trust are standardised with a mean of 0 and a standard deviation of 1.

vote for the right. Le Pen and Mélenchon voters share a low level of life satisfaction but differ in terms of trust. The centre electorate is embodied by Emmanuel Macron, with high life satisfaction and high trust, which correspond to wealthier people with liberal and pro-European values. In contrast, those who vote for centre-right parties are satisfied with life (mainly due to their high income) but are less trusting. We find the same patterns among anti-system parties in Germany (Die Linke versus AfD) and Spain (Podemos versus Vox).

United States

Our framework is more difficult to apply in the US bi-partisan system (Democrats versus Republicans) that protects the parties from splintering. When applying our framework, we uncover a great schism, not only between political parties but also between voters and abstainers, where the latter display the lowest levels of life satisfaction and interpersonal trust. In the US, this group of citizens, who can be considered as the anomic, are not represented by a party.

We focus specifically on the two presidential elections in 2016 and 2020 which are associated with the upsurge of votes for Trump. Although survey data on the 2024 elections are not available yet, we use data on voters' preferences across candidates in the two previous primaries and general presidential elections.

In a multi-party system, voters can choose – or create – a party that corresponds to their preferences over several important dimensions. In a two-party system, the parties often take opposite positions on each salient policy issue, which eventually shapes the nature of polarisation: politicians and voters are divided along a single dimension which aggregates several issues. On the other hand, each of the two parties effectively represents a coalition of voters who support different policy

Figure 7.8A: Life satisfaction, trust, and voting behaviour in US primary elections
ANES (2016, 2020)

stances on various dimensions. In our analysis, we try to identify such subgroups to understand their characteristics and their voting behaviour.

Figures 7.8A and 7.8B illustrate votes in 2016 and 2020 for primary and presidential elections.

In 2016, votes for Hillary Clinton were votes for the status quo, coming from people with high trust and high life satisfaction. Clinton voters are wealthier than the average (which is highly correlated with their life satisfaction) and their high level of trust makes them more favourable to immigration and globalisation. Sanders voters also express a higher level of trust than the average, but a lower level of life satisfaction. This pattern explains why they vote for a candidate whose main campaign motto was about the redistribution of income and taxation of the wealthiest. Trump voters displayed different features. Their level of life satisfaction is slightly higher than the national average (this is due to the composition of this electorate, combining both the white working and middle class and the wealthy), but a low level of trust. This is consistent with both the anti-tax and anti-immigration platform of Trump.

Figure 7.8B: Life satisfaction, trust, and voting behaviour in US presidential elections
ANES (2016, 2020)

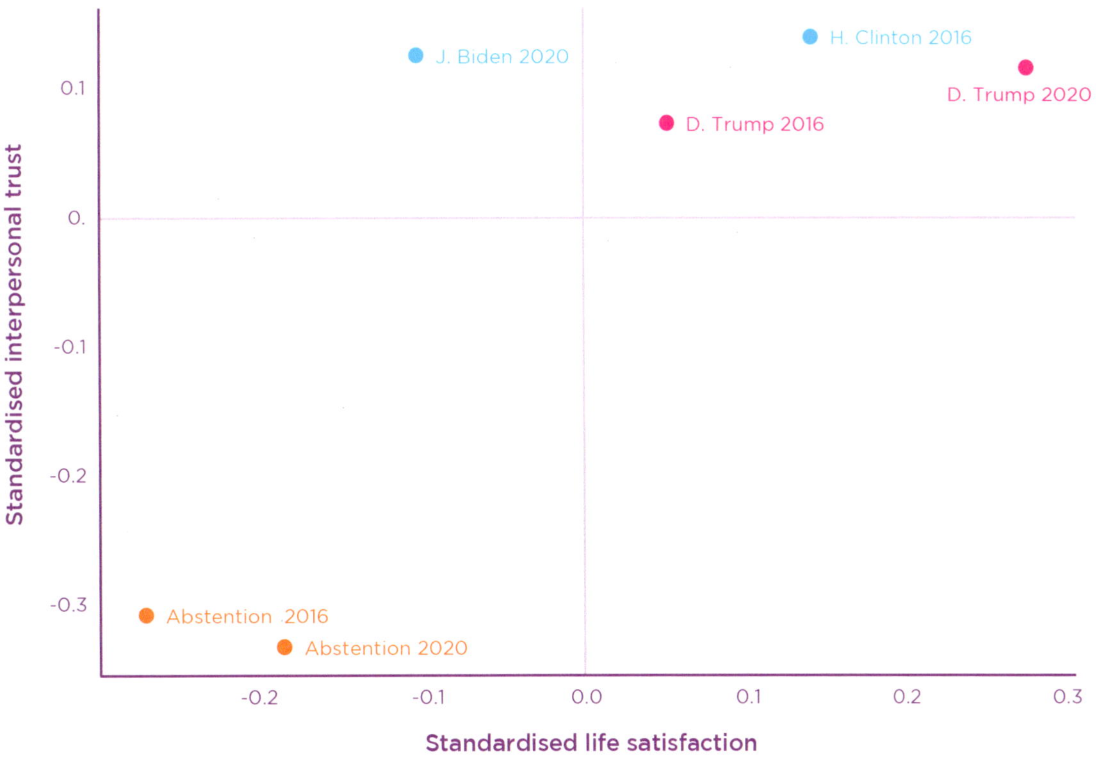

Note: Weighted mean and standardised levels of life satisfaction and interpersonal trust by votes in the presidential elections. Life satisfaction and interpersonal trust are yearly standardised with a mean of 0 and a standard deviation of 1. The normalisation is performed separately for each wave, using the respective mean and standard deviation for 2016 and 2020.

In 2020, Donald Trump was the only Republican candidate. The rally of traditional wealthy Republican voters (who voted for other Republican candidates in the 2016 presidential primaries) explains this group's above-average life satisfaction and average level of trust. By contrast, all Democrat voters displayed low life satisfaction and average or above-average levels of trust.

The most striking feature in the 2016 and 2020 US elections is the low levels of life satisfaction and trust of abstainers compared to voters. While in Europe, the 'anomics' turn to the far-right parties, in the US they withdraw from public life.

To dig deeper, we use a unique 2021 survey conducted by the market research company, Bilendi, on a representative sample of 15,000 US citizens. The survey asked about voting behaviour during the 2016 and 2020 presidential elections alongside specific questions on life satisfaction and various dimensions of trust and ideology.[19]

Figure 7.9A illustrates the schism between abstainers and voters across four dimensions of life satisfaction. Abstainers report higher solitude both at work and in life, lower life satisfaction, and less optimism for the future. The situation of abstentionists worsens between the two elections. Their life satisfaction is 31 percentage points lower than average in 2016, and this increases to 39 percentage points in 2020.

Figure 7.9B shows that abstainers have consistently lower trust than voters in every dimension. They tend to distrust their own family or friends (30 percentage points lower than the average) almost as much as strangers (36 percentage points) and others in general (41 percentage points).

Figure 7.9C shows that abstainers display a specific distrust in institutions. In particular, they had lower than average trust in courts (39 percentage points), police (37 percentage points), and government (30 percentage points) in 2020. This dramatic low level of institutional trust has deteriorated between 2016 and 2020.

Figure 7.9A: Dimensions of life satisfaction for abstainers and voters in US elections
Bilendi (2021)

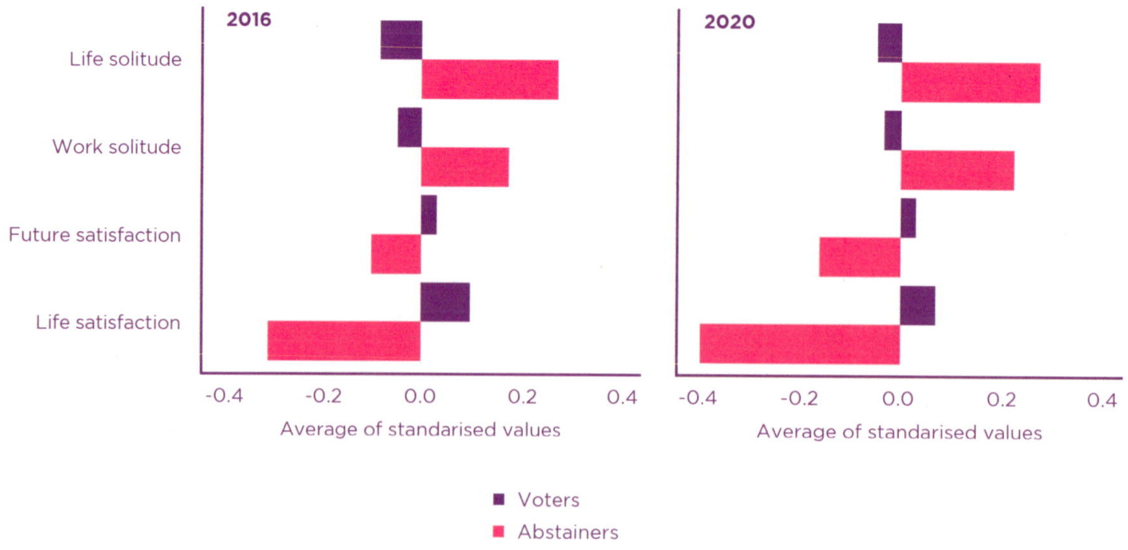

Figure 7.9B: Dimensions of interpersonal trust for abstainers and voters in US elections
Bilendi (2021)

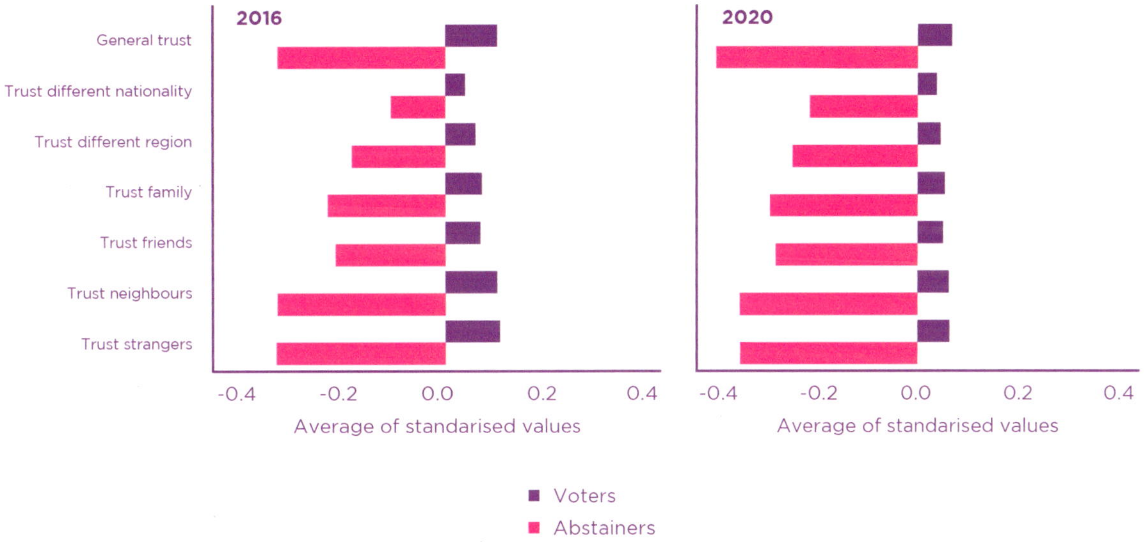

Figure 7.9C: Dimensions of institutional trust for abstainers and voters in US elections
Bilendi (2021)

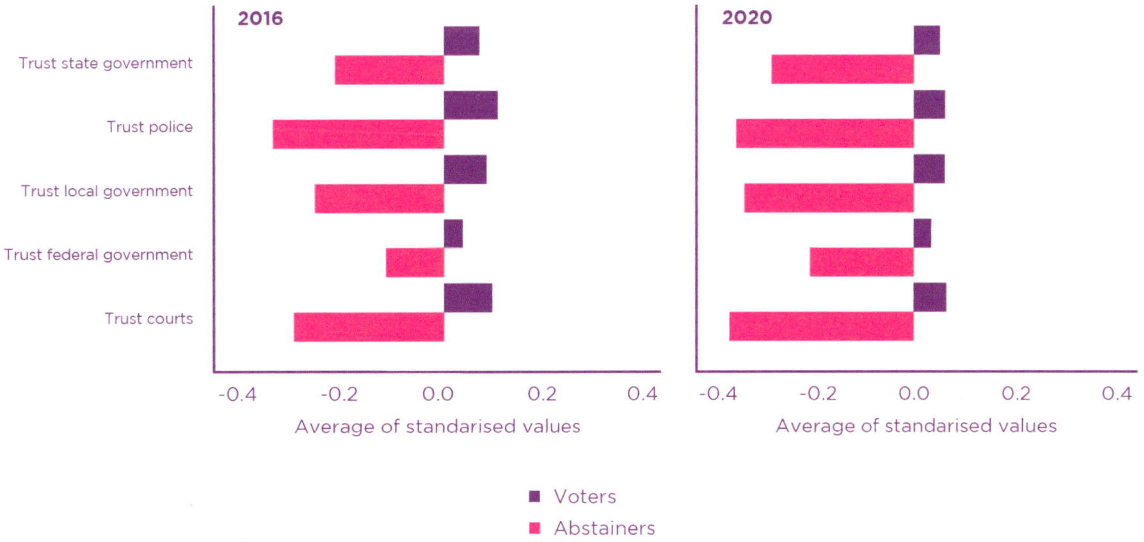

Note: US election representative survey in 2021 on voting behaviour in the primaries and general elections of 2016 and 2020, 15,000 respondents.

This pattern is quite different from the divide between Democrat and Republican voters. In Figure 7.10, we break down their level of trust by education (non-college versus college degree) and age (under 45 versus over 45). The most striking result concerns the differences between trust in one's private circle (family, friends, and neighbours) and trust in open society. Democrat voters have a lower-than-average level of parochial trust, but a much higher-than-average level of trust in strangers. The picture is reversed among Trump voters. Strikingly, this result applies to all generations and education levels.

Values and political behaviour

Finally, we examine the link between economic or cultural values and political behaviour (the third arrow of Figure 7.2A).

Figure 7.10: Dimensions of trust by vote, age, and education in the 2020 US primaries
Bilendi (2021)

Table 7.2A: Political trust and voting behaviour in Europe
ESS (2018–2023)

	Dependent variable				
	Satisfaction democracy (1)	Trust politicians (2)	Trust legal system (3)	EU integration (4)	Trust parliament (5)
Far left	-1.337**	-0.589*	-0.787*	-0.285	-0.836**
	(0.432)	(0.298)	(0.343)	(0.333)	(0.307)
Left	-0.153	0.100	-0.102	0.224	0.109
	(0.435)	(0.317)	(0.270)	(0.323)	(0.327)
Centre (Baseline)					
Right	-0.290	-0.121	-0.277	-0.491	-0.151
	(0.349)	(0.261)	(0.234)	(0.395)	(0.319)
Far right	-1.780***	-1.286***	-1.589***	-2.101***	-1.558***
	(0.477)	(0.322)	(0.209)	(0.154)	(0.378)
Observations	31,606	31,657	31,648	31,067	31,616
R^2	0.165	0.176	0.176	0.158	0.163
Adjusted R^2	0.164	0.176	0.175	0.158	0.162
Residual Std. Error	2.354 (df = 31573)	2.212 (df = 31624)	2.406 (df = 31615)	2.593 (df = 31034)	2.408 (df = 31583)

Note: Weighted least squares. Additional controls: gender, income, age, education level, country and year fixed effects. Standard errors clustered at the country level.

*p<0.1; **p<0.05; ***p<0.01

Western Europe

Table 7.2A shows that all measures of political trust are negatively correlated with far left and far right positioning. Table 7.2B displays the expected relationship between support for income redistribution and equal opportunities versus political positioning and votes. In particular, the far left is more in favour of redistribution and the far right is less in favour compared to the centre. The table also shows the expected relationship between cultural attitudes and voting behaviour with the far right much less tolerant towards immigration and same-sex couples. In terms of magnitude, income redistribution matters much more for the low-satisfaction, high-trust voters of the far left and immigration is the obsession of the low-satisfaction, low-trust voters of the far right.

United States

Similarly in the US case, Trump voters oppose income redistribution and are generally against government intervention. They are also much less supportive of immigration and much less tolerant towards homosexuality (Table 7.3).

Table 7.2B: Ideology and voting behaviour in Europe

ESS (2018–2023)

	Dependent variable				
	Gays and lesbians free to live (1)	Immigration good for economy (2)	Cultural life enriched by immigrants (3)	Government should reduce income inequalities (4)	Equality of opportunities (5)
Far left	0.318**	0.672*	0.993***	0.607***	0.196***
	(0.109)	(0.295)	(0.266)	(0.073)	(0.037)
Left	0.182**	0.329***	0.555*	0.355***	0.158***
	(0.057)	(0.087)	(0.273)	(0.065)	(0.035)
Centre (Baseline)					
Right	-0.454***	-0.486**	-0.352	-0.135	-0.133**
	(0.094)	(0.162)	(0.295)	(0.119)	(0.047)
Far right	-0.671***	-2.188***	-2.180***	-0.006	-0.216***
	(0.119)	(0.193)	(0.124)	(0.073)	(0.041)
Observations	31,647	31,780	31,780	31,633	22,768
R^2	0.160	0.016	0.031	0.112	0.068
Adjusted R^2	0.159	0.015	0.030	0.111	0.067
Residual Std. Error	1.910 (df = 31614)	9.046 (df = 31747)	7.955 (df = 31747)	0.988 (df = 31600)	1.026 (df = 22735)

Note: Weighted least squares. Additional controls: gender, income, age, education level, country and year fixed effects. Standard errors clustered at the country level.

*p<0.1; **p<0.05; ***p<0.01

Table 7.3: Ideology and voting behaviour in the US

ANES (2020)

	Dependent variable				
	Satisfaction with democracy (1)	Immigration is good for US economy (2)	America's culture is not harmed by immigrants (3)	LBGT couples should be allowed to adopt (4)	Government should reduce income inequalities (5)
Trump	0.162*	-0.875***	-1.025***	-1.281***	-2.602***
	(0.093)	(0.091)	(0.098)	(0.159)	(0.141)
Constant	4.847***	6.469***	6.634***	10.258***	5.832***
	(0.208)	(0.203)	(0.220)	(0.355)	(0.316)
Observations	3,604	3,611	3,613	3,604	3,624
R^2	0.019	0.051	0.069	0.050	0.131
Adjusted R^2	0.017	0.049	0.067	0.048	0.130
Residual Std. Error	2.676 (df = 3595)	2.612 (df = 3602)	2.826 (df = 3604)	4.567 (df = 3595)	4.056 (df = 3615)
F Statistic	8.932*** (df = 8; 3595)	24.011*** (df = 8; 3602)	33.610*** (df = 8; 3604)	23.565*** (df = 8; 3595)	68.409*** (df = 8; 3615)

Note: Weighted least squares. Additional controls: gender, income, age, education level. "Trump" is a binary variable equal to 1 when an individual voted for Trump in the 2020 presidential primary election and 0 otherwise.

*p<0.1; **p<0.05; ***p<0.01

Figure 7.11 details these ideological patterns for the 2020 US primary elections across six ideological dimensions: **nationalist** is measured with questions about anti-immigration and protectionism; **progressive** measures support for homosexuality, affirmative actions, and gun controls; **security** measures support for spending on police and defence; **anti-elite** measures negative attitudes towards experts, scientists, MPs, and judges; **government spending** measures attitudes towards taxes on millionaires, income redistribution, and reduction of income inequalities; and **institutional trust** measures trust in institutions and science and whether elections are fair. Democrat voters are much more in favor of immigration, sexual minorities, redistribution, and much less anti-elite and pro-security than the average. Trump voters display the exact opposite cultural and economic values. This pattern of attitudes holds irrespective of socio-demographics and are well captured by life satisfaction and interpersonal trust.

Figure 7.11: Dimensions of ideology by vote, age, and education in the 2020 US primaries
Bilendi (2021)

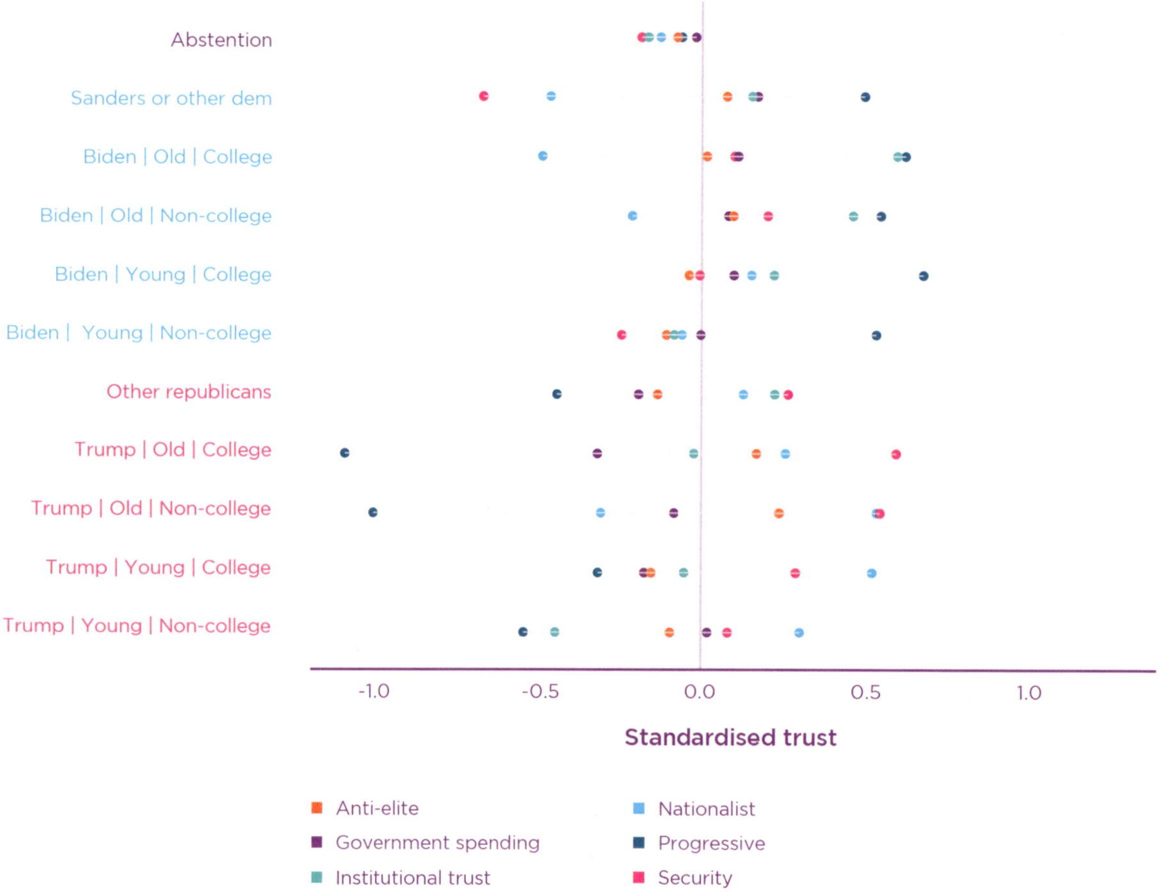

Time trends

We started this chapter by noting the rising support for far-right and far-left parties and then sketching the patterns relating subjective attitudes and economic and cultural values with political votes. Therefore, we expect to observe parallel trends in the evolution of these variables over time. Figure 7.12 shows a sharp fall in life satisfaction in the US, already documented in *World Happiness Report 2023*. The trend is less clear in Western Europe where the initial level of life satisfaction is lower.

It is important to underline that this fall in life satisfaction is not driven by income. To illustrate this, each dot in Figure 7.13 plots life satisfaction and GDP per capita (adjusted for inflation) since 2006. The association between income and life satisfaction is usually found to be positive in the short run, but better represented by a flat trend in the long run.[20] But here, in the 2000s, the association is negative. The time trend in life satisfaction is clearly decreasing in the United States, although GDP per capita is rising over time. The picture is similar in European countries,

Figure 7.12: Life satisfaction in the US and Western Europe
Gallup World Poll (2006–2023)

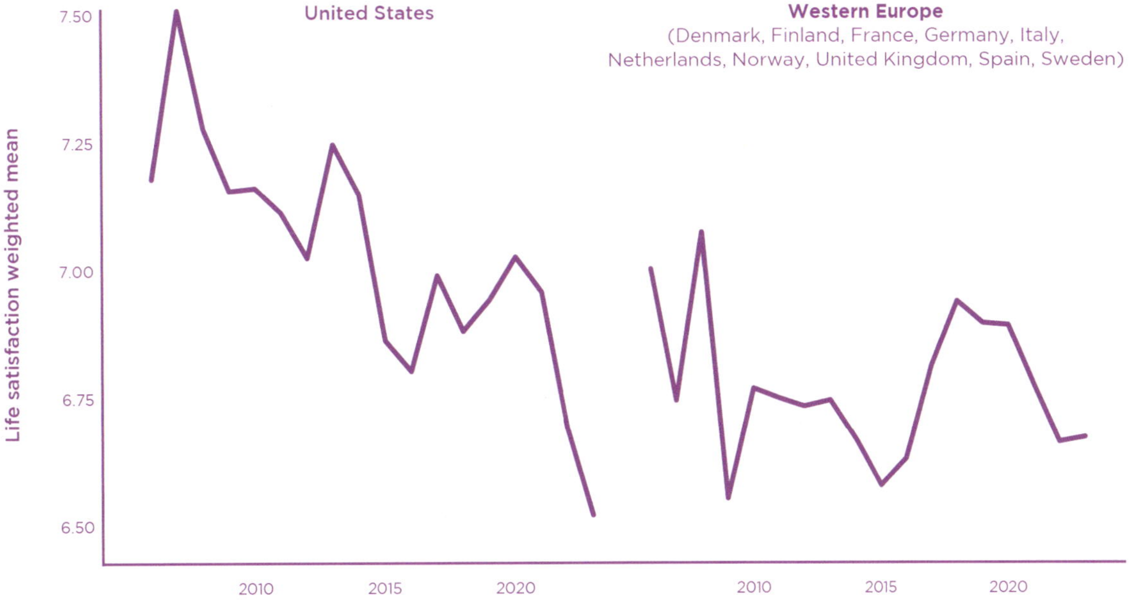

Note: Country-weighted average computed using design weights. Population weights used for region-level average.

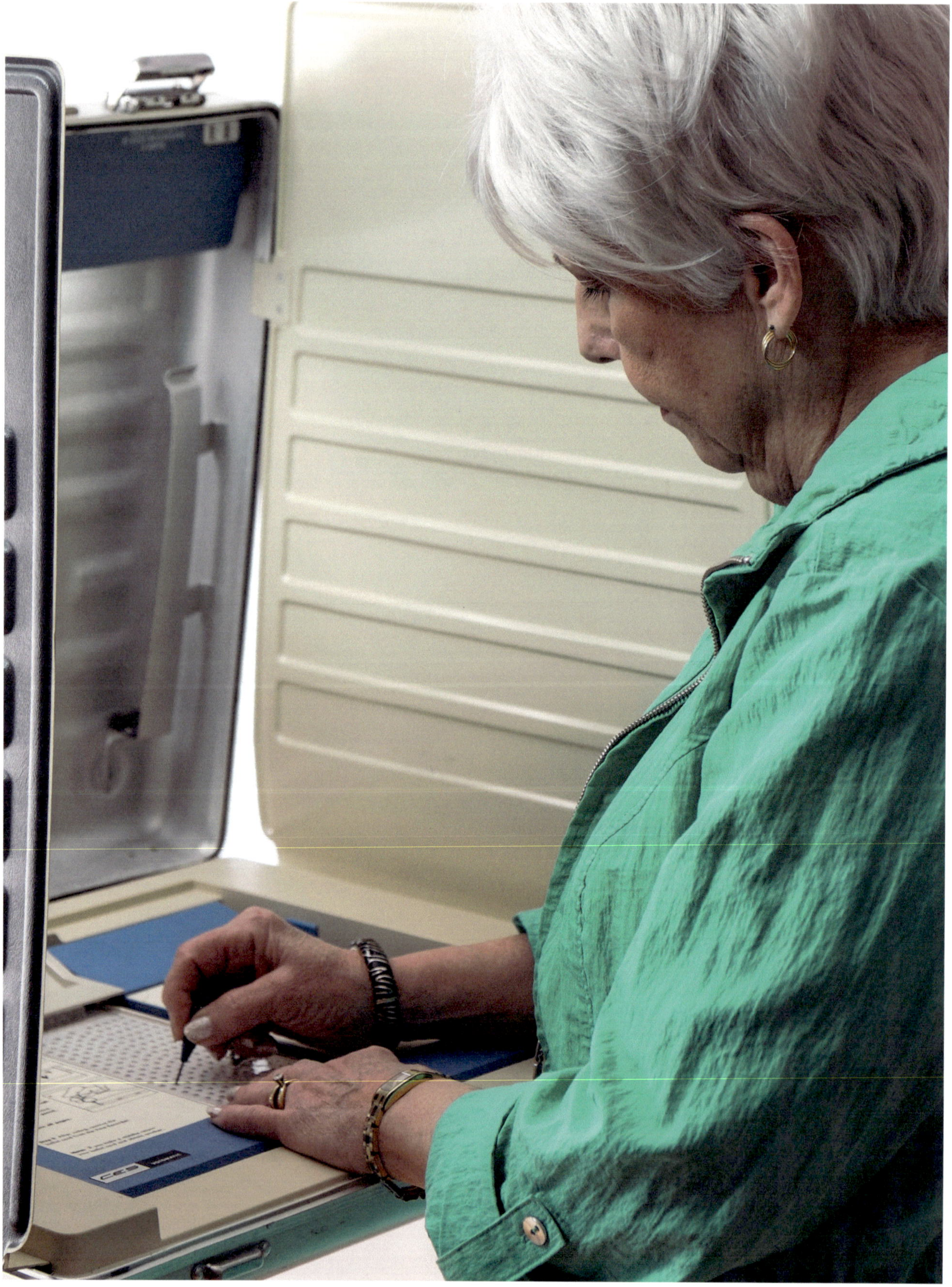

Figure 7.13: Life satisfaction and GDP per capita in the US and Western Europe
Gallup World Poll & WDI (2006–2023)

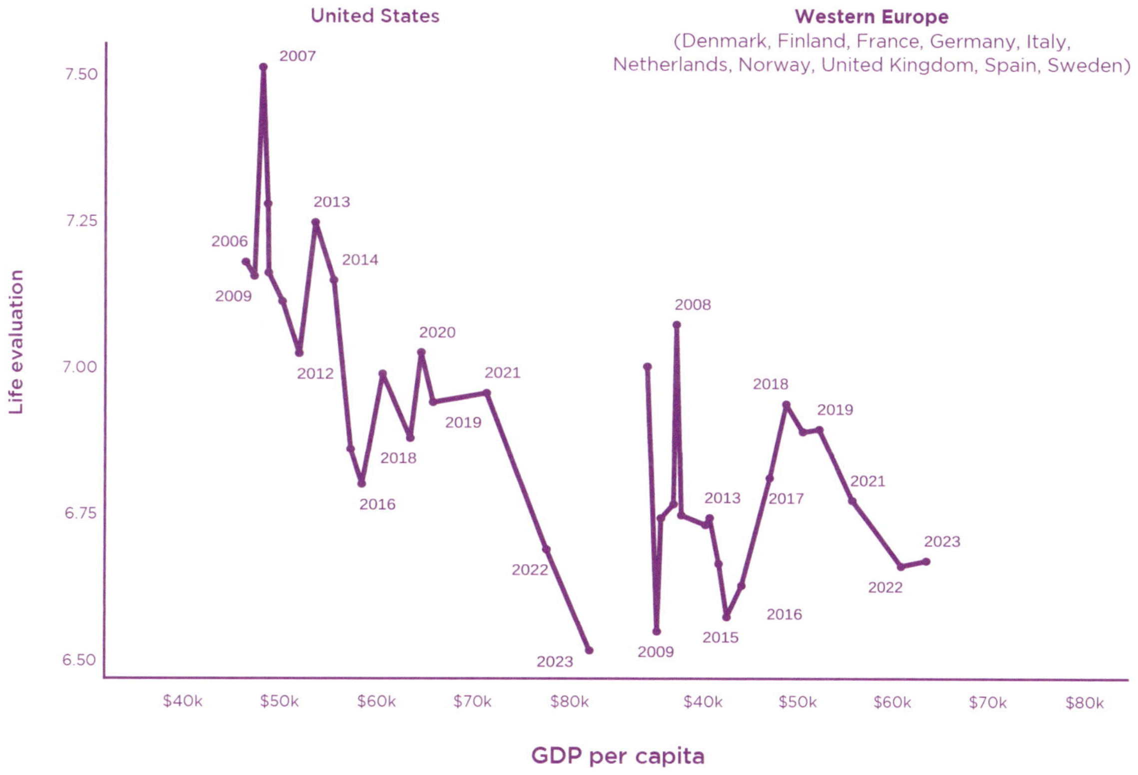

Note: Countries weighted average computed using design weights. Population weights used for region-level average.

especially since the 2020s, where GDP per capita is rising, but happiness is decreasing on average and in most European countries.

Figure 7.14A shows that social trust is also clearly decreasing in the US and Western Europe, with a drop of 10 percentage points in the two continents over the period. Even more strikingly, in Figure 7.14B we see that the percentage of people in the US who trust others has decreased by 20 percentage points since the early 1970s, from 50% to 30%. The evidence is more mixed in Europe, pointing to overall stability.

The sharp drop in social trust in the US is consistent with the fact that rising unhappiness has shifted voters almost exclusively towards the populist right (Donald Trump), and not towards the left (Bernie Sanders). The relative decline or stability in social trust in Europe is consistent with the fact that the decline in life satisfaction has split the unsatisfied electorate between the two opposite extremes of the political spectrum, depending on their level of social trust. Therefore, the next section investigates which groups of citizens have experienced the most significant variation in life satisfaction and social trust, in order to understand the evolution of anti-system voting behaviour.

Figure 7.14A: Trust in the US and Western Europe: Someone to count on
Gallup World Poll (2006-2023)

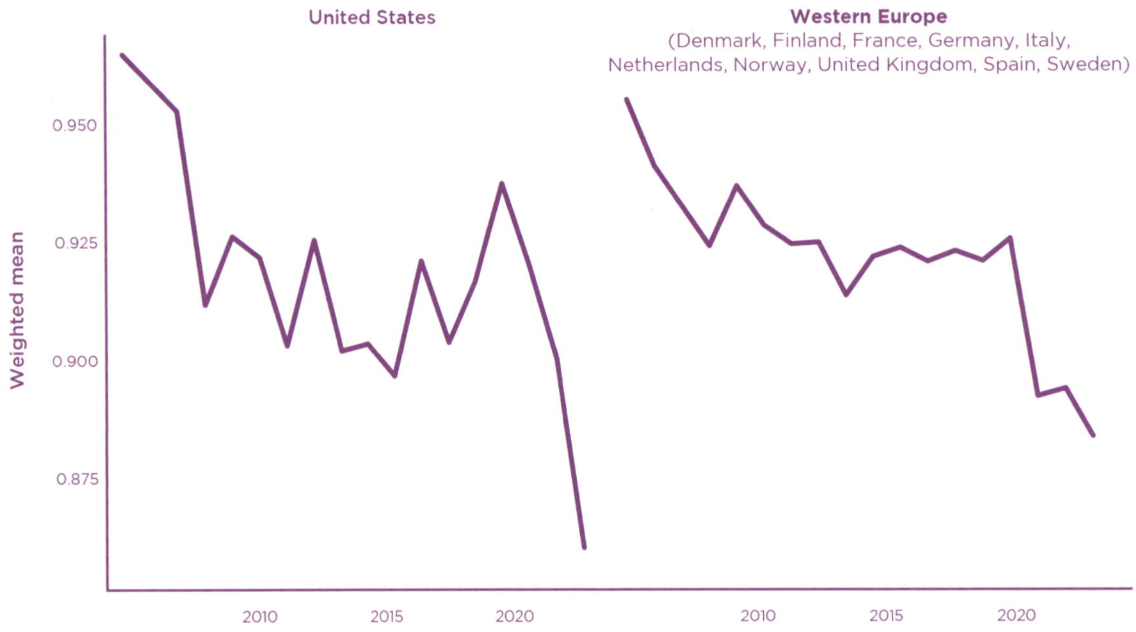

Note: Weighted average computed using design weights. Population weights used for region-level average.

Figure 7.14B: Trust in the US and Western Europe: Interpersonal trust
GSS (1972-2022), ESS (2002-2023)

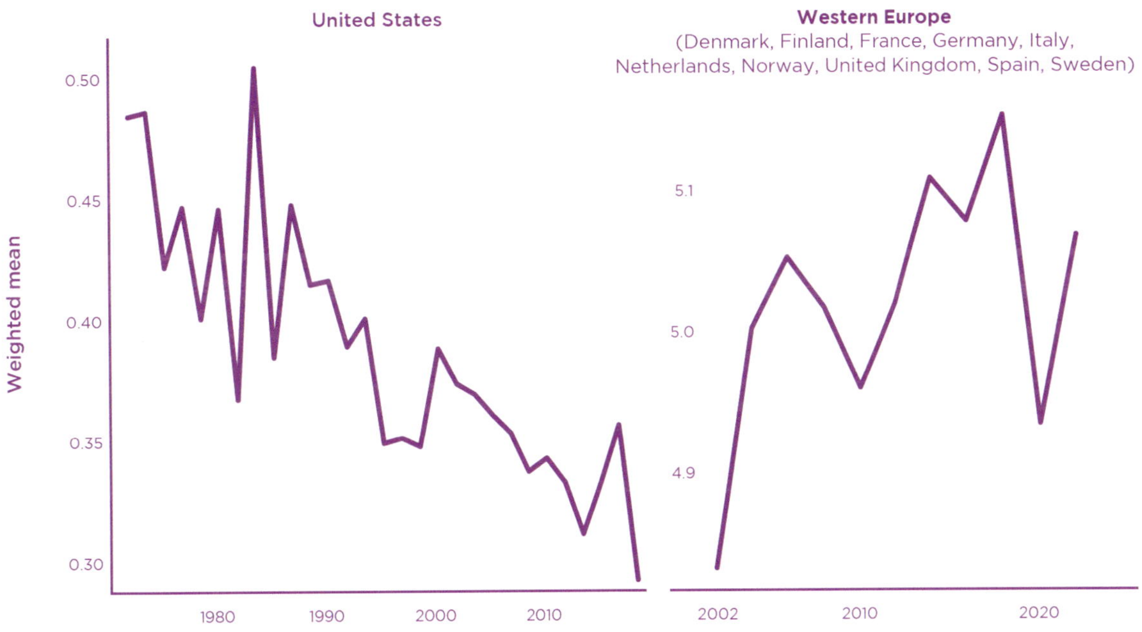

Note: Weighted average computed using design weights. Population weights used for region-level average.

Population sub-groups

The trends in attitudes, values, and voting behaviour that we have described are particularly pronounced among certain groups of the population. This sheds additional light on the underlying factors shaping the rise in anti-system votes.

Figure 7.15 displays the average change in life satisfaction for different population groups based on age, gender, education, economic status, and geography. The groups with increasing life satisfaction stand to the right of the dotted vertical bar and the groups with decreasing satisfaction stand to the left. The vertical axis shows the initial level of life satisfaction measured in 2006. The downward trend in life satisfaction is particularly steep among young people under 30, especially women, both in Western Europe and the United States, as already underlined in *World Happiness Report 2024*.

In terms of social trust, Figure 7.16 shows that, in the US, the average yearly change is negative for all groups, especially for people aged 30–44.

Figures 7.15 and 7.16 also illustrate the importance of financial difficulties. A very strong gradient opposes those who live comfortably on their income and those who find it difficult or very difficult. Education level also makes a difference. People with primary or secondary education endure a larger fall in life satisfaction than those with tertiary education. These are the groups that have shifted to anti-system votes.

Figure 7.15: Life satisfaction for population groups
Gallup World Poll (2006–2023)

Note: Countries weighted average computed using design weights. Population weights used for region-level average.

Figure 7.16: Social trust for population groups
Gallup World Poll (2006-2023)

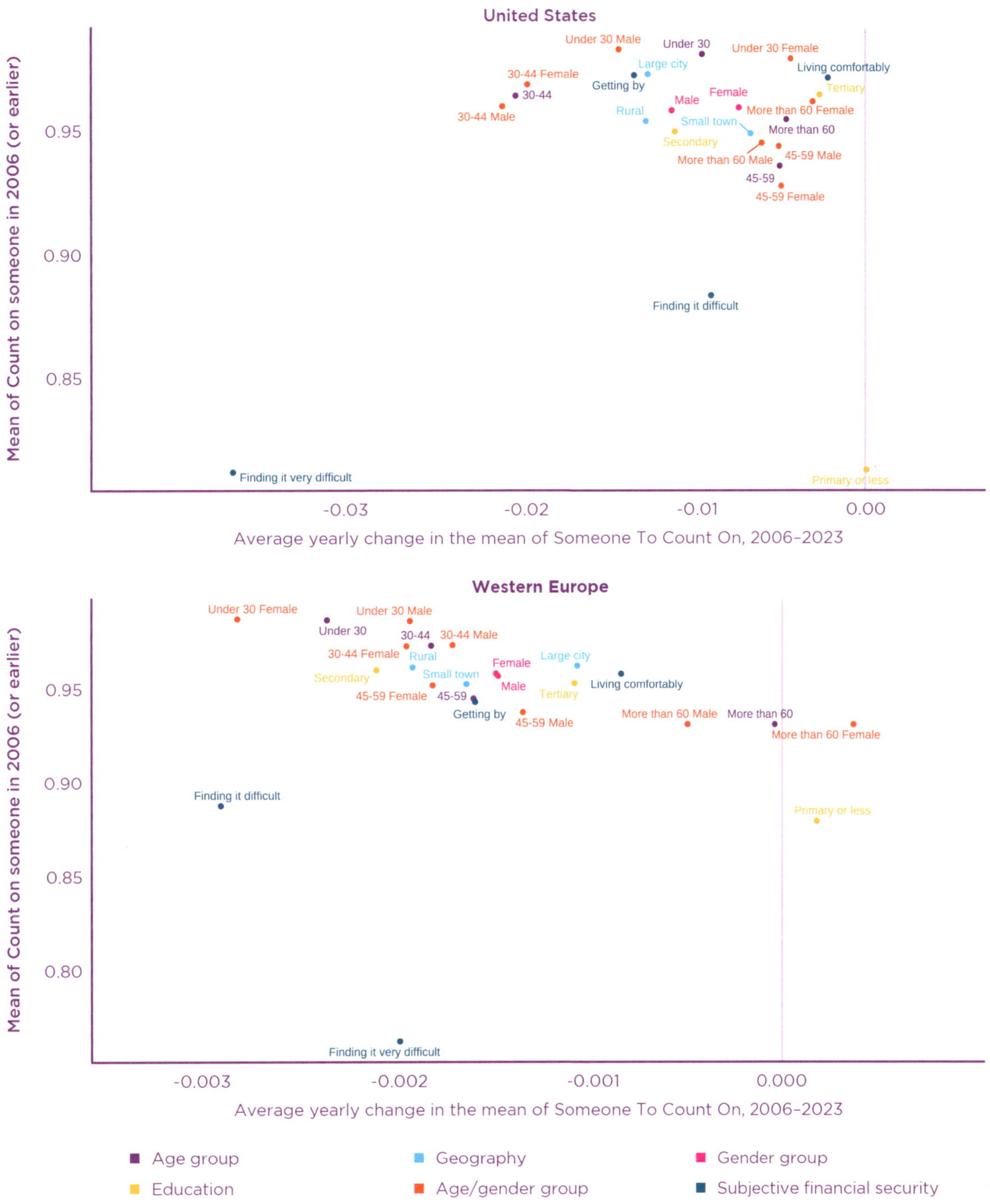

Note: Countries weighted average computed using design weights. Population weights used for region-level average.

Conclusion

In this chapter, we have documented a parallel fall in life satisfaction and social trust and its relationship with the rise in anti-system votes since the early 2000s. We propose a model of the relationships between three blocs of attitudes: (i) life satisfaction and social trust, (ii) cultural, political, and economic values, and (iii) voting behaviour. We show that low life satisfaction comes with anti-system attitudes while people's level of social trust determines their orientation towards anti-system parties at the left or the right. In Europe, citizens with low life satisfaction and low social trust, the 'anomics', tend to vote for far-right parties. In the context of the US two-party system, they tend to abstain and withdraw from public life.

The fall in life satisfaction cannot be explained by economic growth, at least not by average national income, as GDP per capita has been on the rise in the US and Western Europe since the mid-2000s. Rather, it could be blamed on the feelings of financial insecurity and loneliness experienced by Americans and Europeans – two symptoms of a damaged social fabric. It is driven by almost all social categories, but in particular, by the rural, the less-educated, and, quite strikingly, by the younger generation. This low level of life satisfaction is a breeding ground for populism and the lack of social trust is behind the political success of the far right.

Endnotes

1 Guriev and Papaioannou (2022).

2 Algan et al. (2017); Autor et al. (2020); Colantone and Stanig (2018).

3 Bonomi et al. (2021); Inglehart and Norris (2017).

4 In this paper, we use the term "populist" to describe the extreme right, such as the AfD in Germany, the Ligua or Fratelli d'Italia in Italy, or the Rassemblement National in France. The extreme left is equally anti-system, but, as we document here, does not share the same nationalistic tendencies and prejudices against immigrants, and supports a different economic and social platform.

5 Huber and Stanig (2007).

6 Algan et al. (2018); Ward (2019, 2020); Ward et al. (2024).

7 Art (2011); Golder (2016); Mudde (2007, 2016).

8 Helliwell et al. (2024).

9 De Neve et al. (2025); Putnam (2020).

10 Alabrese et al. (2018); Liberini et al. (2017a,b); Ward (2019, 2020); Ward et al. (2024).

11 Ward et al. (2020).

12 Algan and Renault (2024); Ward et al. (2024).

13 Enke (2020); Haidt (2017).

14 Inglehart and Norris (2017).

15 Algan et al. (2016); Coleman (1996); Fehr and Gachter (2000).

16 Algan et al. (2019).

17 See the online appendix for a detailed description of the data and variables.

18 Algan et al. (2016); Coleman (1996); Fehr and Gachter (2000).

19 See the online appendix for more details.

20 Easterlin (1974, 1995).

References

Alabrese, E., Becker, S. O., Fetzer, T., & Novy, D. (2018). Who voted for Brexit? Individual and regional data combined. *European Journal of Political Economy*, In press (available online August 2018).

Algan, Y., Cahuc, P., & Sangnier, M. (2016). Trust and the Welfare State: The Twin Peaks curve. *The Economic Journal*, *126*(593), 861–883.

Algan, Y., Guriev, S., Papaioannou, E., & Passari, E. (2017). The European Trust Crisis and the Rise of Populism. *Brookings Papers on Economic Activity*, 309–382.

Algan, Y., Beasley, E., Cohen, D., & Foucault, M. (2018) *The Rise of Populism and the Collapse of the Left-Right Paradigm: Lessons from the 2017 French Presidential Election*. CEPR Discussion Paper No. DP13103.

Algan, Y., Beasley, E., Cohen, D. & Foucault, M. (2019). *Les Origines du Populisme*, Editions du Seuil

Algan, Y. & Renault, T. (2024). La France sous nos tweets. Cepremap working paper.

Autor, D., Dorn, D., Hanson, G., & Majlesi, K. (2020). Importing Political Polarization? The Electoral Consequences of Rising Trade Exposure. *American Economic Review*, *110*(10), 3139–3183.

Art, D. (2011). *Inside the radical right: The development of anti-immigrant parties in Europe*. Cambridge University Press.

Bonomi, G, Gennaioli, N. & Tabellini, G. (2021). Identity, Beliefs and Political Conflicts, *The Quarterly Journal of Economics*

Colantone, I., & Stanig, P. (2018). Global Competition and Brexit. *American Political Science Review*, *112*(2), 201–218.

Coleman, S. (1996). *The Minnesota Income Tax Compliance Experiment State Tax Results*. Minnesota Department of Revenue.

De Neve, J.-E., Dugan, A., Prati, A., & Kaats, M. (2025). Sharing Meals, Social Connections, and Happiness Around the World. World Happiness Report 2025.

Easterlin, R. A. (1974). Does economic growth improve the human lot? Some empirical evidence. In P. A. David & M. W. Reder (Eds.), *Nations and Households in Economic Growth: Essays in Honor of Moses Abramowitz* (pp. 89-125). New York: Academic Press, Inc.

Easterlin, R. A. (1995). Will Raising the Incomes of All Increase the Happiness of All? *Journal of Economic Behavior and Organization*, *27*(1) 35-48.

Enke, B. (2020). Moral Values and Voting. *Journal of Political Economy*, *128*(10), 3679–3729. https://doi.org/10.1086/708857

Fehr, E. & Gachter, S. (2000). Cooperation and punishment in public goods experiments. *American Economic Review*, *90*(4), 980-994.

Funke, M., Schularick, M., & Trebesch, C. (2023). Populist Leaders and the Economy. *American Economic Review*, *113*(12), 3249–3288.

Golder, M. (2016). Far right parties in Europe. *Annual Review of Political Science*, *19*, 477-497.

Guriev, S., & Papaioannou, E. (2022). The political economy of populism. *Journal of Economic Literature*, *60*(3), 753-832.

Inglehart, R., & Norris, P. (2017). Trump and the populist authoritarian parties: the silent revolution in reverse. *Perspectives on Politics*, *15*(2), 443-453.

Helliwell, J. F., Layard, R., Sachs, J. D., De Neve, J.-E., Aknin, L. B., & Wang, S. (Eds.). (2024). World Happiness Report 2024. University of Oxford: Wellbeing Research Centre.

Huber, J. D. & Stanig, P. (2007). *Why do the poor support right-wing parties? A cross-national analysis*. Unpublished manuscript. Columbia University, New York.

Liberini, F., Redoano, M., & Proto, E. (2017a). Happy voters. *Journal of Public Economics*, *146*, 41-57.

Liberini, F., Oswald, A. J., Proto, E., & Redoano, M. (2017b). Was Brexit caused by the unhappy and the old? *IZA Institute of Labor Economics, Discussion Paper No. 11059*. Retrieved from http://ftp.iza.org/dp11059.pdf

Mudde, C. (2007). *Populist Radical Right Parties in Europe*. Cambridge, UK: Cambridge University Press.

Mudde, C. (2016). Europe's Populist Surge: A Long Time in the Making. *Foreign Affairs*, *95*, 25.

Putnam, R. (2001). *Bowling Alone*. Simon & Schuster.

Ward, G. (2019). Happiness and Voting Behavior. *World Happiness Report 2019*.

Ward, G. (2020). Happiness and Voting: Evidence from Four Decades of Elections in Europe. *American Journal of Political Science*. *64*(3), 504-518.

Ward G., De Neve, J.-E., Ungar, L. H., & Eichstaedt, J. C. (2020). (Un)happiness and voting in U.S. presidential elections. *Journal of Personality and Social Psychology*. *120*(2), 370-383.

Ward, G., Schwartz, H. A., Giorgi, S., Menges, J. I., & Matz, S. C. (2024). The role of negative affect in shaping populist support: Converging field evidence from across the globe. *American Psychologist*.

Chapter 8

Giving to others

How to convert your money into greater happiness for others

Michael Plant
Happier Lives Institute; University of Oxford

Joel McGuire
Happier Lives Institute

Samuel Dupret
Happier Lives Institute

Ryan Dwyer
Happier Lives Institute

Ben Stewart
Happier Lives Institute

From the *World Happiness Report*, we thank Barry Grimes, Lara B. Aknin, John F. Helliwell, and Richard Layard for their helpful feedback. We also appreciate the comments from the other chapter writers. Lastly, we thank Maggie Moss, Chris Conway, and Jim McGuire for their reviews of the chapter to check whether our work was at least somewhat accessible to a lay audience. All the feedback has much improved the chapter.

doi.org/10.18724/whr-e0cy-0r69

This version was last updated on 13 March 2025. Please see worldhappiness.report for the latest data.

Money *can*
buy happiness
for other
people.

8

Key Insights

The cost-effectiveness of charities can be measured and compared with a standardised metric of value: wellbeing-years (WELLBYs).

In the first global review of published evidence, we find that the cost-effectiveness of charities varies dramatically. The best charities in our sample are hundreds of times better at increasing happiness per dollar than others. Therefore, you can multiply your impact at no extra cost by funding more cost-effective charities.

There are no published evaluations of large, well-known charities or typical acts of charitable giving, such as helping the homeless, using a wellbeing approach. We present some rough evaluations of these cases but find them to be less cost-effective than nearly all the charities in our sample.

Introduction

We hope that if you're reading this, you're not just *interested* in world happiness, you want to *do something* about it. But, what can you do? This chapter focuses on something many of us already do and nearly all of us can do: give to charity. Each year, over a billion people donate more than $500 billion to charity,[1] driven in large part by a desire to help others.[2]

However, there are many worthy problems in the world, our resources are limited, and we don't want to waste our money. So, how can you get real change for your dollar? More specifically, how can you make the biggest difference to world happiness with what you have to spare?

People say "money can't buy happiness". At the Happier Lives Institute, we reject that claim but with a twist. We show that money *can* buy happiness *for other people* and we highlight the 'best buys' that have been identified so far. To do this, we compare the impact of charities using wellbeing-years (WELLBYs) per dollar, a method we will explain in due course.

In the first global review of published evidence, we find the best charities are hundreds of times better than others. This means you have an opportunity to make a far greater difference to world happiness, at no extra cost to yourself, simply by changing where you donate. If a friend told you they gave $200,000 to a charity, you'd probably be extremely impressed – that could be their life savings! However, it's possible to have that sort of *impact* for a fraction of the cost: giving $1,000 to the best charities may do just as much good as $200,000 to a randomly selected one.

You may be familiar with the claim – widely made in the effective altruism movement[3] and endorsed in a survey of charity experts[4] – that the top charities are a hundred times more impactful than the average charity. While this claim is believable, we are unaware of any research that demonstrates, or even tests this, with evidence. Indeed, we cannot think of any attempts to assess the cost-effectiveness of a representative sample of charities – a necessary first step for comparing 'the average' to 'the best'. The most relevant work we know of finds that health interventions can differ dramatically in cost-effectiveness (when measured with a standard health metric).[5] This suggests, but doesn't demonstrate, that the same may be true for charities. Our analysis provides the most direct test (we know of) for the claim that the best charities can be a hundred times more cost-effective than others.

We begin the chapter by introducing some key ideas behind our empirical comparison of charities. This includes brief discussions of effective giving, the focus on wellbeing, the WELLBY, and assessing charity cost-effectiveness.

We then turn to the main part of the chapter: a global comparison of charity cost-effectiveness. Scientific research into happiness has been growing for decades. In recent years, around 4,000 papers are published annually on the topic[6] and at least 24 countries now measure subjective wellbeing routinely.[7] However, efforts to find the most cost-effective ways to improve happiness are only getting started. A handful of WELLBYs per dollar estimates for charities and policies have been produced in the last 10 years. We found 24 estimates of different charities from four different evaluators. While this is neither a large nor representative sample of evaluations, it does cover a variety of charitable activities across the world. Our emphasis is on the variability in cost-effectiveness and the ability of research to reveal this, rather than the specific 'winners' and 'losers'. Given the newness of this field, we want to spark interest from donors and researchers, not draw definite conclusions.

In the first global review of published evidence, we find the best charities are hundreds of times better than others. This means you have an opportunity to make a far greater difference to world happiness, at no extra cost to yourself, simply by changing where you donate.

This sample of pre-existing estimates has two key gaps: it doesn't include any of the biggest and most well-known NGOs or any typical acts of charity, such as helping the homeless. How do these two options compare to the charities in our sample? We explain why it's hard to estimate the cost-effectiveness of many charities using WELLBYs, particularly for Multi-Armed NGOs (MANGOs) that run many programmes. We attempt to partially fill the gap by providing back-of-the-envelope calculations for a couple of well-known NGOs and for helping the homeless in wealthy countries. Our tentative conclusion is that the top charities in our sample are likely to be considerably more impactful per dollar than the missing options.

In the final parts of the chapter, we anticipate questions and objections and set out directions for further work.

Effective giving: a wellbeing approach

We expect many readers of the *World Happiness Report* will like the idea of finding and supporting the charities that make the most difference, per dollar, to world happiness. But, we don't want to assume readers have thought about why and how to do this. Before we get to the empirical analysis, we motivate and explain the project. For ease of reading, we have split up the various ideas. Readers familiar with these should feel free to skip over them.

Why give at all

The classic argument for giving to charity comes from the philosopher Peter Singer. He asks us to imagine we are walking past a shallow pond when we suddenly see that a child is drowning.[8] We can jump in and save that child, but this will ruin our expensive new suit. Are we morally required to save the child at some cost to ourselves? The reaction most people have is that we must wade in. The principle that appears to explain this reaction is that, as Singer puts it, "if it is in our power to prevent something bad from happening without thereby sacrificing anything of comparable importance, we ought, morally, to do it".[9]

This dilemma is not merely theoretical. People are suffering all around the world. Even if we can't help them directly, we could give some share of our income – perhaps 1% – to charities that can. This may not feel the same as doing it ourselves when the person is right in front of us, but the outcome may be identical. It seems that, if we have a moral duty to rescue the child, we have a moral duty to give something to charity.[10]

You may not feel wealthy enough to give, but you're probably wealthier than you think.

You may not feel wealthy enough to give, but you're probably wealthier than you think. If you earn the median US salary ($42k), you are in the top 2% of the global income distribution.

If you earn the median United States (US) salary ($42k), you are in the top 2% of the global income distribution.[11] What's more, humans have existed for a few hundred thousand years and the world has never been wealthier.[12] Believe it or not, you may be one of the richest people who has ever lived.

That's the stern, 'bad cop' argument for giving, but we can offer a 'good cop' one too: altruism is its own reward. Research shows that prosocial behaviour – and charitable giving in particular – improve self-reported wellbeing.[13] This is no surprise if you've ever felt the warm glow of giving a gift to family or showing kindness to a stranger. Other chapters in this edition of the *World Happiness Report* also show the importance of prosocial behaviour. Chapter 2 highlights how having a clear sense of one's positive impact increases its reward to well-being. This chapter will give you clear information about how impactful your charitable donations can be. If you are not sure if giving will make you happier, why not try it and find out?

Why to give effectively

This chapter will be most useful for those who want to give effectively i.e., to make the biggest (or at least, a bigger) difference with their donations based on evidence.

The argument for effective giving is simple: if you can make a bigger difference to others without a significant extra cost to yourself, you should do so. As we've already said and later show, this is not merely hypothetical: some charities are much more impactful than others, in terms of the happiness they create per dollar.

Not everyone is, or wants to be, an 'effective giver'.[14] Research has found that people are not effective givers due to: (A) information, they don't know how or where to give effectively, and (B) motivation, they prefer to support causes they are attached to, even if this would have less impact.[15]

At first glance, it's understandable that few donors seek information on charity cost-effectiveness. According to recent research, donors predict that the best charities helping the global poor are only 1.5 times better than the average ones.[16] If you

We may feel a greater bond with those who are close to us, but we may want to look beyond that if we can make a much bigger difference to those further away.

believe that charities don't differ much by cost-effectiveness, it doesn't make sense to invest time looking for a slightly better option. However, the evidence in this chapter shows that this belief is misplaced.

Research has also found that people focus too much on overheads (i.e., non-programme expenses like office costs and management salaries) and mistakenly think that charities with higher over-heads must be less cost-effective.[17] When donors realise that overheads and cost-effectiveness can diverge, they give more to charities with higher cost-effectiveness. To make this point, consider a hypothetical charity, Donuts for Billionaires, which uses volunteers and spends 100% of its money on snacks for the world's wealthiest people: no overheads, but not cost-effective. All the estimates we present in this chapter integrate the overheads when calculating cost-effectiveness.

Regarding motivation, it's understandable that people want to support projects close to their hearts - particularly if you think all charities are about as cost-effective as each other, or it's hard to compare them. Someone might want to support a charity that works on a particular health condition because a family member suffers from it. If this sounds like you, ask yourself this: do you care about that specific health problem, or do you care about it because of the impact it has, for instance the suffering it causes? If your ultimate goal is to have an impact, you may want to choose another charity that is better at achieving that objective.

People say that charity "begins at home", but we don't think it should end there. We may feel a greater bond with those who are close to us, but we may want to look beyond that if we can make a much bigger difference to those further away. We encourage readers to consider the global impact they could have.

Finally, the choice between giving to what makes you feel good and what does the most good doesn't have to be all or nothing. A middle option here is to split your donations, something we return to in the section on objections.

Assessing charity impact on wellbeing

Our analysis defines 'impact' in terms of changes to overall wellbeing. More conventional approaches might focus on poverty or health. However, we don't believe that improving poverty or health is the ultimate goal. Rather, the ultimate goal is to help people live happier lives. We should think of health and wealth as *means to an end*: the end being happiness.[18] Here's a test. If you gave to charity and it had no impact on anyone's happiness, nor reduced any suffering, would you be disappointed? If you would, that suggests you believe happiness is what ultimately matters.

Quantifying charity impact in WELLBYs

We quantify impact using wellbeing-years (WELLBYs).[19] One WELLBY is equivalent to a 1-point increase on a 0–10 self-reported wellbeing scale (typically life satisfaction) for 1 person for 1 year. So, if your wellbeing went from 6/10 to 7/10 for two years, that would be a gain of 2 WELLBYs.

Data from self-reported wellbeing questions, such as life satisfaction, are increasingly common[20] and widely viewed as meaningful.[21] The *World Happiness Report* uses a life evaluation question for its global ranking of countries. One of the main benefits of using self-reported wellbeing is that we can see, from the data, how much things like wealth and health really matter to people's lives, rather than assuming we know.

How significant is a 1-point change in life satisfaction? In high-income countries, being depressed is associated with a 1.3-point decrease in life satisfaction,[22] being unemployed is about a 0.5-point decrease,[23] a doubling of income is about a 0.2-point increase,[24] and marriage is associated with a 0.3-point increase a year after getting married.[25]

Quantifying impact in WELLBYs is a recent research area. It's a simple and powerful approach and we are not the only people to propose or use

it. The WELLBY was proposed as an alternative or complement to measures of health and wealth in *World Happiness Report 2021*,[26] and has been discussed in mainstream academia as a method for evaluating public policy.[27] In 2021, the WELLBY methodology was adopted by the United Kingdom (UK) Treasury as an official way of evaluating the impact of government policies.[28]

We see no good reason not to apply WELLBYs to charities: it gives us a scientifically credible, evidence-based way to work out how much good we do, per dollar, by giving to different charities. So, how can we assess charity impact in WELLBYs, and what's been found so far?

A review of the current literature

In this section, we discuss and compare the previous work that has analysed the cost-effectiveness of life-improving charities in WELLBYs. We offer some context first before presenting a table of the results and a few case studies to illustrate how the evaluations were done. Finally, we present a figure indicating the differences in cost-effectiveness and discuss how to interpret these results.

Context for WELLBY charity evaluation

Evaluations of charity cost-effectiveness in WELLBYs only started in the last few years. The first ones we know of were in Plant (2019) and the first explicitly couched in WELLBYs were all published in 2021: by Frijters and Krekel, State of Life, and the Happier Lives Institute (where the authors work).[29]

At the time of writing, we found 24 charity evaluations from four evaluators:[30]

- State of Life: 3 charities
- Pro Bono Economics: 3 charities
- Happier Lives Institute: 14 charities
- Krekel and colleagues: 4 charities

The first three are organisations. (Christian) Krekel is an academic who has produced estimates with different colleagues so, for simplicity, we say "Krekel and colleagues".[31]

At its simplest, a charity evaluation might look at the total amount an organisation spends to provide a service, how many people it provides that service to, and then estimate the average benefit per person reached. So, if a charity spent $1 million, reached 50,000 people, and they each got a 1 WELLBY benefit, that's 50,000 WELLBYs for $1 million, a cost-effectiveness of $1 million / 50,000 = $20 per WELLBY.

Getting sensible estimates for these numbers can be a resource-intensive process. The charity evaluations we draw on usually consist of a technical report at least as long as this chapter. We summarise each evaluation in a few paragraphs in the online appendix. For brevity and readability, we do not describe every charity evaluation in the main text.

All the evaluations produce a similar *output*: a cost-per-WELLBY figure for each charity (or estimates easily convertible to these terms). However, they differ in terms of their *inputs*: the evaluations are not all done in the same way. The main differences are:

1) The depth of the analyses and the quality and quantity of the evidence used.[32]
2) Modelling choices, such as adjusting for internal and external validity, and whether researchers tried to include longer-term and societal effects in addition to the short-term impacts on the direct beneficiaries.[33]

For cost-effectiveness estimates to be credible, they need good data and analysis. If we want perfect data and analysis, we will be waiting forever. The estimates we present below are informed by data, not 'facts', and we hope further work will refine them. Nevertheless, as the alternative to explicit, quantitative estimates is to rely on intuitive judgments, we see real value in producing and using estimates for decision-making purposes.

We take the evaluators' assessments at face value, rather than critiquing or adjusting them. Further work could attempt to 'harmonise' the estimates, but this wasn't essential to draw our main conclusion that cost-effectiveness differs radically between charities.[34] We do, however, include our subjective assessments of the relevance of the evidence and depth of analysis: these can be understood as indicating uncertainty. These estimates will also need to be updated in the future. They reflect the charities' programmes at a particular point in time and will naturally become less realistic as programmes and operating conditions change.

The charity evaluations

With those caveats out of the way, we present our results. In Table 8.1 we summarise the 24 evaluations (including unpublished ones) we have collected for this chapter. The columns are largely self-explanatory, but we provide further details for interested readers in an endnote.[35]

Table 8.1: Wellbeing cost-effectiveness estimates for 24 charities
(ordered by 'Cost per WELLBY', lowest to highest)

Charity	What does it do?	Cost per WELLBY	Duration of effect (years)	Country income	Total sample	Total studies	Causal evidence	Evidence relevance	Depth of analysis	Evaluator
Pure Earth	Advocacy for reducing lead exposure (Advocacy campaign in Ghana)	$9.23	lifetime	LMICs	947	2	No (longitudinal associative studies with adjustment)	low	Medium	Happier Lives Institute
Taimaka	Treating acute malnutrition (2 months of therapeutic food)	$15.15	lifetime	LMICs	118,370	18	Yes (RCTs)	low	Medium	Happier Lives Institute
Icddr,b	Early childhood psychosocial stimulation (32 sessions of educational play)	$19.95	32	LMICs	2,928	4	Yes (RCTs)	medium	Medium	Happier Lives Institute
Friendship Bench	Psychotherapy (6 sessions)	$20.61	4	LMICs	35,854	95	Yes (meta-analysis of RCTs)	medium	In-depth	Happier Lives Institute
StrongMinds	Psychotherapy (6 sessions)	$24.77	4	LMICs	35,739	92	Yes (meta-analysis of RCTs)	medium	In-depth	Happier Lives Institute
Earthenable	Upgrading flooring (1 new earthen floor)	$34.06	8	LMICs	2,742	1	Yes (RCT)	medium	Shallow (unpub-lished)	Happier Lives Institute
Tearfund	Multifaceted, religious (community engagement)	$39.33	1	LMICs	7,212	1	No (comparing to non-randomised group without treatment)	high	Medium	State of Life
NEPI	CBT and cash transfers for crime reduction (12 sessions + $300)	$46.34	10	LMICs	15,899	2	Mixed (RCT and associative study)	medium	Medium	Happier Lives Institute
Fortify Health	Fortifying wheat flour with iron (1 year of fortified wheat)	$46.19	1	LMICs	1,002,135	25	No (associative studies with adjustment)	medium	Medium	Happier Lives Institute
TECHO	Emergency housing (1 new small house)	$70.11	3	LMICs	2,203	4	Yes (RCTs)	high	Shallow (unpub-lished)	Happier Lives Institute
Royal Volun-tary Service	Volunteering (15 tasks to help with COVID crisis)	$81.99	1	HIC (UK)	4,033	1	Yes (RCT)	high	Medium	Krekel et al. (2024)
Action for Happiness	Happiness courses (6 sessions)	$100.00	1	HIC (UK)	146	1	Yes (RCT)	medium	Shallow	Frijters and Krekel (2021)
GiveDirectly	Cash transfers ($1,000)	$132.40	8	LMICs	35,961	12	Yes (meta-analysis of RCTs)	high	In-depth	Happier Lives Institute
Parkrun	Exercise, volunteer (going on more runs – unclear)	$205.67	1	HIC (UK)	567	2	No (associative studies with adjustment)	high	Medium	State of Life
London Youth Rowing's Active Row	Sport, exercise (unclear amount of sports training)	$500.00	Unclear (assume 1)	HIC (UK)	525	1	No (associative study)	high	Medium	State of Life
Walking with the Wounded Head Start	Therapy (number of sessions is unclear)	$1,674.81	Unclear (assume 1)	HIC (UK)	118	1	No (matched comparison group with differ-ence-in-difference)	high	Medium	Pro Bono Econom-ics
Restoration Trust: Human Henge	Mental health support (several months of mental health activities)	$3,568.93	1	HIC (UK)	20	1	No (associative study)	high	Shallow	Frijters and Krekel (2021)

Table 8.1: Wellbeing cost-effectiveness estimates for 24 charities (continued) (ordered by 'Cost per WELLBY', lowest to highest)

Charity	What does it do?	Cost per WELLBY	Duration of effect (years)	Country income	Total sample	Total studies	Causal evidence	Evidence relevance	Depth of analysis	Evaluator
Royal National Lifeboat Institution	Search and rescue, volunteering (a year of search and rescue)	$6,385.66	lifetime	HIC (UK)	NA	0	No (associative study)	high	Shallow (BOTEC for this chapter)	Happier Lives Institute
Walking with the Wounded Employment	Employment (unclear amount of employment aid)	$5,601.19	Unclear (assume 1)	HIC (UK)	92	1	No (matched comparison group with difference-in-difference)	high	Medium	Pro Bono Economics
Football Beyond Borders	Sports, education and counselling (1 year of sports training, tutoring, counsel)	$8,690.85	1	HIC (UK)	153	1	No (matched comparison group with difference-in-difference)	high	Medium	Pro Bono Economics
Hypothetical charity	Cash transfers to unhoused people (lump-sum unconditional 7,500 CAD)	$19,994.12	2	HIC (Canada)	115	1	Yes (RCT)	unclear	Shallow (BOTEC for this chapter)	Happier Lives Institute
Hypothetical charity	Housing and support for unhoused	$35,027.50	1	HIC (UK and Canada)	2,148	1	Yes (RCT)	unclear	Shallow	Frijters and Krekel (2021)
Guide Dogs	Providing guide dogs for the blind (one guide dog companion)	$40,766.67	7	HIC (UK)	87	1	No (associative study)	medium	Shallow (BOTEC for this chapter)	Happier Lives Institute
Deworm the World	Mass deworming (1 year of deworming)	Unclear	Unclear	LMICs	5,094	1	Yes (RCT)	low	In-depth	Happier Lives Institute

In the following sections, we highlight one analysis from each evaluator for illustrative purposes. A longer description of each charity and its evaluation is provided in the online appendix. Some readers may want to skip over these details to see the visual comparisons of cost-effectiveness in the figures below – and return here afterwards.

Psychotherapy in Sub-Saharan Africa (Happier Lives Institute)

StrongMinds[36] (in Uganda and Zambia) and Friendship Bench[37] (in Zimbabwe) are two charities scaling access to basic mental healthcare in Sub-Saharan Africa using lay practitioners to deliver psychotherapy to people with depression or anxiety.[38]

To estimate the effect of these charities we combined evidence from several sources. For the effect on the direct recipient, we drew on three types of evidence.

- First, we performed a systematic search and collected 84 randomised control trials (RCTs) of psychotherapy delivered in low- and middle-income countries.
- Second, we used RCTs associated with the charities' programmes (1 for StrongMinds and 4 for Friendship Bench).
- Third, we used the monitoring and evaluation pre-post data collected by the charities themselves.

We then combined the estimates from these different evidence sources based on our judgement of the relative statistical uncertainty, quality, and relevance of each estimate. To estimate the effect on other members of the household, we combined data from six studies[39] and adjusted them for validity concerns (e.g., adjustments for publication bias[40]).

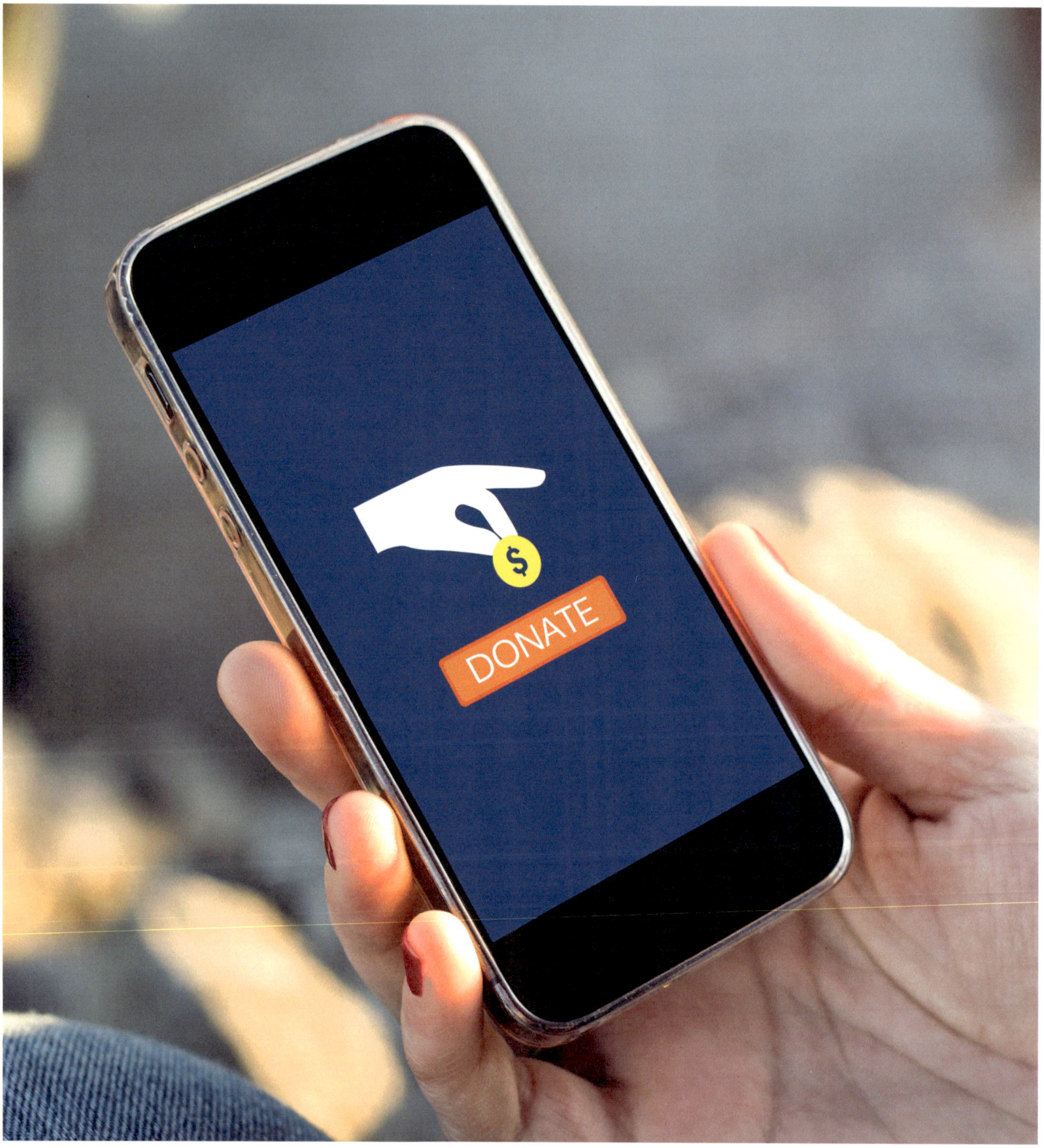

Based on these evidence sources and our methodology, we estimated an overall effect of 0.80 WELLBYs for Friendship Bench and 1.80 WELLBYs for StrongMinds per person treated (including the effects over time and the impact on the beneficiary's household). The cost of these programmes is very low as they work in low-income countries and use trained volunteers. The cost to treat an individual for therapy is $17 through Friendship Bench and $45 for StrongMinds. This results in a cost per WELLBY of $21 for Friendship Bench and $25 for StrongMinds. We consider this an in-depth analysis.

Parkrun (State of Life)

Parkrun[41] is a UK charity that supports free community runs delivered by volunteers every weekend. State of Life estimates its effects using correlational evidence.[42] First, they use a larger, but less relevant, dataset from the UK (n ≈ 60,000) that shows the relationship between running, volunteering, and wellbeing more broadly. Second, they use a longitudinal study that followed participants before and after they participated in a Parkrun event in the UK (n = 576).

They apply these effects to the total number of walks or runs (8,590,393) and volunteering for Parkrun instances (974,255) in 2019 (see their Table 8 and Approach 3). They then discount the effects in both cases to account for the fact they are missing a randomly assigned control group and so can't be sure how much of the correlation is due to the effect of related activities.

They did not provide a central estimate between the two estimates, so we took the liberty of averaging them together. The result is Parkrun producing 27,651 WELLBYs in 2019. The total operational cost of Parkrun in 2019 was £4.5 million, resulting in a cost per WELLBY of £165 ($206).[43] We consider this analysis to be of medium depth.

Action for Happiness (Frijters and Krekel)

Action for Happiness[44] is a charity that delivers courses and promotes wellbeing skills. The analysis[45] is based on a randomised control trial which found that a six-week course teaching wellbeing skills raised life satisfaction by around 1 point on a 0-10 scale at the two-month follow-up.

To get the total effects over time the authors assumed the effects remained constant for a year before dropping off completely. They used the course data that suggests it costs £80 per participant and assumed these costs and randomised control trial results are representative of the charity in general. Overall, Action for Happiness has a cost per WELLBY of £80 (or $100).[46]

We consider this to be a shallow analysis. Note that Action for Happiness has since switched from an in-person to a virtual model, meaning the evidence is now less relevant to the delivery in practice. We are uncertain how this would change its cost-effectiveness.

Football Beyond Borders (Pro Bono Economics)

Football Beyond Borders[47] is a UK charity dedicated to improving school outcomes for students through a combination of tutoring with a trusted adult, football practice, and therapeutic support. Their cost-effectiveness was evaluated by Pro Bono Economics.[48]

The study was not a randomised control trial, but they used a matched control method and a difference-in-difference estimator, which is better than relying on pre-post changes. 153 Football Beyond Borders participants were statistically matched with individuals drawn from the Manchester BeeWell dataset. They found that participants' wellbeing improved by 0.15 points on a 0–10 life-satisfaction scale[49] and assumed the effects lasted for one year.

During the 2022–23 school year, 2,401 students benefited from the Football Beyond Borders programme, representing a 360 WELLBY benefit. It cost £2.5 million to run the programme. This results in a cost of £6,953 ($8,691)[50] per WELLBY. We consider this a medium-depth analysis.

Results and interpretation

Distribution of cost-effectiveness

In Figure 8.1, we present the cost-effectiveness estimates for 19 of the 24 analyses.[51] We identify the different evaluators with unique colours and indicate the depth of analysis with different circle sizes. We use report length as a proxy for depth: in-depth evaluations have multiple analyses that could each be a separate report, medium-depth evaluations have a standalone report, and shallow evaluations are rough, brief analyses that are not presented in standalone reports.

We do not include confidence intervals around the central estimate – which is a typical way of representing uncertainty – for reasons explained in an endnote.[52]

Figure 8.2: The cost-effectiveness of 19 charities (WELLBYs per $1,000)

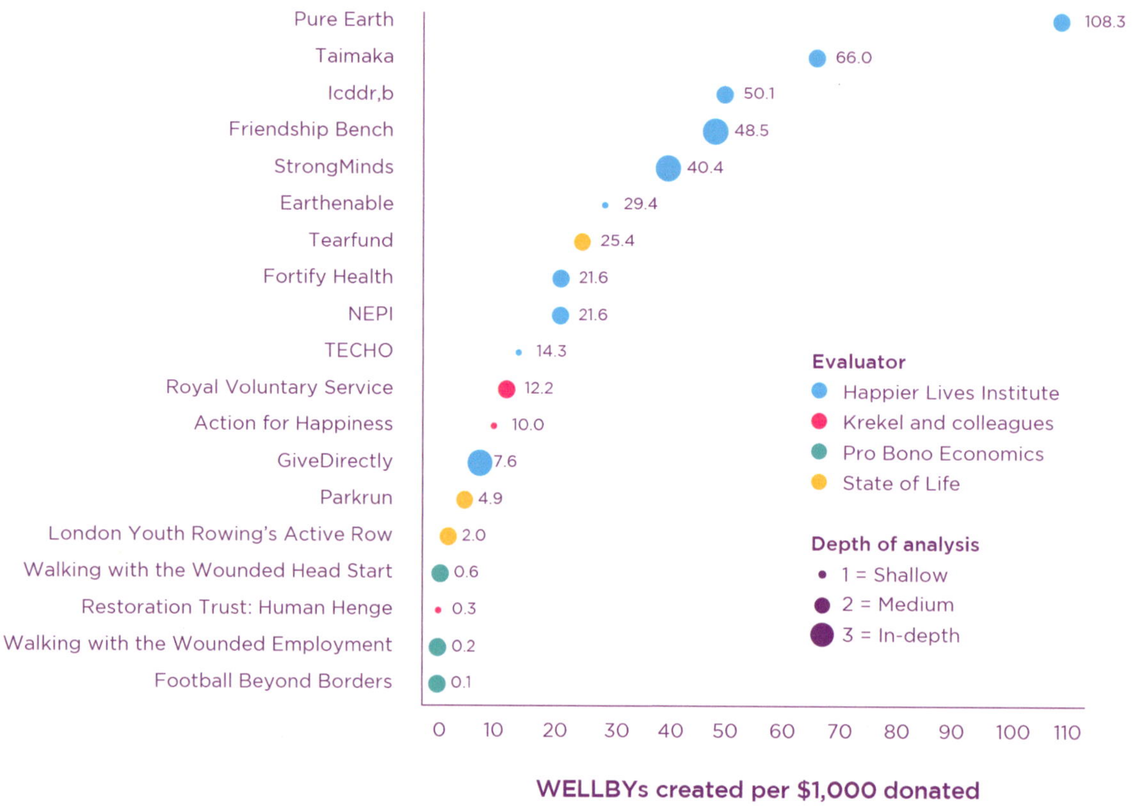

The figure shows that the cost-effectiveness of the charities varies substantially from $9 to $8,691 per WELLBY – a 942 fold difference in cost-effectiveness. In Figure 8.2, we present the same results, but this time in *WELLBYs per $1,000*, rather than *cost per WELLBY*. This is the same information, presented differently. The advantage of showing WELLBYs per $1,000 is that it does not compress the top options in the way cost per WELLBY did above. This presentation shows how much more cost-effective the top options are compared to the middle and bottom ones.

Explaining the distribution

These estimates show that charities differ radically in how much happiness they create per dollar.

Why is this? The natural explanation is that the top charities are providing cheap and impactful interventions in low- and middle-income countries (LMICs). In contrast, the less cost-effective charities are working in high-income countries in ways that are much more expensive.

This difference is starkly presented in Figure 8.3. The top five charities in our sample[53] all operate in LMICs and have an average cost per WELLBY of $18. This is 142 times more cost-effective than the seven UK charities in our sample[54] which have an average cost per WELLBY of $2,553.[55] However, those seven charities are still cost-effective in the UK context, where governmental guidelines value 1 WELLBY at £13,000.[56]

Figure 8.1: The cost-effectiveness of 19 charities ($ per WELLBY)

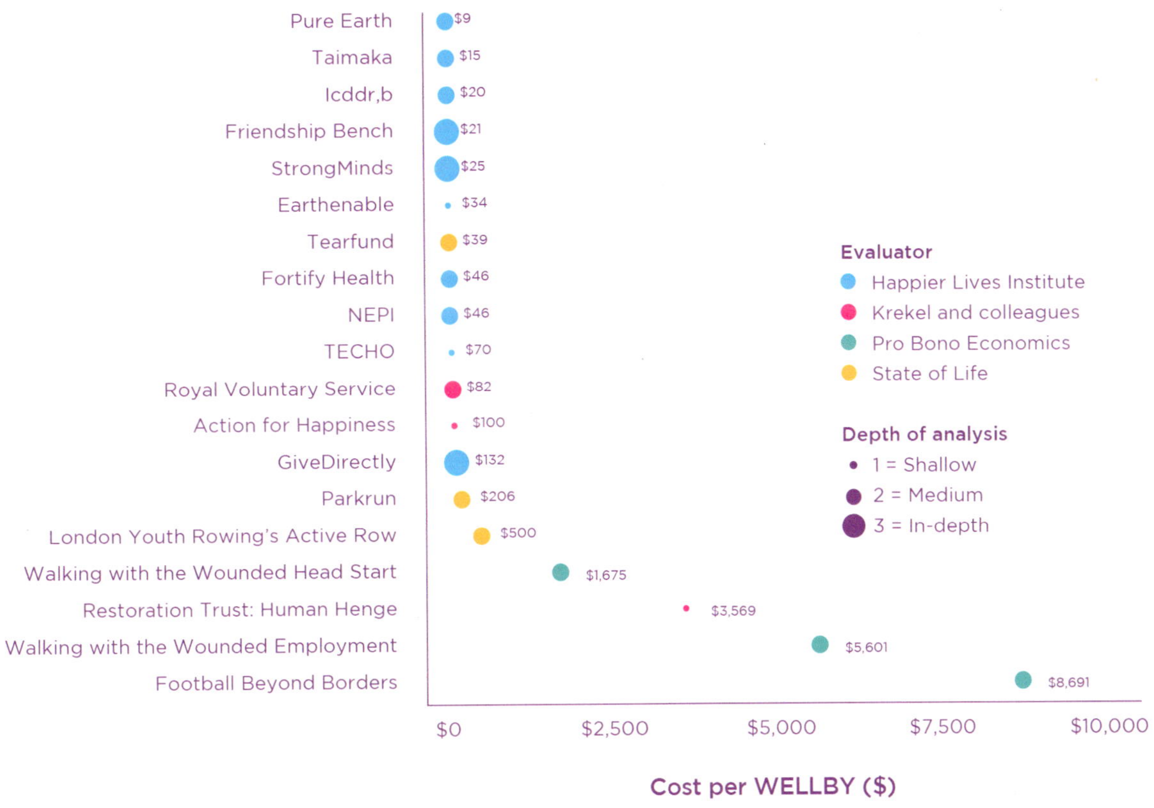

Figure 8.3: The average cost-effectiveness of UK and LMIC charities

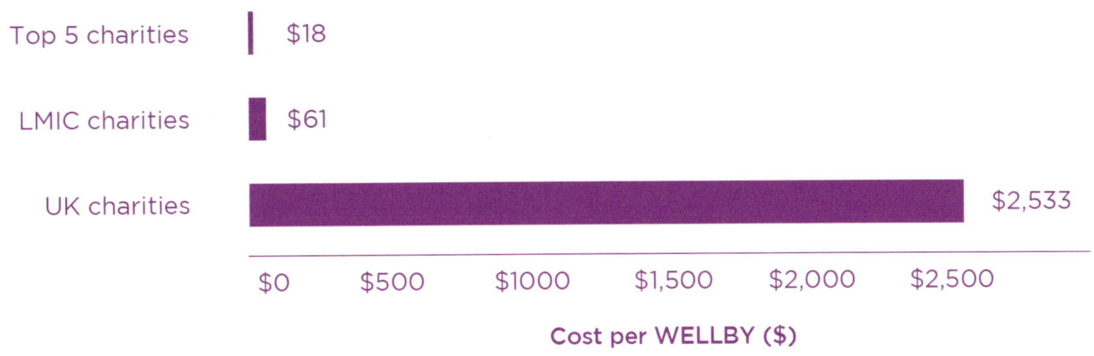

Figure 8.4: The average cost-effectiveness of charities by evaluator

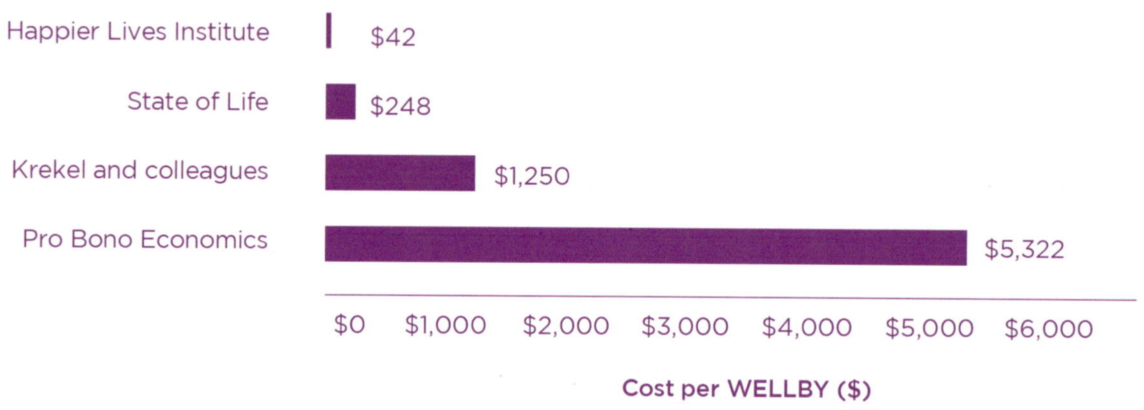

That said, could the differences in cost-effectiveness be due to some evaluators being more conservative or generous in their analysis than others? At first glance, this may seem like an important factor since the differences in average cost-effectiveness across evaluators are also substantial (see Figure 8.4).

There are differences in methodology that may explain some of the variation we see between evaluators. We discuss these in more detail in the online appendix but highlight the more important points below.

1) Only a few evaluations have been done per evaluator: four or less for State of Life, Pro Bono Economics, and Krekel and colleagues, and just over 10 for the Happier Lives Institute (HLI). As these are such small samples, a single charity evaluation can have a large impact on the average.

2) The evaluators have different sampling processes. State of Life, Pro Bono Economics, and Krekel and colleagues are, to a large extent, analysing UK charities that they were asked to analyse or which were convenient to analyse. In contrast, HLI explicitly set out to look for the most cost-effective charities and focuses on charities working in low-income countries.

It seems reasonable to assume each evaluator is using similar levels of conservatism over time. The main result – charity cost-effectiveness differs substantially – is true even if we look within each evaluator. For the Happier Lives Institute, the best charity is 14x more cost-effective than the least cost-effective charity; for State of Life it's 13x; for Krekel and colleagues it's 44x; and for Pro Bono Economics it's 5x. If charity cost-effectiveness differs considerably *within* each evaluator, it is unsurprising that cost-effectiveness differs considerably *between* evaluators with different sampling methods.

A final reason to be reassured comes from analogous WELLBY analyses for policies which also find that the cost-effectiveness of policy initiatives varies to a large degree – even more so than the charities discussed here.[57] On the low end, some policies, such as government-funded psychotherapy in the UK, are estimated to have a *negative* cost per WELLBY as they save the government money.[58] On the higher end, some policies appear highly ineffective. For example, extending the winter fuel allowance in the UK to help older people cover heating costs (a policy the Labour party ended while we were writing this) is estimated to cost $100,000 per WELLBY.

In summary, we find that charities differ substantially in their impact per dollar, with some hundreds of times better than others. This finding is in sharp contrast to the seemingly common but mistaken belief that the best charities are less than two times better than average charities.[59]

What's missing from the sample?

Our review has two obvious gaps. First, it doesn't include any large and well-known NGOs, such as Oxfam or Save the Children. Second, it doesn't include typical acts of charity, like giving to the homeless within one's own country. How do these compare to the charities in our sample? In this section, we try to answer that question but we're unable to draw strong conclusions and more work is needed.

Multi-armed NGOS

The immediate challenge with evaluating large charities is that they often run tens, even hundreds, of programmes. We call these types of charities 'MANGOs', standing for 'multi-armed NGOs'. An example of a MANGO is Oxfam. In 2023, it reported running 727 programmes across diverse areas such as economic and gender justice, climate action, and humanitarian response.[60]

The charities in our sample only run a single (or a few) programmes. Assessing one programme carefully in terms of WELLBYs per dollar is a considerable task. Assessing a MANGO requires an evaluation of every individual programme and weighing them by budget allocation to calculate average cost-effectiveness. This is out of scope. The alternative would be guesswork, which would not be informative. Leaving aside the focus of WELLBYs per dollar, it is very hard to find cost-effectiveness assessments by charities of any kind. Out of the largest 100 charities in the UK,[61] we can only provide a ballpark cost-effectiveness estimate for two, and this is because they have a clear primary output.

The Royal National Lifeboat Institute (RNLI) is a UK charity that rescues people in danger at sea. In 2023, they report saving a total of 269 people[62] and a total expenditure of £242.6m[63] ($303.3m).[64] These figures imply a cost of $902k per life saved. We estimate that each life they save produces an average of 177 WELLBYs.[65] This results in a cost per WELLBY of $6,386, not accounting for any psychological benefits for the wider population.[66]

The Guide Dogs for the Blind Association spent around £74m on guide dogs in 2023[67] and reported that they formed 469 new guide dog partnerships. We estimate, based on three limited studies, that having a guide dog leads to a 0.7-point increase in life satisfaction which is a very large effect per year.[68] Guide dogs work for seven years on average.[69] Therefore, we estimate that one extra guide dog leads to 4.83 WELLBYs (counting just the impact on humans). This means the total impact in 2023 was 2,266 WELLBYs, costing $40,767 per WELLBY.

While these two organisations are clearly making a difference to people's lives, our very shallow estimates suggest they are substantially less cost-effective than the top charities in our sample which cost around $18 per WELLBY.

Our second concern about MANGOs – on top of the assessment challenge – is the 'dilution effect'. A MANGO's cost-effectiveness will be the combined cost-effectiveness of its individual programmes. If an organisation has one very cost-effective programme, adding a low cost-effectiveness programme will *reduce* its average impact per dollar. To remain as cost-effective, it has to add a programme that is just as efficient. For a MANGO to be more cost-effective than the top charities in our sample, their average (budget-weighted) programme would need to be as good as the best single programme NGOs. This means some of their programmes need to be better than anything we've observed so far. This isn't impossible, but it seems unlikely. As a result, we expect MANGOs to be less cost-effective than charities that do one thing well.

By reallocating resources from the least to the most impactful options, MANGOs could make a far greater difference.

Considering the funding allocated to MANGOs and how little is known about their cost-effectiveness, we encourage further work on this by researchers and the charities themselves. Indeed, we hope MANGOs see the value of the WELLBY approach. It makes it possible to compare otherwise incomparable programmes. This means charities can see what works well and what does not. By reallocating resources from the least to the most impactful options, MANGOs could make a far greater difference.

Homelessness interventions

As well as donating to large, well-known charities, many people donate to small, local causes or even individuals. When people think of 'charity', we expect that helping the poor and homeless in their local area is one of the first things that comes to mind.[70] For the purposes of comparison, we briefly try to estimate the impact of helping the homeless using two analyses of homelessness interventions.

Frijters and Krekel (2021) estimate, based on an RCT, that providing "housing first" for people experiencing homelessness is reasonably effective, providing 0.67 WELLBYs per person.[71] However, they also found it was very expensive, costing around $23k per person. Thus, they concluded that providing a housing first intervention costs $35k per WELLBY,[72] which is about 2,000 times less cost-effective than the top charities in our sample.

However, economic theory suggests that cash is (typically) better as it allows you to buy whatever you want.[73] We found one study looking at cash transfers to homeless people living in Vancouver, Canada.[74] This RCT looked at unconditional, lump-sum cash transfers of 7,500 CAD (6,637 USD). Based on the results of the trial, we model that the impact lasts for two years, leading to an overall effect of 0.33 WELLBYs costing $6,667 per person.[75] This leads to a cost per WELLBY of $19,994. This is more promising than 'housing first', but still hundreds of times less impactful per dollar than the best-performing alternatives.

An obvious limitation of our estimates is the exclusion of possible 'spillover' benefits, for instance, that homeless people, once housed, require fewer emergency services.[76] Nevertheless, we think these estimates provide at least some evidence to suggest that a very common charitable act, helping the homeless, is difficult to do cost-effectively.[77]

These estimates provide at least some evidence to suggest that a very common charitable act, helping the homeless, is difficult to do cost-effectively.

Questions and objections

In this section, we anticipate some questions and concerns that we haven't addressed so far and provide brief responses. We end with some concluding remarks for donors and researchers.

How much should I give?

As a rule of thumb, the right amount to give is the largest amount that *you can sustain*. Peter Singer (the originator of the Shallow Pond thought experiment) proposes a stepped scale. It starts at 1% for those earning $40k-$80k, is 5% between $80k-$120k, and eventually rises to 50% for those earning over $50m.[78]

Will giving make me happier?

As we mentioned earlier, the evidence indicates it does.[79] However, this is presumably only up to a limit. We know of no research on the tipping point of when giving reduces our wellbeing.

What about causes you haven't discussed?

WELLBY cost-effectiveness allows us to shine a bright, scientifically credible, and evidence-based light in a narrow spot: the topics where we have good self-reported wellbeing data. This is both a strength and a limitation. Our exclusion of charities from other areas, such as climate change and animal welfare, doesn't mean those topics have no value; only that the more we want to go beyond the existing evidence, the more we have to peer into the darkness and speculate. For instance, there is no clear theoretical problem with applying the WELLBY approach to climate change interventions - at base, it involves estimating the impacts on wellbeing now versus wellbeing later - but any analysis will rest heavily on assumptions about long-term effects.[80] We hope that readers will be interested in the evidence we've assembled here, but not especially interested in our opinions, which is why we don't extend the analysis beyond the available data.

I already give to [insert charity]. Is it wrong to switch my donations to a more cost-effective charity?

In short: no. Some people are understandably reluctant to switch because it feels like they are letting down the organisation they've been supporting and the people it benefits. But if you think an alternative charity will do more good per dollar, it helps to keep in mind the greater total benefit you'll cause by supporting that alternative.

Should I split my donations?

A simple answer is that, if you want to maximise your impact, you should only give to the charity you think is most cost-effective. However, that organisation could reach 'diminishing marginal returns'.[81] For instance, their budget gets filled and they can't usefully spend the money, so they stop being as cost-effective. At that point, you should switch to the next best option, and so on. This implies small donors shouldn't switch, but perhaps large donors should as they have enough money to fill one organisation's budget and then move to the next.[82]

However, this simple answer isn't very satisfying. One problem is that it's often very hard to compare the options. For instance, you might care about helping people today and helping future generations (or helping globally and helping locally, and so on). Lots of us have the intuition we don't really know how to compare these, but we want to do *something* about both. Faced with this problem, you might have a current people 'bucket' and a future generations 'bucket', then somehow split your budget between these, and then try to identify the best thing(s) in each. The authors share the intuition that this is the right approach, but it's unclear how to justify it or make the details precise, and very little seems to have been written on it in philosophy.[83]

Another issue is that many people have a strong desire to split. Perhaps you're not motivated to give everything to a 'boring-but-effective' charity (even if you think that would make the most difference) and you want to give something to a cause that tugs on your heartstrings. In this case, it's clearly better to split your donations, rather than give nothing. Perhaps you give 80% to the boring-but-effective option and 20% to your 'heartstrings' project to stay motivated. Again, the best giving is sustainable.

What about inequality?

Lots of people have the intuition that it's more important to increase the happiness of someone who's at 3/10 than someone who's at 8/10, even if the costs and all other factors are the same. This can be understood as a concern for equality, not merely efficiency. None of the estimates in our sample account for equality: they treat a 1-point increase as having the same value for both the 3/10 and the 8/10 person.[84]

Further work could examine this, but we don't think accounting for equality would change the rankings. The most cost-effective charities tend to target people with low happiness anyway - what makes them so efficient is that they address serious problems. For instance, Strong-Minds and Friendship Bench help depressed people in low-income countries, some of the worst-off people on the planet. In contrast, the less cost-effective options in the sample are often targeting better-off people in a wealthy country. So, adding in an equality factor would probably exaggerate the differences, making the most cost-effective options look more valuable, rather than reverse the results.

Concluding remarks

For donors

The main takeaway for donors is that charities vary hugely in terms of cost-effectiveness. By picking carefully, we can have vastly more impact on happiness for the same monetary cost. We have to emphasise how extraordinary the differences in charity cost-effectiveness are. When we are buying items for ourselves we are typically delighted when there's a sale and we discover we can get 20% more for our money. But if you want to buy happiness *for other people*, you can potentially get 100x, or more, by donating to the most cost-effective charities. It would be like a store running a secret campaign where you spend $10 on a phone charger and receive a complimentary $1,000 laptop.

Research on WELLBY cost-effectiveness is in its early days, for both policies and charities. We hope more research will be done, even better charities will be revealed, that the estimates we have presented will be updated, and therefore this chapter will quickly become outdated.

For those wanting to see the latest findings and recommendations, we advise you to visit www.happierlivesinstitute.org which acts as a living literature review for the cost-effectiveness of (top) charities.

From a global and historical perspective, you are probably much wealthier than you realise. The difference you can make to people's happiness globally is probably far greater than you ever thought – if you follow the evidence and support the best charities. We often want to help other people and make the world better, but feel like there's nothing we can do. We hope we've shown that's not true. We can do a great deal. It's up to each of us to decide what to do.

For researchers

This chapter has only scratched the surface of applying wellbeing cost-effectiveness analysis to maximising philanthropic returns. There are many ways for researchers to contribute to this new, and important field.

First and foremost, we need more wellbeing cost-effectiveness analyses in general. We know so little about the huge variety of things that could be funded or done. Particularly, we need more analyses to be published in academia to stress test the methodology and develop best practices.

In our research so far, we have found there's often information on the direct, immediate impact on recipients, but very sparse data on all the other aspects. There is practically no information on household spillovers. This is crucial given that in some cases, as we have argued,[85] household members as a whole will get a greater total benefit than the direct recipient.

Evidence on the duration of benefits is also often missing. Again, it seems very plausible that some interventions will last much longer than others, so long-term data will have a large influence on effectiveness estimates.[86]

Lastly, the benefits of some interventions are inherently more difficult to quantify because they have a particularly long-term or diffuse group of beneficiaries. Examples of this include climate change,[87] children (who can't report their wellbeing reliably),[88] education,[89] cultural activities,[90] research,[91] and religious activities.[92] However, even for these, we believe that it is possible to use existing evidence to get a better sense of their effect on wellbeing.

There are many other methodological questions, such as empirical and philosophical issues relating to extending versus improving lives,[93] or how to trade-off between internal and external validity.[94]

We believe these are all rich veins of inquiry waiting to be mined. Wellbeing cost-effectiveness is an unusual area. Barely any work has been done but it has huge and direct practical implications. We hope researchers take up the challenge and use their skills to make a difference.

Endnotes

1 Nonprofits Source (2024).

2 See Bekkers and Wiepking (2010) for a review of the mechanisms (including altruism) supporting charitable giving and Ottoni-Wilhem et al. (2017) for a discussion of how one may distinguish between true altruism and warm glow motivations.

3 Giving What We Can, an organisation which encourages people to pledge 10% of their income to the most effective charities, and an original, central organisation of the Effective Altruism movement, claims "you can have 100x more impact by donating to the most effective charities". https://www.givingwhatwecan.org/ Accessed 03/01/2025. See MacAskill (2017) for an introduction to the philosophy of Effective Altruism, or visit https://www.effectivealtruism.org/ for a more up to date and accessible overview.

4 Caviola et al. (2020).

5 See Ord (2019) for health interventions. Todd (2023) examines how much cost-effectiveness differs for various health and policy interventions, but has no data on charities.

6 Barrington-Leigh (2022).

7 OECD (2024).

8 Singer (1972).

9 Singer (1972, p. 231).

10 Singer's thought experiment of whether we ought to save a drowning child from a shallow pond has since generated a huge literature in philosophy, one we cannot hope to summarise here. Interested readers can find articles on this by going to https://philpapers.org/ (a philosophy database) and searching for "shallow pond" and/or "drowning child".

11 The median individual income figure of $42,220 comes from the Federal Reserve Economic Data (2024), based on data from the U.S. Census Bureau. The income percentile this falls into comes from Gapminder's (2023) data on global income distribution after converting the yearly income to $115 dollars a day.

12 See Our World in Data (Roser et al., 2024) for many striking presentations of this economic fact.

13 See Aknin et al. (2020) and Aknin et al. (2022) for pre-registered reports tackling this the topic of the personal wellbeing benefits of beneficence.

14 See Caviola et al. (2021) or Berman et al. (2018) for a paper length discussion or Schubert and Caviola (2024) for an extensive discussion in Part 1 of their book of why people do not prefer giving to effective charities.

15 Caviola et al. (2020a).

16 Caviola et al. (2020b).

17 See the references given on page 159 of Caviola et al. (2020a).

18 We use 'happiness' and 'wellbeing' as synonyms here and throughout. This is, admittedly, an abuse of terminology, though not one we expect readers to mind. In philosophy, 'wellbeing' refers to what makes life ultimately good for us. Philosophers have three accounts of wellbeing. The first is hedonism, on which wellbeing consists in happiness (a positive balance of enjoyment over suffering). The second are desire theories, on which wellbeing consists in having one's desires satisfied. The third is the objective list, which claims there can be things that make a person's life go better which are neither pleasurable to nor desired by them. Classic items for the objective list include success, friendship, knowledge, and virtuous behaviour. Hence, wellbeing refers to happiness only on one of these accounts. That said, the debate amongst philosophers is generally about whether happiness is the only thing that matters for our wellbeing. Those holding desire theory or objective lists accounts will tend to say that happiness matters for wellbeing somehow, either because we desire happiness, or because it is one of the goods that matter intrinsically. A view of wellbeing that held that our experiences, including suffering, did not matter in any way would be a peculiar, implausible view of wellbeing. We direct readers to Crisp (2021) for the best introduction to the topic; it is open access.

19 See Brazier and Tsuchiya (2015); Layard and Oparina (2021).

20 The national statistics offices of over 37 countries collect this sort of data about their citizens (OECD, 2023).

21 OECD (2013, Chapter 1).

22 Happiness Research Institute (2020).

23 Clark et al. (2018, Chapter 4, p. 63).

24 Clark et al. (2018, Chapter 2); Frijters and Krekel (2021, Table 2.2).

25 Clark et al. (2008).

26 Layard and Oparina (2021).

27 Frijters et al. (2020); Frijters et al. (2024).

28 HM Treasury (2022).

29 Frijters and Krekel (2021); McGuire and Plant (2021b); State of Life (2021a).

30 As most of this work is not published in academic journals, we generally found these evaluations by word of mouth. While it's possible we have missed something, the world of WELLBY research is small (all of the researchers seem to be associated with UK organisations) so we don't think this is a large concern.

31 Frijters and Krekel (2021); Krekel et al. (2024).

32 It is widely accepted that there is a 'hierarchy of evidence', where research that uses evidence that is higher up is considered more reliable and better at demonstrating causality than those which use data from further down. At the top of the hierarchy are studies of groups of causal studies (meta-analyses), then singular causal studies, then groups of studies with non-causal methods, etc. One difference between the charity evaluations in our dataset is that they do not all rely on evidence at the same level of the hierarchy. Beyond the quantity of studies and whether they demonstrate causality or not, academics often use frameworks like 'GRADE' which assess the quality and relevance of study to whatever question is at hand. See

Guyatt et al. (2011) for a discussion of the role of indirectness in GRADE assessments and Guyatt et al. (2008) or Siemieniuk and Guyatt (n.d.) for more about GRADE generally.

33 The estimates here generally focus on directly-measurable improvements to quality of life over a few years or less. It's possible, in principle, to extend the WELLBY approach to incorporate, for instance, (A) interventions with very long-run, indirect effects, such as those from climate policy, or (B) comparing life-improving and life-extending interventions. But these end up being far more speculative. For (A) there's an empirical challenge about predicting the future. For (B) there are difficult philosophical questions, such as when a life has negative wellbeing, that apply whatever measure of impact is used. See Plant et al. (2022) for a discussion of the empirical and philosophical difficulties with comparing the wellbeing value of improving and extending lives. We lack the space to do justice to these more complex topics here, so we restrict our attention to the most data-based 'apples-to-apples' comparisons.

34 This conclusion would only be under threat if there was an unlikely pattern of error: all the more cost-effective charities were too high, all the less cost-effective charities were too low, and coincidentally the correct estimates yield nearly identical cost-effectiveness numbers. As we say later, the differences in cost-effectiveness is most easily explained by what the charities do and where they work.

35 Each row in the table represents a charity evaluation. It summarises what the charity does, who evaluated it, and its cost-effectiveness, expressed as cost per WELLBY.

The **duration** column reflects how many years the effect of the intervention is expected to last.

The **total sample** refers to the number of participants included in the analysis. The **total studies** column shows the number of datasets or interventions analysed. This is not always the same as the number of published papers because a single intervention might appear in multiple papers (e.g., follow-ups). Analyses often rely on diverse evidence sources that are sometimes difficult to quantify. We aimed to ensure consistency in how these totals are reported across charity evaluations.

We also provide brief assessments of the **causality** of the evidence. Evidence is ranked on a hierarchy, with meta-analyses of randomised control trials (RCTs) being the most robust and associative studies being the least. Methods such as matched groups with difference-in-difference fall somewhere in between.

Similarly, we offer brief assessments of the **relevance** of the evidence to the specific charity being evaluated. When evidence is rated as highly relevant, it often comes directly from studies of the charity's own participants. Lower relevance ratings apply when the evidence is drawn from studies of similar interventions (e.g., psychotherapy) rather than the specific charity.

Finally, we provide brief assessments of how **in-depth** the analysis is. These assessments are relative. In this case, we view report length as a proxy for depth. In-depth evaluations may have multiple analyses that could each be a separate report, medium depth evaluations provide a standalone analyses, and shallow depth evaluations are rough, brief analyses that are not presented in standalone reports.

36 https://strongminds.org.

37 https://www.friendshipbenchzimbabwe.org.

38 McGuire et al. (2024b).

39 Note that there is no spillover data directly related to the charities programme. We use wider sources of data to estimate a spillover ratio (5 RCTs and 1 controlled trial). Namely, we estimate that non-recipient members of the household experience 16% of the wellbeing benefit that the psychotherapy recipient experiences. We apply this to the average household in the countries where these charities operate, which is between 3 and 4 other household members.

40 We discounted this estimate for a range of internal validity (i.e., is the data accurate) and external validity (i.e., does the data we use generalise to the case we are estimating) concerns. For example, for internal validity, we multiply the effect by 0.69 (a 31% discount) for publication bias. For example, for external validity, we multiply the effect by 0.90 (a 10% discount) for the use of lay therapists. The charities rely on lay deliverers of manualised psychotherapy instead of experts because there are too few experts in low-income countries. This is often called 'task-shifting'. While our modelling suggests it reduces the effect a little bit, it also allows for a lot more people to be treated and at lower costs; hence, it improves the cost-effectiveness of the charities.

41 https://www.parkrun.org.uk.

42 State of Life (2021a).

43 Charities evaluated in the UK generally have results reported in pound sterling. For consistency, we convert these results to USD, using the average conversion rate reported by the World Bank (2023): $1 = £0.8, so we convert results with £X * 1/0.8.

44 https://actionforhappiness.org.

45 In Frijters and Krekel (2021) but based on an RCT by Krekel et al. (2021).

46 Charities evaluated in the UK generally have results reported in pound sterling. For consistency, we convert these results to USD, using the average conversion rate reported by the World Bank (2023): $1 = £0.8, so we convert results with £X * 1/0.8.

47 https://www.footballbeyondborders.org.

48 Pro Bono Economics (Franklin, 2024) and Cheng and Humphrey (n.d).

49 Pro Bono Economics (Franklin, 2024; see also Cheng & Humphrey (n.d)) used the effect on students considered to be 'at risk'. They found an improvement of 2.4 points on the SWEMWBS measure of mental wellbeing, which they convert at a 0.24 rate to life satisfaction on a 0–10 scale; namely, an effect of 0.576. They assume benefits only occur for the 'at risk' children. They adjust the impact by the proportion of 'at risk' student in the total sample of Football Beyond Borders in the 2022–23 school year (26%), resulting in an average effect per student of 0.26*0.576 = 0.15 points.

50 Charities evaluated in the UK generally have results reported in pound sterling. For consistency, we convert these results to USD, using the average conversion rate reported by the World Bank (2023): $1 = £0.8, so we convert results with £X * 1/0.8.

51 We restrict it to pre-existing estimates at the time of writing for charities that readers could fund; so, we removed hypothetical charities and the estimates we produced for this chapter to fill in gaps. We discuss the hypothetical and new analyses later in the chapter or in the online appendix.

52 We do not include confidence intervals for two reasons. One is that the evaluations themselves often don't include them. The other, more conceptual issue, is that confidence intervals only capture statistical uncertainty. There are other, harder-to-quantify uncertainties to pay attention to. For example, is the quality of the data sufficient to determine a causal effect? Is the data relevant to the context being evaluated? Has the analysis considered all the relevant parameters? Hence, providing confidence intervals would create false precision.

53 Pure Earth, Taimaka, Icddr,b, Friendship Bench, and StrongMinds.

54 Royal Voluntary Service, Action for Happiness, Parkrun, London Youth Rowing's Active Row, Walking with the Wounded (Head Start and Employment programmes), and Football Beyond Borders.

55 Note that averaging ratios introduces a mathematical inconsistency: the average 'cost per WELLBY' does not necessarily align with the inverse of the average 'WELLBYs created per $1,000 donated.' This discrepancy arises because the average of a ratio is not equivalent to the ratio of averages when values vary across analyses. As a result, the direction of calculation can yield different outcomes. These comparisons are intended for illustrative purposes rather than precise calculations.

56 HM Treasury (2022).

57 See Frayman et al. (2024); Fritjers and Krekel (2021); State of Life (2023b).

58 Policies, unlike charities, can have zero cost per WELLBY if they increase future tax revenues, and thus have no net cost to governments over the long run.

59 Caviola et al. (2020). This raises a puzzle: if some charities are so much better than others, why would people think they are all about as good? Schubert and Caviola (2024, pp 33) propose the underlying cause is that donors are not motivated by efficiency and often lack good metrics to compare charities' impact. They contrast this to consumer goods. We want good deals for ourselves and can tell if we're getting them, so competitive pressures mean that overpriced goods are driven out of the market, and companies that produce better goods can charge more for them. Hence for consumer goods, it is reasonable to expect equally priced products are equally good. As we are used to buying goods for ourselves, it's natural for donors to (mistakenly) apply the same thinking to charities. Schubert and Caviloa (2024, pp 68) also point to an illustrative disanalogy between donations and investments. Investors are generally rational and seeking the best returns for themselves, so share prices approximate companies' true value – which is why it is difficult to 'beat the market'. In contrast, because so few donors are seeking to maximise their philanthropic returns, it should be relatively easy for impact-minded donors to beat the *philanthropic* market and have outsized returns. We say 'relatively easy' as there *are* impact-minded donors, and occasionally there is a scramble to fund the best charities – just as there is to invest in the most promising companies – such that you can't give to the best options because their budgets are full.

60 See Oxfam's (2023) annual report.

61 See the data collected by findthatcharity.uk, which compiles the open source data from the UK charity regulators for England and Wales (https://register-of-charities.charitycommission.gov.uk/en/sector-data/top-10-charities), Northern Ireland (https://www.charitycommissionni.org.uk/), and Scotland (https://www.oscr.org.uk/).

62 See the RNLI's (2023) annual report.

63 See the 2023 expense information from the Charity Commission for England and Wales.

64 Charities evaluated in the UK generally have results reported in pound sterling. For consistency, we convert these results to USD, using the average conversion rate reported by the World Bank (2023): $1 = £0.8, so we convert results with £X * 1/0.8.

65 We use a standard (but disputable) method to calculate the value of extending a life. We explain this further in the online appendix.

66 The RNLI arguably provides a psychological benefit to those that don't directly use its service, but *could* do – the feeling that someone would save you, if you needed it. However, any attempt to estimate the size of this benefit would be very speculative. Further, many charities also have a "it's good to know it's there if you need it" factor, so it's not obvious how accounting for this factor would change the numbers across the board. Hence, we don't try to account for it here. We thank the editors for drawing this issue to our attention.

67 See Guide Dogs' (2023) annual report.

68 We take a sample-weighted average of Refson et al. (1999, Scotland, effect is 0.80 on a 0–10 scale converted linearly from the SWLS; n = 117), Yarmolkevich (2017, USA, effect is 1.08 on a 0–10 scale converted linearly from the SWLS; n = 58), and Glenk et al. (2019, Austria, effect is -0.30 on a 0–10 scale converted from the WHOQOL-BREF psychological subscale, n = 36). These studies are comparing visually impaired individuals with and without guide dogs, they are associative and not causal. We do not apply an adjustment for the lack of causality. This is a very shallow analysis.

69 See Guide Dogs' FAQ.

70 According to the Charities Aid Foundation (2024), giving to homelessness is one of the largest charitable causes in the UK, absorbing around 7% of all donations in the UK. The homelessness share of donations is comparable to everything that went towards international aid (7%) but below the share to religious organisations (13%).

71 This was based on Stergiopoulos et al. (2015), a well-powered (n = 2,148) trial in Canada that compared the effect of housing and social assistance to the typical assistance provided. They estimated the costs based on studies of similar interventions in the UK. See Frijters and Krekel (2021, p. 210) for their discussion of it.

72 Frijters and Krekel (2021).

73 Thurow (1974).

74 Dwyer et al. (2023).

75 This cost consists of the $5,555 USD cash transfer and assuming it costs 20% of the value of the transfer to deliver it. We assume this figure because it's a similar value to what we estimated as the overhead share of the cost of a cash transfer delivered by GiveDirectly, a well run organization (McGuire & Plant, 2021a).

76 Dwyer et al. (2023) report that recipients of the cash transfers spend less time in shelters which resulted in net cost-savings for society.

77 The authors feel the intuitive pull of doing something to help those in our local communities - but not of allocating the lion's share of our (spare) resources here if there are better ways to help people. See the comments on donation splitting later.

78 Singer (2009, p. 221).

79 See Aknin et al. (2020) and Aknin et al. (2022) for pre-registered reports on the personal wellbeing benefits of beneficence.

80 For animal welfare, it's not obvious what we would compare the self-reported wellbeing scores of humans to, or how to form an evidence-based rate of exchange.

81 None of the estimates above - by us or the other researchers - have tried to account for diminishing marginal returns. This is both because it's difficult to do so and it's unusual for organisations to suddenly be offered far more money than they can spend.

82 See Snowden (2019) for elucidation of the standard argument on why giving to one charity maximises expected utility, absent concerns about diminishing marginal returns.

83 For instance, in a blog post, Karnofsky (2016) endorses 'worldview diversification', that is, "putting significant resources behind each worldview that [one finds] highly plausible" but does not provide a fully-developed argument for this; keys terms, such as 'worldview' and 'strong uncertainty' are undefined. Kaczmarek, Lloyd and Plant (forthcoming) observe that none of the standard philosophical theories of *moral* uncertainty provide independent grounds for diversification (moral uncertainty is distinct from empirical uncertainty). They offer a novel, bargaining-based theory of moral uncertainty, the 'moral marketplace', on which diversification is *sometimes* appropriate. We know of no other work that argues, on grounds of moral uncertainty, that diversification is ever appropriate.

84 The philosophical topic here is the ethics of aggregation, sometimes known as distributional ethics. The implicit approach taken in WELLBY cost-effectiveness is an additive aggregation function, where each 1-point change has the same value, regardless of who accrues it or how well-off they are (an additive aggregation function is necessary but not sufficient for utilitarianism). There are alternative options here, e.g., prioritarian or egalitarian functions, both of which can capture the intuition it's better to help the worse off, even if the change in total wellbeing is the same. See Holtug (2015) and reference therein for discussion.

85 See McGuire et al. (2022).

86 See McGuire et al. (2024a) Section 4.1 for a discussion.

87 We did not find evidence establishing a link between global carbon dioxide emissions and subjective wellbeing. However, the consequences of climate change have a clear relationship to mental wellbeing. For example, increases in ambient temperature are related to lower wellbeing (Liu et al., 2021; Noelke et al. 2016), hurricane risk is related to lower life satisfaction (Berlemann, 2016), and storms and floods also are related to lower wellbeing (Sekulova & van den Bergh, 2016; von Möllendorff & Hirschfeld, 2016).

88 For more discussion of this see Little and Parkes (2024), McGuire et al. (2024b), and McGuire et al. (2024d).

89 While the correlational relationship between years of education and mental wellbeing is positive (Bücker et al., 2018; Clark et al., 2018), the causal evidence is more mixed with some studies finding positive effects (Chevalier and Feinstein, 2006; Oreopoulos, 2007; Oreopoulos & Salvanes, 2011; Powdthavee et al., 2013) other null (Avendano et al., 2020; Dahmann & Schnitzlein, 2019; Davies et al., 2019; Viinikainen et al., 2018) or in some cases a negative effect (Courtin et al. 2019).

90 Frijters and Krekel (2021) evaluate the wellbeing cost-effectiveness of two cultural programmes from a policy perspective: the city of culture programme in the UK and the London Olympics.

91 This isn't true for other outcomes, which gives some reason for hope. Several studies find a high ROI for research (Jones & Summers, 2020; Kremer et al., 2021; Pardey et al., 2016). Unger et al. (2023) is optimistic about the cost-effectiveness of cancer research on DALYs, estimating it costs $326 per year of life gained through investment in research.

92 Zotti et al. (2016) estimates that, in the UK, the causal effect of being religious is around 0.03 points on a 0–10 life-satisfaction scale.

93 See Plant et al. (2022) for a discussion of the challenges in comparing the wellbeing value of improving and extending lives.

94 See McGuire et al. (2024a, Section 5) for an extensive example.

References

Aknin, L. B., Dunn, E. W., Proulx, J., Lok, I., & Norton, M. I. (2020). Does spending money on others promote happiness?: A registered replication report. *Journal of Personality and Social Psychology*, *119*(2), e15–e26. https://doi.org/10.1037/pspa0000191

Aknin, L. B., Dunn, E. W., & Whillans, A. V. (2022). The Emotional Rewards of Prosocial Spending Are Robust and Replicable in Large Samples. *Current Directions in Psychological Science*, *31*(6), 536–545. https://doi.org/10.1177/09637214221121100

Avendano, M., de Coulon, A., & Nafilyan, V. (2020). Does longer compulsory schooling affect mental health? Evidence from a British reform. *Journal of Public Economics*, *183*, 104137. https://doi.org/10.1016/j.jpubeco.2020.104137

Bekkers, R., & Wiepking, P. (2010). A Literature Review of Empirical Studies of Philanthropy: Eight Mechanisms That Drive Charitable Giving. *Nonprofit and Voluntary Sector Quarterly*, *40*(5), 924–973. https://doi.org/10.1177/0899764010380927

Berlemann, M. (2016). Does hurricane risk affect individual well-being? Empirical evidence on the indirect effects of natural disasters. *Ecological Economics*, *124*, 99–113. https://doi.org/10.1016/j.ecolecon.2016.01.020

Berman, J., Barasch, A., Levine, E., & Small, D. A. (2018). *Impediments to Effective Altruism: The Role of Subjective Preferences in Charitable Giving* (SSRN Scholarly Paper No. 3224734). https://papers.ssrn.com/abstract=3224734

Blattman, C., Chaskel, S., Jamison, J. C., & Sheridan, M. (2023). Cognitive Behavioral Therapy Reduces Crime and Violence over Ten Years: Experimental Evidence. *American Economic Review: Insights*, *5*(4), 527–545. https://doi.org/10.1257/aeri.20220427

Brazier, J., & Tsuchiya, A. (2015). Improving Cross-Sector Comparisons: Going Beyond the Health-Related QALY. *Applied Health Economics and Health Policy*, *13*(6), 557–565. https://doi.org/10.1007/s40258-015-0194-1

Bücker, S., Nuraydin, S., Simonsmeier, B. A., Schneider, M., & Luhmann, M. (2018). Subjective well-being and academic achievement: A meta-analysis. *Journal of Research in Personality*, *74*, 83–94. https://doi.org/10.1016/j.jrp.2018.02.007

Cattaneo, M. D., Galiani, S., Gertler, P. J., Martinez, S., & Titiunik, R. (2009). Housing, Health, and Happiness. *American Economic Journal: Economic Policy*, *1*(1), 75–105. https://doi.org/10.1257/pol.1.1.75

Caviola, L., Schubert, S., & Greene, J. D. (2021). The Psychology of (In)Effective Altruism. *Trends in Cognitive Sciences*, *25*(7), 596–607. https://doi.org/10.1016/j.tics.2021.03.015

Caviola, L., Schubert, S., & Nemirow, J. (2020). The many obstacles to effective giving. *Judgment and Decision Making*, *15*(2), 159–172. https://doi.org/10.1017/S1930297500007312

Caviola, L., Schubert, S., Teperman, E., Moss, D., Greenberg, S., & Faber, N. S. (2020). Donors vastly underestimate differences in charities' effectiveness. *Judgment and Decision Making*, *15*(4), 509–516. https://doi.org/10.1017/S1930297500007452

Charities Aid Foundation. (2024). *UK Giving: Mapping Generosity Across the Country*. https://www.cafonline.org/docs/default-source/uk-giving-reports/uk_giving_report_2024.pdf

Charity Commission for England and Wales. (2023). *The Royal Lifeboat Institution—Charity 209603*. Register of Charities. https://register-of-charities.charitycommission.gov.uk/en/charity-search/?p_p_id=uk_gov_ccew_onereg_charitydetails_web_portlet_CharityDetailsPortlet&p_p_lifecycle=0&p_p_state=maximized&p_p_mode=view&_uk_gov_ccew_onereg_charitydetails_web_portlet_CharityDetailsPortlet_regId=209603&_uk_gov_ccew_onereg_charitydetails_web_portlet_CharityDetailsPortlet_subId=0

Cheng, Q., & Humphrey, N. (n.d.). *Preliminary Evaluation of a Targeted, School-Based Social and Emotional Learning Intervention for At Risk Youth: Football Beyond Borders*. PsyArXiv. https://osf.io/preprints/psyarxiv/ph3vs

Chevalier, A., & Feinstein, L. (2006). *Sheepskin or Prozac: The Causal Effect of Education on Mental Health* (SSRN Scholarly Paper No. 923530). https://doi.org/10.2139/ssrn.923530

Clark, A. E., Diener, E., Georgellis, Y., & Lucas, R. E. (2008). Lags and Leads in Life Satisfaction: A Test of the Baseline Hypothesis. *The Economic Journal*, *118*(529), F222–F243. https://doi.org/10.1111/j.1468-0297.2008.02150.x

Clark, A. E., Flèche, S., Layard, R., Powdthavee, N., & Ward, G. (2018). *The Origins of Happiness*. Princeton University Press. https://press.princeton.edu/books/hardcover/9780691177892/the-origins-of-happiness

Courtin, E., Nafilyan, V., Glymour, M., Goldberg, M., Berr, C., Berkman, L. F., Zins, M., & Avendano, M. (2019). Long-term effects of compulsory schooling on physical, mental and cognitive ageing: A natural experiment. *J Epidemiol Community Health*, *73*(4), 370–376. https://doi.org/10.1136/jech-2018-211746

Crisp, R. (2021). *Well-being*. The Stanford Encyclopedia of Philosophy. https://plato.stanford.edu/cgi-bin/encyclopedia/archinfo.cgi?entry=well-being

Dahmann, S. C., & Schnitzlein, D. D. (2019). No evidence for a protective effect of education on mental health. *Social Science & Medicine*, *241*, 112584. https://doi.org/10.1016/j.socscimed.2019.112584

Davies, N. M., Dickson, M., Davey Smith, G., van den Berg, G. J., & Windmeijer, F. (2018). The causal effects of education on health outcomes in the UK Biobank. *Nature Human Behaviour*, *2*(2), 117–125. https://doi.org/10.1038/s41562-017-0279-y

Drysdale, L. (2018). *Human Henge Evaluation Report*. Human Henge. https://humanhenge.org/2019/01/30/human-henge-evaluation-report/

Dupret, S., McGuire, J., Dwyer, R., & Plant, M. (2022). *A can of worms: The non-significant effect of deworming on happiness*. Happier Lives Institute. https://www.happierlivesinstitute.org/report/a-can-of-worms/

Dupret, S., McGuire, J., Dwyer, R., & Plant, M. (2024). *Converting measures of mental health and wellbeing into WELLBYs*. Happier Lives Institute. https://www.happierlivesinstitute.org/report/converting-measures-of-mental-health-and-wellbeing-into-wellbys/

Dwyer, R., Palepu, A., Williams, C., Daly-Grafstein, D., & Zhao, J. (2023). Unconditional cash transfers reduce homelessness. *Proceedings of the National Academy of Sciences*, *120*(36), e2222103120. https://doi.org/10.1073/pnas.2222103120

Federal Reserve Economic Data. (2024, September 10). *Median Personal Income in the United States*. https://fred.stlouisfed.org/series/MEPAINUSA646N

Field, M. S., Mithra, P., & Peña-Rosas, J. P. (2021). Wheat flour fortification with iron and other micronutrients for reducing anaemia and improving iron status in populations. *Cochrane Database of Systematic Reviews*, *1*. https://doi.org/10.1002/14651858.CD011302.pub3

Franklin, J. (2024). *Investing in trusted relationships: The economic value of Football Beyond Borders' impact on children's wellbeing*. https://www.probonoeconomics.com/Handlers/Download.ashx?IDMF=c7989d9a-133b-4fe9-91da-378511099018

Frayman, D., Krekel, C., Layard, R., MacLennan, S., & Parkes, I. (2024). Value for money: How to improve wellbeing and reduce misery. *CEP Reports*. https://ideas.repec.org//p/cep/cepsps/44.html

Frijters, P., Clark, A. E., Krekel, C., & Layard, R. (2020). A happy choice: Wellbeing as the goal of government. *Behavioural Public Policy*, *4*(2), 126–165. https://doi.org/10.1017/bpp.2019.39

Frijters, P., & Krekel, C. (2021). *A Handbook for Wellbeing Policy-Making: History, Theory, Measurement, Implementation, and Examples* (1st ed.). Oxford University PressOxford. https://doi.org/10.1093/oso/9780192896803.001.0001

Frijters, P., Krekel, C., Sanchis, R., & Santini, Z. I. (2024). The WELLBY: A new measure of social value and progress. *Humanities and Social Sciences Communications*, *11*(1), 1–12. https://doi.org/10.1057/s41599-024-03229-5

Galiani, S., Gertler, P. J., & Undurraga, R. (2018). The Half-Life of Happiness: Hedonic Adaptation in the Subjective Well-Being of Poor Slum Dwellers to the Satisfaction of Basic Housing Needs. *Journal of the European Economic Association*, *16*(4), 1189–1233. https://doi.org/10.1093/jeea/jvx042

Gapminder Tools. (2023). *Gapminder Tools*. https://www.gapminder.org/tools/#$model$markers$billy$encoding$selected$data$;;;;;&chart-type=mountain&url=v2

Glenk, L. M., Pˇribylová, L., Stetina, B. U., Demirel, S., & Weissenbacher, K. (2019). Perceptions on Health Benefits of Guide Dog Ownership in an Austrian Population of Blind People with and without a Guide Dog. *Animals*, *9*(7), Article 7. https://doi.org/10.3390/ani9070428

Guide Dogs. (n.d.). *What does a guide dog do?* Retrieved 21 December 2024, from https://www.guidedogs.org.uk/getting-support/guide-dogs/what-a-guide-dog-does/#what-happens-at-the-end-of-a-guide-dog-partnership

Guide Dogs. (2023). *Annual report & Accounts 2023*. https://gd-prod.azureedge.net/-/media/project/guidedogs/guidedogsdotorg/files/about-us/finance-and-governance/reports-and-accounts/annual-report-2023.pdf

Guyatt, G. H., Oxman, A. D., Kunz, R., Woodcock, J., Brozek, J., Helfand, M., Alonso-Coello, P., Falck-Ytter, Y., Jaeschke, R., Vist, G., Akl, E. A., Post, P. N., Norris, S., Meerpohl, J., Shukla, V. K., Nasser, M., & Schünemann, H. J. (2011). GRADE guidelines: 8. Rating the quality of evidence—indirectness. *Journal of Clinical Epidemiology*, *64*(12), 1303–1310. https://doi.org/10.1016/j.jclinepi.2011.04.014

Guyatt, G. H., Oxman, A. D., Vist, G. E., Kunz, R., Falck-Ytter, Y., Alonso-Coello, P., & Schünemann, H. J. (2008). GRADE: An emerging consensus on rating quality of evidence and strength of recommendations. *BMJ*, *336*(7650), 924–926. https://doi.org/10.1136/bmj.39489.470347.AD

Hamory, J., Miguel, E., Walker, M., Kremer, M., & Baird, S. (2021). Twenty-year economic impacts of deworming. *Proceedings of the National Academy of Sciences*, *118*(14), e2023185118. https://doi.org/10.1073/pnas.2023185118

Haushofer, J., & Shapiro, J. (2016). The Short-term Impact of Unconditional Cash Transfers to the Poor: Experimental Evidence from Kenya. *The Quarterly Journal of Economics*, *131*(4), 1973–2042. https://doi.org/10.1093/qje/qjw025

Heaslip, V., & Darvill, T. (2017). *Human Henge Wellbeing Research*. Human Henge. https://humanhenge.org/2019/01/30/human-henge-evaluation-report/

HM Treasury. (2022, November 18). *Green Book supplementary guidance: Wellbeing*. GOV.UK. https://www.gov.uk/government/publications/green-book-supplementary-guidance-wellbeing

Holtug, N. (2015). Theories of Value Aggregation: Utilitarianism, Egalitarianism, Prioritarianism. In I. Hirose & J. Olson (Eds.), *The Oxford Handbook of Value Theory* (p. 0). Oxford University Press. https://doi.org/10.1093/oxfordhb/9780199959303.013.0015

Jones, B. F., & Summers, L. (2020). *A Calculation of the Social Returns to Innovation* (SSRN Scholarly Paper No. 3700691). https://papers.ssrn.com/abstract=3700691

Kaczmarek, P., Lloyd, H. R., & Plant, M. (forthcoming). Moral Uncertainty, Proportionality and Bargaining. *Ergo: An Open Access Journal of Philosophy*. https://philarchive.org/rec/KACMUP

Karnofsky, H. (2016). Worldview Diversification. *Open Philanthropy*. https://www.openphilanthropy.org/research/worldview-diversification/

Kramer, S. (2020, March 31). With billions confined to their homes worldwide, which living arrangements are most common? *Pew Research Center*. https://www.pewresearch.org/short-reads/2020/03/31/with-billions-confined-to-their-homes-worldwide-which-living-arrangements-are-most-common/

Krekel, C., De Neve, J.-E., Fancourt, D., & Layard, R. (2021). A local community course that raises wellbeing and pro-sociality: Evidence from a randomised controlled trial. *Journal of Economic Behavior & Organization*, *188*, 322–336. https://doi.org/10.1016/j.jebo.2021.05.021

Krekel, C., Shreedhar, G., Lee, H., Marshall, C., Boler, A., Smith, A., & Dolan, P. (2024). Happy to Help: Welfare Effects of a Nationwide Volunteering Programme. *Review of Economics and Statistics*, 1–64. https://doi.org/10.1162/rest_a_01533

Kremer, M., Thomas, M., Gallant, S., & Rostapshova, O. (2021). *Is Development Innovation a Good Investment? Evidence on Scaling and Social Returns from USAID's Innovation Fund*. https://bpb-us-w2.wpmucdn.com/voices.uchicago.edu/dist/0/2830/files/2022/04/SROR-21.11.04_clean-2.pdf

Layard, R., & Oparina, E. (2021). Living long and living well: The WELLBY approach. In *World Happiness Report*. https://worldhappiness.report/ed/2021/living-long-and-living-well-the-wellby-approach/

Lee, Y.-J., & Kim, H.-B. (2020). Association between anaemia and adult depression: A systematic review and meta-analysis of observational studies. *J Epidemiol Community Health*, *74*(7), 565–572. https://doi.org/10.1136/jech-2020-213842

Little, A., & Parkes, I. (2024). *Child Wellbeing-Years (C-WELLBYs)*. State of Life. https://drive.google.com/file/d/1_hwHw4_j4M-6CybdpMihNMtQulKAvIfeQ/view?usp=embed_facebook

Liu, J., Varghese, B. M., Hansen, A., Xiang, J., Zhang, Y., Dear, K., Gourley, M., Driscoll, T., Morgan, G., Capon, A., & Bi, P. (2021). Is there an association between hot weather and poor mental health outcomes? A systematic review and meta-analysis. *Environment International*, *153*, 106533. https://doi.org/10.1016/j.envint.2021.106533

MacAskill, W. (2017). Effective Altruism: Introduction. *Essays in Philosophy*, *18*(1), 1–5. https://doi.org/10.7710/1526-0569.1580

McGuire, J., Dupret, S., Dwyer, R., Plant, M., Stewart, B., Goddard, J., Klapow, M., Giraldi, D., Olshin, B., Michelet, J., & Beuchot, T. (2024a). *The wellbeing cost-effectiveness of StrongMinds and Friendship Bench: Combining a systematic review and meta-analysis with charity-related data (Nov 2024 Update)*. Happier Lives Institute. https://www.happierlivesinstitute.org/report/the-wellbeing-cost-effectiveness-of-strong-minds-and-friendship-bench-combining-a-systematic-review-and-meta-analysis-with-charity-related-data-nov-2024-update/

McGuire, J., Dupret, S., & Plant, M. (2022). *Happiness for the whole family*. Happier Lives Institute. https://www.happierlivesinstitute.org/report/happiness-for-the-whole-family/

McGuire, J., Kaiser, C., & Bach-Mortensen, A. M. (2020). *Cash transfers: Systematic review and meta-analysis*. Happier Lives Institute. https://www.happierlivesinstitute.org/report/cash-transfers-systematic-review-and-meta-analysis/

McGuire, J., & Plant, M. (2021a). *Cash transfers: Cost-effectiveness analysis*. Happier Lives Institute. https://www.happierlivesinstitute.org/report/cash-transfers-cost-effectiveness-analysis/

McGuire, J., & Plant, M. (2021b). *Donating money, buying happiness*. Happier Lives Institute. https://www.happierlivesinstitute.org/report/donating-money-buying-happiness/

McGuire, J., Dwyer, R., & Plant, M. (2024e). *Stealing Happiness? The wellbeing cost-effectiveness of NEPI, preventing crime with cash and cognitive behavioural therapy. Happier Lives Institute*. https://www.happierlivesinstitute.org/report/stealing-happiness-the-wellbeing-cost-effectiveness-of-nepi-preventing-crime-with-cash-and-cognitive-behavioural-therapy/

McGuire, J., Stewart, B., Dupret, S., Dwyer, R., & Plant, M. (2024b). *Does a well-fed infancy make for a more felicitous life? How treating malnutrition impacts happiness: a charity evaluation of Taimaka*. Happier Lives Institute. https://www.happierlivesinstitute.org/report/how-treating-malnutrition-impacts-happiness-a-charity-evaluation-of-taimaka/

McGuire, J., Stewart, B., Dupret, S., Dwyer, R., & Plant, M. (2024d). *Does improving parenting practices in childhood lead to happier adults?* Happier Lives Institute. https://www.happierlivesinstitute.org/report/does-improving-parenting-practices-in-childhood-lead-to-happier-adults/

McGuire, J., Stewart, B., Dwyer, R., & Plant, M. (2024c). *Ironing out wellbeing: A shallow exploration into the wellbeing cost-effectiveness of fortifying wheat with iron*. Happier Lives Institute. https://www.happierlivesinstitute.org/report/how-iron-fortification-impacts-happiness-a-charity-evaluation-of-fortify-health/

Mitchell, A., Macció, J., & Mariño Fages, D. (2019). The Effects of Emergency Housing on Wellbeing: Evidence from Argentina's Informal Settlements. *The European Journal of Development Research*, *31*(3), 504–529. https://doi.org/10.1057/s41287-018-0166-z

Noelke, C., McGovern, M., Corsi, D. J., Jimenez, M. P., Stern, A., Wing, I. S., & Berkman, L. (2016). Increasing ambient temperature reduces emotional well-being. *Environmental Research*, *151*, 124–129. https://doi.org/10.1016/j.envres.2016.06.045

Nonprofits Source. (2024). *2024 Charitable Giving Statistics*. https://nonprofitssource.com/online-giving-statistics/#Charitable

OECD. (2013). OECD Guidelines on Measuring Subjective Well-being. *OECD Publishing*. https://doi.org/10.1787/9789264191655-en

OECD. (2023). Subjective well-being measurement: Current practice and new frontiers. *OECD Publishing*. https://doi.org/10.1787/4e180f51-en

Ord, T. (2019). The Moral Imperative Toward Cost-Effectiveness in Global Health. In H. Greaves & T. Pummer (Eds.), *Effective Altruism* (1st ed., pp. 29–36). Oxford University PressOxford. https://doi.org/10.1093/oso/9780198841364.003.0002

Oreopoulos, P. (2007). Do dropouts drop out too soon? Wealth, health and happiness from compulsory schooling. *Journal of Public Economics*, *91*(11), 2213–2229. https://doi.org/10.1016/j.jpubeco.2007.02.002

Oreopoulos, P., & Salvanes, K. G. (2011). Priceless: The Nonpecuniary Benefits of Schooling. *Journal of Economic Perspectives*, *25*(1), 159–184. https://doi.org/10.1257/jep.25.1.159

Ottoni-Wilhelm, M., Vesterlund, L., & Xie, H. (2017). Why Do People Give? Testing Pure and Impure Altruism. *American Economic Review*, *107*(11), 3617–3633. https://doi.org/10.1257/aer.20141222

Oxfam. (2023, November 8). *Annual Report 2023*. https://www.oxfamamerica.org/explore/research-publications/annual-report-2023/

Pardey, P. G., Andrade, R. S., Hurley, T. M., Rao, X., & Liebenberg, F. G. (2016). Returns to food and agricultural R&D investments in Sub-Saharan Africa, 1975–2014. *Food Policy*, *65*, 1–8. https://doi.org/10.1016/j.foodpol.2016.09.009

Plant, M. (2019). *Doing good badly? Philosophical issues related to effective altruism* [University of Oxford]. https://ora.ox.ac.uk/objects/uuid:59a79c4f-2a25-4ea7-afa6-aa04c1537acf

Plant, M., McGuire, J., & Dupret, S. (2022). *The elephant in the bednet: The importance of philosophy when choosing between extending and improving lives*. Happier Lives Institute. https://www.happierlivesinstitute.org/report/the-elephant-in-the-bednet/

Powdthavee, N., Lekfuangfu, W. N., & Wooden, M. (2013). The Marginal Income Effect of Education on Happiness: Estimating the Direct and Indirect Effects of Compulsory Schooling on Well-Being in Australia. *SSRN Electronic Journal*. https://doi.org/10.2139/ssrn.2259323

Pro Bono Economics. (2022). *The wellbeing impacts of Walking With The Wounded programmes*. https://www.probonoeconomics.com/the-wellbeing-impacts-of-walking-with-the-wounded

Pure Earth. (2024). *Global Lead Program: Results, 2020-2023*. https://www.pureearth.org/wp-content/uploads/2024/01/Global-Lead-Program-achievements-2020-2023-Jan_2024.pdf

Refson, K., Jackson, A. J., Dusoir, A. E., & Archer, D. B. (1999). The health and social status of guide dog owners and other visually impaired adults in Scotland. *Visual Impairment Research*, *1*(2), 95–109. https://doi.org/10.1076/vimr.1.2.95.4411

RNLI. (2023). *Annual Reports, Accounts And Future Plans*. https://rnli.org/about-us/how-the-rnli-is-run/annual-reports

Roser, M., Arriagada, P., Hasell, J., Ritchie, H., & Ortiz-Ospina, E. (2024). *Global GDP over the long run*. Our World in Data. Retrieved 20 December 2024, from https://ourworldindata.org/grapher/global-gdp-over-the-long-run

Samuelsson, C., Dupret, S., Plant, M., & Kaiser, C. (2023). *Can we trust wellbeing surveys? A pilot study of comparability, linearity, and neutrality*. Happier Lives Institute. https://www.happierlivesinstitute.org/report/can-we-trust-wellbeing-surveys-a-pilot-study-of-comparability-linearity-and-neutrality/

Schubert, S., & Caviola, L. (2024). *Effective Altruism and the Human Mind: The Clash Between Impact and Intuition* (1st ed.). Oxford University Press. https://doi.org/10.1093/oso/9780197757376.001.0001

Sekulova, F., & van den Bergh, J. C. J. M. (2016). Floods and happiness: Empirical evidence from Bulgaria. *Ecological Economics*, *126*, 51–57. https://doi.org/10.1016/j.ecolecon.2016.02.014

Siemieniuk, R., & Guyatt, G. (n.d.). *What is GRADE?* BMJ Best Practice. Retrieved 20 December 2024, from https://bestpractice.bmj.com/info/us/toolkit/learn-ebm/what-is-grade/

Singer, P. (1972). Famine, Affluence, and Morality. *Philosophy & Public Affairs*, *1*(3), 229–243.

Singer, P. (2009). *The Life You Can Save*. Pan Macmillan. https://www.thelifeyoucansave.org/the-book/

Snowden, J. (2019). Should We Give to More Than One Charity? In H. Greaves & T. Pummer (Eds.), *Effective Altruism: Philosophical Issues* (p. 0). Oxford University Press. https://doi.org/10.1093/oso/9780198841364.003.0005

State of Life. (2017). *If you could bottle it: A wellbeing and human capital value for money analysis of the NCS 2015 programme*. NCS. https://wearencs.com/sites/default/files/2020-09/NCS%20Wellbeing%20and%20Human%20Capital%20Valuation%20-%20Jump.pdf

State of Life. (2021a). *Parkrun*. State of Life. https://drive.google.com/file/d/1imx6DJ0-xehaMgDBpxBLiTdzO2QJVLUI/view?usp=embed_facebook

State of Life. (2023a). *Social Impact and Value Assessment of Tideway's Legacy Programme*. Tideway. https://www.tideway.london/media/6069/tideway_social-impact-technical-report-1.pdf

State of Life. (2023b). *Active Row Social Value Study*. https://heyzine.com/flip-book/907baf2538.html

State of Life. (2023c). *Tearfund, Church Transformation in Africa*.

https://learn.tearfund.org/en/resources/series/cct-impact-study-series/2022---local-church-lasting-transformation

Stergiopoulos, V., Gozdzik, A., Misir, V., Skosireva, A., Connelly, J., Sarang, A., Whisler, A., Hwang, S. W., O'Campo, P., & McKenzie, K. (2015). Effectiveness of Housing First with Intensive Case Management in an Ethnically Diverse Sample of Homeless Adults with Mental Illness: A Randomized Controlled Trial. *PLOS ONE*, *10*(7), e0130281. https://doi.org/10.1371/journal.pone.0130281

Stewart, B., Dupret, S., McGuire, J., Dwyer, R., & Plant, M. (2024). *Toxic Cosmetics: A shallow evaluation of Pure Earth advocacy against leaded cosmetics in Ghana*. Happier Lives Institute. https://www.happierlivesinstitute.org/report/toxic-cosmetics-a-shallow-evaluation-of-pure-earth-advocacy-against-leaded-cosmetics-in-ghana/

Thurow, L. C. (1974). Cash Versus In-Kind Transfers. *The American Economic Review*, *64*(2), 190–195.

Todd, B. (2023, February 14). *How much do solutions to social problems differ in their effectiveness? A collection of all the studies we could find*. 80,000 Hours. https://80000hours.org/2023/02/how-much-do-solutions-differ-in-effectiveness/

Unger, J. M., LeBlanc, M., George, S., Wolmark, N., Curran, W. J., O'Dwyer, P. J., Schnall, M. D., Mannel, R. S., Mandrekar, S. J., Dignam, J. J., Gray, R. J., Zhao, F., Bah, M., Vaidya, R., & Blanke, C. D. (2023). Population, Clinical, and Scientific Impact of National Cancer Institute's National Clinical Trials Network Treatment Studies. *Journal of Clinical Oncology*, *41*(11), 2020–2028. https://doi.org/10.1200/JCO.22.01826

Viinikainen, J., Bryson, A., Böckerman, P., Elovainio, M., Pitkänen, N., Pulkki-Råback, L., Lehtimäki, T., Raitakari, O., & Pehkonen, J. (2018). Does education protect against depression? Evidence from the Young Finns Study using Mendelian randomization. *Preventive Medicine*, *115*, 134–139. https://doi.org/10.1016/j.ypmed.2018.08.026

von Möllendorff, C., & Hirschfeld, J. (2016). Measuring impacts of extreme weather events using the life satisfaction approach. *Ecological Economics*, *121*, 108–116. https://doi.org/10.1016/j.ecolecon.2015.11.013

Walker, S. P., Chang, S. M., Wright, A. S., Pinto, R., Heckman, J. J., & Grantham-McGregor, S. M. (2022). Cognitive, psychosocial, and behaviour gains at age 31 years from the Jamaica early childhood stimulation trial. *Journal of Child Psychology and Psychiatry*, *63*(6), 626–635. https://doi.org/10.1111/jcpp.13499

World Bank. (2023). *Official exchange rate*. World Bank Open Data. https://data.worldbank.org

World Bank Open Data. (n.d.). World Bank Open Data. Retrieved 20 December 2024, from https://data.worldbank.org

Yarmolkevich, N. (2017). *Bonds Beyond Time: Are There Differences In Well-Being, Autonomy, And Bond Between Visually Impaired Individuals With Guide Dogs Versus Pet Dogs?* [MS, Illinois State University]. https://doi.org/10.30707/ETD2017.Yarmolkevich.N

Zotti, R., Speziale, N., & Barra, C. (2016). On the causal effect of religiosity on life satisfaction using a propensity score matching technique. *International Journal of Social Economics*, *43*(10), 1031–1048. https://doi.org/10.1108/IJSE-12-2014-0262

Made in the USA
Las Vegas, NV
03 November 2026

57422015R00155